HEATING AND AIR-CONDITIONING OF BUILDINGS

Plate I. An Oil-fired Boilerhouse serving an industrial estate (see p. 96)

HEATING AND AIR-CONDITIONING OF BUILDINGS

(BY OSCAR FABER AND J. R. KELL)

FIFTH REVISED EDITION (IN SI UNITS)

BY

J. R. KELL

C.B.E., C.ENG., F.I.MECH.E., PP.I.H.V.E., F.INST.F., M.CONS.E.
HOLDER OF THE I.H.V.E. GOLD MEDAL

AND

P. L. MARTIN

C.ENG., F.I.MECH.E., P.I.H.V.E., M.INST.F., A.R.AE.S., M.CONS.E.

THE ARCHITECTURAL PRESS: LONDON

First published 1936
Second and revised edition 1943
Reprinted 1945
Reprinted 1948
Reprinted 1951
Third and revised edition 1957
Reprinted 1958
Reprinted 1961
Fourth and revised edition 1966
Fifth and revised edition 1971

Errata

Page 92 In the formula (at foot of page), for '75' read '0·75'.

Page 119 Line 11 should read 'velocity of 3·0 to 5·0 m/s'.

Page 119 Line 13, for 'gas' read 'gas flues'.

Page 145 Omit the dagger preceding the *Note* below Table 7·3.

Page 169 13th line from foot of page, omit the words 'and radiator'.

Page 215 In the example under *Sizing by Pressure Drop*, alter the following to read:

Latent heat $= 2121\,\text{kJ/kg}$

Steam flow $\dfrac{1000\,\text{kJ/s}}{2121} = 0\cdot47\,\text{kg/s}$

$= 470\,\text{g/s}$

Page 306 Fig. 14.4: vertical scale reading 'Water flow rate in energy units kW/°C' has been omitted.

PRINTED IN GREAT BRITAIN BY
ROBERT MACLEHOSE AND CO. LTD, THE UNIVERSITY PRESS, GLASGOW

PREFACE TO FIFTH EDITION

THE CHANGE OVER FROM IMPERIAL UNITS to *Système International d'Unités* (SI) renders all existing text books on engineering subjects out of date. Thus, when the fourth edition of this book ran out of print, it was obvious that it must be revised in the new units. To have perpetuated a re-print in the old terms might have met a demand for collectors' pieces but it would have afforded no guidance or instruction for readers of the future.

The task of conversion has, however, been a mammoth one due to fundamental changes in the basis of measurement of heat, pressure and so on. Old idols such as the Btu are relegated to the scrap-heap! Thus, a complete re-write has been necessary in most sections, with a corre-sponding use of unfamiliar factors for heat exchange, energy flow and hygrometric differences, to mention only a few. The change to metric for weights and measures being a thing in itself is not by any means the whole story, and the resultant for this branch of engineering is probably more far-reaching than for any other.

In the past, this book has endeavoured to follow fairly straight-forward lines, leaving a multitude of by-ways and side-tracks to be followed by those so inclined. The present edition, notwithstanding the change in units, follows the same plan with the object of being an intro-duction to the subjects treated, rather than being exhaustive in any sense.

Two new chapters have been added dealing with District Heating and Total Energy. The latter was, in fact, covered to some extent in the first three editions of this book, but under the heading of *Combined Electrical Generating Stations*. The new title extends the conception beyond what was previously possible by making use of modern developments in plant and fuel sources. District Heating, whilst standing on its own feet, may be com-bined with Total Energy—so opening up a great new field of possibilities. Thus, the two additional chapters are, to some extent, complementary.

In presenting this new edition, the Authors have leant heavily on the *I.H.V.E. Guide*, 1970, as will be seen from the numerous references. They would like to acknowledge with thanks the permission of the Institution of Heating and Ventilating Engineers to quote from this source. In most cases, extracts only from Tables are given and, for fuller data, the reader is referred to the *Guide* itself.* In other cases—notably in pipe sizing—the

---

* The *I.H.V.E. Guide*, 1970. Published by the Institution of Heating and Ventilating Engi-neers, 49 Cadogan Square, London, S.W.1. In three volumes (Volumes A, B and C). Prices: £8 per volume for A and B; £6 for Volume C.

Guide formulae have been presented graphically in energy units, thereby overcoming a disadvantage of SI under which specific heat of water is no longer unity. There are various other Tables presented in a form perhaps more convenient for use in arriving at a quick approximate answer.

Figures by way of illustration have, of necessity, been largely redrawn due to the change in units and, for this task, the Authors wish to express their thanks to Mr John Freeman, chief draughtsman (services) of Oscar Faber and Partners. They would also like to express their gratitude to that firm for making available their computer for deriving certain of the graphical material. To various manufacturers who have provided illustrations and information they would also like to convey their thanks.

Thanks are also due to Mrs Pamela Kell for the unremitting toil of typing, checking and indexing, and above all to Mr Alan Hastings of the Publishers for his unbelievable patience throughout.

St Albans, 1971   J. R. K.
                                                                                                P. L. M.

CONTENTS

LIST OF PLATES

ACKNOWLEDGMENTS

The Authors wish to express their thanks to the following for their kind permission to reproduce illustrations:

Plate III: Riley Products (I.C.) Ltd.
Plate V: Reznor Nesbitt, London.
Plate VII: Frenger Heating Ltd.
Plate IX: The Imperial Cancer Research Fund, London.
Plates XI and XII: *Power and Works Engineering Journal.*
Plate XV: Drake and Gorham (Contractors) Ltd.
Plate XVII: Colt Heating and Ventilation Ltd.
Plate XXV, and Jacket photo: Laing Development Co. Ltd. (Photograph by courtesy of G. N. Haden and Sons Ltd.).
Plates XIX, XX, XXII, XXIV, XXVII and XXVIII: Barclays Bank Ltd. (Photographs by courtesy of Benham and Sons Ltd.).

THE TAKE-OVER AT THE TEMPLE!

The old household gods have been deposed. Of the new gods, Kelvin and Celsius vie for the same niche. The British Standards Institute is reverently inserting the last statue. The initials in the pediment have simply been reversed—IS (Imperial System) to SI (Système International). Two of the twelve steps are being cut away, to the discomfort of the gentleman on the left who is used to taking them three at a time.

The Old System

IN THE PAST, ENGINEERS have been accustomed to using a variety of disconnected and unrelated units. Thus we had temperatures measured in degrees Fahrenheit, which started from the freezing point of water at 32 degrees and divided the space between this and boiling point into 180 degrees. Heat quantities were based on the properties of water—the *British Thermal Unit* in the British and American system, or the *Calorie* in the Continental system. But the properties of water vary according to temperature, so this basis had to be defined in relation to temperature.

Energy was measured in horsepower using an odd figure of 33 000 foot pounds a minute to one h.p. Its equivalent in terms of heat, originating in the experiments of Joule (the son of a brewer) was the even odder figure of 2544 Btu/hour. For pressures there were various measurements to choose from—pounds per square inch or square foot, inches of water gauge, inches of mercury (or their metric equivalents: millimetres of mercury).

This whole edifice of engineering data was built up on the foundations of the inch, foot, yard and rod, pole or perch for linear dimensions; the square foot, square yard, pole, rood and three varieties of acre for areas; and the cubic inch, cubic foot, gallon and so on for volumes. For weight (avoirdupois) we had the pound, hundredweight, and ton, and for small weights the grain, of which 7000 went to the pound. A gallon of water weighed 10 lb which was convenient enough, but confusion with the American gallon at $8\frac{1}{3}$ lb had to be avoided, as likewise the American short ton of 2000 lb compared with the British 2240 lb.

This multifarious array of archaic tools—for means of measurement and calculation are no more than tools—has been likened to the Chinese alphabet. Those who mastered it over the years were proud of their great feat of memory, but it could not be regarded as suited to the present age of rapid scientific and technical development. Indeed, the scientists had long abandoned it, so that a student changing from the laboratory to applied engineering was faced with a completely new set of unfamiliar symbols and values to be learnt.

The New System

In the United Kingdom it has been decided to adopt the *Système International*, or SI. This is based on the metric system commonly in use on the Continent in so far as it is decimal, but it differs in that it is coherent throughout, which the old metric is not. For instance, in the metric system the unit for measurement of heat is, as explained, calorimetric—relying on the properties of water as in the British Thermal Unit. In the SI system the basis is energy.

SI units have already been accepted by a large number of other countries throughout the world and in course of time may well become completely international.

It is not proposed here to enter into a complete dissertation on SI, which can be found elsewhere,* but to summarize its principal features as they affect the subject of this book.

There are, in fact, only six basic units:

Length	the metre	(m)
Mass	the kilogramme	(kg)
Time	the second	(s)
Electric current	the ampere	(A)
Absolute temperature	deg. Kelvin	(K)
Luminous intensity	the candela	(cd)

All other units are derived from these: such as the units of force, energy, pressure, volume, area, motion, viscosity, etc.

Force

The unit of force is the *newton* (N), which is the force necessary to accelerate a mass of one kilogramme to a speed of one metre per second in one second,

$$\text{or } 1 \text{ N} = 1 \text{ kg m/s}^2.$$

When a mass of one kilogramme is subjected to acceleration due to gravity (which is 9·81 m/s²), the 'force' or 'weight' exerted is 9·81 N. Weight is not otherwise referred to in SI terms.

Energy: Heat

Energy may take a number of forms which are mutually convertible; for instance, chemical energy in a battery may produce electrical energy which in turn, in passing through a resistance, will produce heat energy.

Heat energy in fuel, in a boiler producing steam, may drive an engine producing mechanical energy, and so on.

The unit of energy, or quantity of heat, is the *joule* (J), which is equal to a force of one newton acting through one metre, or

$$1 \text{ J} = 1 \frac{\text{kg m}^2}{\text{s}^2}.$$

* *I.H.V.E.: Change to Metric* and the *I.H.V.E. Guide*, 1970.

Heat-Flow Rate

The unit for rate of heat flow or power is the *watt* (W). One watt is equal to one joule produced or expended in one second:

$$1 \text{ W} = 1\frac{\text{J}}{\text{s}} = 1\frac{\text{Nm}}{\text{s}}.$$

The watt is already familiar in its narrower electrical context:

$$1 \text{ watt} = 1 \text{ ampere} \times 1 \text{ volt}.$$

Thus the volt can be derived from the newton.

The amount of energy flowing through unit area in unit time is to be known as *the density of heat-flow rate*—W/m².

Temperature

The absolute Kelvin scale of temperature will be little used in practical engineering work, but the *Celsius* scale (°C) is the same in that the difference between freezing point and boiling point of water is divided into 100 equal parts.

Absolute zero is 0 deg. K, or $-273 \cdot 15°$ C.
Freezing point of water is $0°$ C.
Boiling point of water is $100°$ C.

Celsius was a Swedish professor and, as the inventor of the scale, his name is used in preference to 'centigrade' to avoid confusion with a French term concerning angular measurement.

A difference of $1°$ Celsius $= 1 \cdot 8°$ Fahrenheit
A temperature of $0°$,, $= 32°$,,
    ,,   ,,   ,, $100°$,, $= 212°$,,

Thus we say goodbye for ever to our old familiar friend, Mr Fahrenheit. The degree Celsius is a coarser unit, but has the great advantage of conforming to the decimal system.

Volume

The cubic metre (m³) is the preferred unit, as is the cubic millimetre (mm³). In view of the disparity in magnitude between these units, however, the cubic decimetre (dm³) and cubic centimetre (cc or cm³) are acceptable. The litre (originally $1 \cdot 000\ 028$ dm³) has been redefined such that

$$1 \text{ dm}^3 = 1 \text{ litre} = 0 \cdot 001 \text{ m}^3.$$

A difficulty arises as to the symbol for *litre* (l), which when typed is indistinguishable from the figure *one*. Hence '1 l' is easily mistaken for eleven. This matter is not finally resolved. 'L' has been suggested, but could be confused with the sign for the Italian Lire. In this book it is intended to use *litre* in full.

At standard atmospheric pressure and 4° C:

> 1 litre of water has a mass of 1 kilogramme
> 1 cc „   „   „ „ „ „ 1 gramme
> 1 m³ „   „   „ „ „ „ 1 tonne (1000 kg)

Time

The *second* is the only acceptable unit in SI. This should avoid error sometimes liable to creep in when using a formula in which some quantities are in one unit and others are in a different unit. The *hour, day, week* and *year* remain, but it will be important in calculations to convert them to seconds:

$$1 \text{ hour} = 3600 \text{ secs.}$$

The *cycle per second* is replaced by the term *hertz* (Hz), named after Heinrich Rudolf Hertz, the pioneer of the theory of radiation.

Pressure

The standard is the *newton per square metre* (N/m²). This is for some purposes an inconveniently small unit, and hence the *bar* is permitted, which is approximately one standard atmosphere.

> 1 bar   =   100 000 N/m²
> 1 pound/sq in =   6895 N/m²
> 1 inch wg   =   249 N/m²

In meteorological usage the *millibar* occurs,

> 1 mb   =   100 N/m².

The unit, N/m², is also to be known as the *pascal*.

Multiples and Sub-multiples

The preferred multiples are in threes:

> kilo   (k), a thousand, 10^3
> mega   (M), a million, 10^6
> giga,* (G), a thousand million, 10^9
> tera, (T), a billion, 10^{12}.

Sub-multiples are:

> milli (one thousandth), 10^{-3}
> micro (one millionth), 10^{-6}.

Other smaller sub-multiples need not concern us. Reference has, however, been made to exceptions where use has had to be made of:

> deci   (one tenth), 10^{-1}
> centi (one hundredth), 10^{-2}.

* Pronounced 'geeger' (hard 'gs').

The purpose of multiples and sub-multiples is to enable the values of units in common use to be whole numbers or single decimals, the preferred range to be o·1 to 1000. For instance, as pointed out, the newton is too small for most purposes, but the *kilonewton* (kN) may be used. The same applies to the watt where, for most purposes, the *kilowatt* (kW), *megawatt* (MW), and *gigawatt* (GW) will be used.*

The Decimal Point

In the British version of SI, the decimal marker remains as the familiar point above the line for writing and printing, and on the line for typescript —until special typewriters have become common.

The use of the comma to mark thousands is not acceptable owing to its liability to confusion with its use in other countries as the decimal marker. Instead, a gap is to be left to indicate thousands, but this will not apply where only four digits occur. Thus, for example,

six thousand will be   6000
sixty thousand will be   60 000.

Fractions will always be preceded by a nought, and thus one-tenth will be 'o·1' (*not* '·1').

Coherence and Consistency

It will be clear that SI achieves the greatest possible measure of co-herence and interchangeability as between one set of units and another. It is unfamiliar ground to regard heat as measurable in terms of force, distance and time, but this is what constitutes work; and work and heat have long had equivalents.

The units may seem strange for a time, but in use this disadvantage will no doubt be overcome. What can never be overcome in the decimal system is, of course, its indivisibility by three. No longer can the yard be divided into feet or the shilling into fourpences. On the other hand, who could divide 2240 by three?

---

* See the list of SI symbols at the end of this book for various multiples and sub-multiples. Also, for conversion factors, Imperial to SI units, see Table 1.4, page 15.

CHAPTER 1

Heat

KEEPING WARM IS A primitive instinct in man, and modern life still depends on warmth for existence.

The more civilized we become, the more sophisticated are the demands—the fashion is to wear lighter clothes and hence to require warmer environments.

Buildings can be warmed by the sun in hot countries and, even in temperate climates, given sufficient insulation, they can be warmed by the internal heat from occupants and lights. But it is necessary, as a rule, to warm buildings by the burning of some sort of fuel, or by consuming electrical energy which may in turn be derived from fuel or from nuclear energy or water power. It is to the means for performing these functions and distributing the heat to places where it is wanted that the heating engineer directs his attention.

But warmth alone is not the sole criterion of comfort. There must also be a pleasant atmosphere, which involves considerations of ventilation for the supply of fresh air and the removal of foul air, and air-conditioning for cooling in summer as well as warming in winter. Modern buildings with their large expanses of glass and high concentration of occupancy, coupled with the noise and dirt of large cities, create an increasing demand for such systems.

Apart from comfort conditioning, there is the further field in industrial applications of providing carefully controlled temperatures and humidities for process requirements and for the storage and manufacture of a wide variety of materials.

Fundamentally, all these and other related problems may be resolved simply into questions of how to add or remove heat, how to add or remove moisture and how to move air. Air will generally be involved somehow or another in all these mechanisms. It will be as well to keep this clear perception in mind in all that follows, but in the first place it is necessary to define some of the terms commonly used, with special reference to the particular meaning which the engineer normally applies to them.

DEFINITIONS

Temperature—Temperature is that state of a substance that determines the direction of heat flow to or from the substance. Heat will flow from a warm body to a cold body, and the rate will be directly proportional to the temperature difference or temperature 'head'. Temperature is akin to potential or pressure and is a relative term. The temperature of boiling water is higher than cold, that of cold water is higher than ice. Ice may be said to be hot, or at a high temperature compared with liquid air at − 190° C.

The scale of temperature now to be used in the United Kingdom is (as explained in the Introduction) the *Celsius* Scale (° C). If a mercury-in-glass thermometer (which measures the relative expansion of mercury and glass) is so calibrated that the freezing point is marked 0° and the boiling point is marked 100°, then any intermediate temperature is given a figure between these two, proportionate to the length of the mercury column.

The *absolute zero* has been established by studying the characteristics of expansion of gases, and other phenomena, and is assumed to be the temperature of space. The absolute scale is referred to in degrees *Kelvin*; absolute zero = − 273·15° C.

The Fahrenheit scale, having been ingrained in the mind for so long, will no doubt die hard; hence conversion tables will be useful and one is produced in Table 1.1, being in whole numbers of degrees Celsius. Fig. 1.1 serves to illustrate the comparisons.

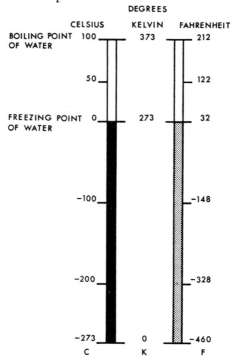

FIG. 1.1.—Temperature Scales.

To convert Celsius degrees into Fahrenheit, multiply by 9, divide by 5 and add 32, and conversely, to convert Fahrenheit into Celsius, deduct

TABLE 1.1
DEGREES CELSIUS TO DEGREES FAHRENHEIT

C	0°	1°	2°	3°	4°	5°	6°	7°	8°	9°
	°F	°F	°F	°F	°F	°F	°F	°F	°F	°F
0°	32·0	33·8	35·6	37·4	39·2	41·0	42·8	44·6	46·4	48·2
10°	50·0	51·8	53·6	55·4	57·2	59·0	60·8	62·6	64·4	66·2
20°	68·0	69·8	71·6	73·4	75·2	77·0	78·8	80·6	82·4	84·2
30°	86·0	87·8	89·6	91·4	93·2	95·0	96·8	98·6	100·4	102·2
40°	104·0	105·8	107·6	109·4	111·2	113·0	114·8	116·6	118·4	120·2
50°	122·0	123·8	125·6	127·4	129·2	131·0	132·8	134·6	136·4	138·2
60°	140·0	141·8	143·6	145·4	147·2	149·0	150·8	152·6	154·4	156·2
70°	158·0	159·8	161·6	163·4	165·2	167·0	168·8	170·6	172·4	174·2
80°	176·0	177·8	179·6	181·4	183·2	185·0	186·8	188·6	190·4	192·2
90°	194·0	195·8	197·6	199·4	201·2	203·0	204·8	206·6	208·4	210·2
100°	212·0	213·8	215·6	217·4	219·2	221·0	222·8	224·6	226·4	228·2
110°	230·0	231·8	233·6	235·4	237·2	239·0	240·8	242·6	244·4	246·2
120°	248·0	249·8	251·6	253·4	255·2	257·0	258·8	260·6	262·4	264·2
130°	266·0	267·8	269·6	271·4	273·2	275·0	276·8	278·6	280·4	282·2
140°	284·0	285·8	287·6	289·4	291·2	293·0	294·8	296·6	298·4	300·2
150°	302·0	303·8	305·6	307·4	309·2	311·0	312·8	314·6	316·4	318·2
160°	320·0	321·8	323·6	325·4	327·2	329·0	330·8	332·6	334·4	336·2
170°	338·0	339·8	341·6	343·4	345·2	347·0	348·8	350·6	352·4	354·2
180°	356·0	357·8	359·6	361·4	363·2	365·0	366·8	368·6	370·4	372·2
190°	374·0	375·8	377·6	379·4	381·2	383·0	384·8	386·6	388·4	390·2
200°	392·0	393·8	395·6	397·4	399·2	401·0	402·8	404·6	406·4	408·2
210°	410·0	411·8	413·6	415·4	417·2	419·0	420·8	422·6	424·4	426·2
220°	428·0	429·8	431·6	433·4	435·2	437·0	438·8	440·6	442·4	444·2
230°	446·0	447·8	449·6	451·4	453·2	455·0	456·8	458·6	460·4	462·2
240°	464·0	465·8	467·6	469·4	471·2	473·0	474·8	476·6	478·4	480·2
250°	482·0	483·8	485·6	487·4	489·2	491·0	492·8	494·6	496·4	498·2
260°	500·0	501·8	503·6	505·4	507·2	509·0	510·8	512·6	514·4	516·2
270°	518·0	519·8	521·6	523·4	525·2	527·0	528·8	530·6	532·4	534·2
280°	536·0	537·8	539·6	541·4	543·2	545·0	546·8	548·6	550·4	552·2
290°	554·0	555·8	557·6	559·4	561·2	563·0	564·8	566·6	568·4	570·2
300°	572·0	573·8	575·6	577·4	579·2	581·0	582·8	584·6	586·4	588·2

32, multiply by 5 and divide by 9. A convenient short cut for mental arithmetic, reasonably accurate in the range of British weather charts, is to multiply degrees Celsius by 2 and add 30 to obtain degrees Fahrenheit.

It is interesting to note that the intermediate temperatures as measured by a mercury-in-glass thermometer do depend a little on the expansion characteristics of mercury and glass, and that thermometers which depend on the expansion of other materials do not necessarily give an exact subdivision between freezing and boiling point when checked against a mercury thermometer. For very accurate scientific work this has to be taken into account, but is not necessary for the purposes of this present treatise.

Fig. 1.2 illustrates various forms of thermometer in common use.

The reading of an ordinary mercury thermometer is said to give the temperature of a gas, liquid or solid in which it is immersed, but in reality what it gives is something considerably more complicated when applied to air in buildings. It gives a temperature at which the heat received from the

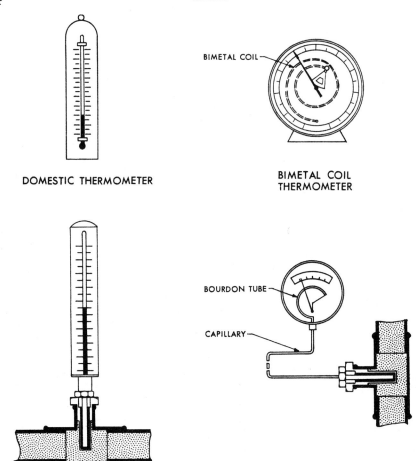

DOMESTIC THERMOMETER

BIMETAL COIL
THERMOMETER

PIPE THERMOMETERS

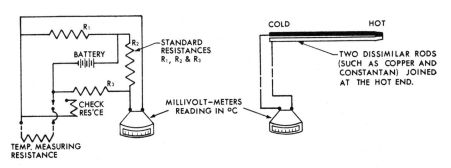

ELECTRICAL RESISTANCE
THERMOMETER

THERMOCOUPLE
THERMOMETER

FIG. 1.2.—Types of Thermometer.

surrounding objects exactly equals the heat given off to other surrounding objects when a perfect balance is obtained, and this heat transference may be partly by conduction, partly by convection and partly by radiation (defined later).

It is a matter of considerable interest that an ordinary mercury thermometer gives approximately the same reading in a room which has been allowed to reach stable conditions as regards temperature, whether it is screened from the source of radiant heat or not, provided that the source of radiant heat is one of relatively low temperature. This is because glass is practically impervious to the radiation from low-temperature sources.

This, however, is not at all true where high-temperature radiation, such as that received from the sun and from incandescent surfaces such as open fires, gas and electric fires is concerned. This is really the explanation of the curious phenomenon of the ordinary glass greenhouse, which, as everyone knows, gets extremely hot inside when the sun is pouring down on it, and retains its heat long after the sun has ceased to shine upon it.

The air outside the greenhouse in the sun receives exactly the same amount of radiation from the sun as the air inside the greenhouse. Why, therefore, does that outside remain, say, at 20° when the air inside may be at 40°? In both cases the temperature of any body subject to the radiation goes up until the total quantity of heat received from the sun and all other surrounding objects is balanced by the heat lost to the other surrounding objects, and at first there appears to be no reason why the temperature reached in the two cases should be in any way different. The explanation of this is to be found in the peculiar property of glass in being pervious to high-temperature radiation and impervious to low-temperature radiation (a phenomenon sometimes referred to as *diathermancy*). The heat from the sun (which is a source of high-temperature radiation) passes through the glass and warms the objects therein contained, which are unable to radiate this heat back through the glass.

A thermometer known as a *solar thermometer* is employed when it is desired to obtain a reading which includes the effect of solar radiation (see Fig. 1.3). This thermometer con-sists of a glass bulb containing a vacuum, in the middle of which is an ordinary thermometer bulb with a blackened surface. The idea is that the thermometer bulb will not be affected by the air tempera-ture, from which it is insulated by the intervening vacuum, but will

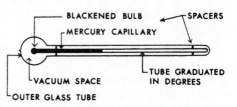

FIG. 1.3.—Solar Thermometer.

receive only radiation, and so will measure the radiant heat only. This gives quite satisfactory readings when applied to the measurement of sun temperatures, but it is of little use when we wish to measure radiation from low-temperature surfaces for the reason stated above.

TABLE 1.2

MOISTURE IN SATURATED AIR AT
VARIOUS TEMPERATURES.

kg per kg dry air

Temperature °C	kg moisture
0	0·0038
5	0·0054
10	0·0076
15	0·0107
20	0·0148
25	0·0202
30	0·0273
35	0·0367
40	0·0491
45	0·0653
50	0·0868

Humidity—The total pressure of the air is the sum of the partial pressures of the separate constituents. Water vapour, or the constituent of air which we describe as *humidity* exists in the space independently of the oxygen, nitrogen and other gases forming the atmosphere. When the highest partial pressure of water vapour appropriate to the air temperature exists, the air is said to be saturated. At lower partial pressures, the air is unsaturated, and the ratio of the two pressures is known as *relative humidity*. The greatest mass of moisture which can obtain in a given space is dependent only on temperature, and Table 1.2 expresses this in terms of kilogrammes per kg of dry air. Such a condition is stated to be saturated. If air which is unsaturated is lowered in temperature, the relative humidity will rise until it reaches 100 per cent. and is saturated. Any further lowering causes condensation. Another term used is *percentage saturation* which for practical purposes is almost the same as relative humidity.

The relative humidity is the condition of air upon which depends, more than anything else, the rate at which evaporation from a moist surface will occur. Thus at 100 per cent. relative humidity no evaporation can occur, while at 50 per cent. relative humidity the evaporation will be rapid. The human body provides such a moist surface, and the evaporation which takes place from it with low relative humidities produces various physiological effects, such as a parchiness in the throat and a cooling effect on the face and hands, especially in the presence of air movements, thus producing discomfort.

Too high a relative humidity produces other signs of discomfort, such as lassitude, caused primarily by the inability of the skin to rid itself of moisture and heat.

The normal evaporation from the skin is accompanied by a cooling effect on the body owing to the great heat required (equivalent to approximately 2300 kJ/kg) to convert water into water vapour (see also *latent heat*, page 14). It is on this that the cooling effect of a current of air principally depends. With excessive humidity, when this evaporation does not occur, the cooling effect clearly disappears also, and discomfort arises at high temperatures.

This explains what travellers in tropical climates have experienced for many years, that a high air temperature with a low relative humidity may be borne more easily than a lower air temperature in very humid conditions.

In cold climates, however, the opposite is the case. Air at low temperatures has a greater feeling of coldness at high relative humidities than at low. The reason for this effect on the sensation of cold in cold weather and of heat in hot weather, being both greatly accentuated in a humid atmosphere, is due to the changes which high humidity induce on the human skin. In a dry atmosphere, the skin dries up and hardens and becomes more insulating, and so feels the cold or heat less.

Conversely, in a moist atmosphere the skin swells, the pores open, the skin becomes more conducting and sensitive. This is an important effect, which no instrument so far devised takes into account.

It therefore appears that high relative humidity causes discomfort both at high and low temperatures, in the former case by producing a sensation of extreme heat, and in the latter case of extreme cold. The temperature at which no change in sensation occurs with change of humidity is stated to be 8° C for still air and 10° to 13° C for air moving at speeds of 0·5 m/s to 2·5 m/s respectively.

Humidification and De-humidifying—From what has been said it will be clear that in hot summer weather the hot air from outside coming into a cool building will have its relative humidity raised, and will therefore produce oppressive conditions.

Comfortable conditions may be produced by some device whereby the relative humidity may be reduced, and this is known as de-humidifying. Most de-humidifying processes depend on allowing the air to come in contact with cold surfaces where the excess moisture is condensed by being lowered below saturation temperature, and then warming it again to a comfortable temperature before distribution.

In cold winter weather the opposite occurs. Cold air from the outside comes into a warm room, and is warmed by the heating system to a higher temperature accompanied by a drop in the relative humidity. In such cases there is need for humidification, i.e. for adding to the moisture content of the air. One method of effecting this is to pass the air through a washer in which it is intimately mixed with water in the form of a very fine spray or mist.

In general it may be said that to most people a comfortable humidity is between 40 and 70 per cent. combined with a temperature of somewhere between 18° and 24° C.

Conduction, Convection and Radiation—It will be important, in the discussions that follow, to understand clearly the difference between the three main methods of transferring heat from one body to another.

Conduction may be described as a heat transfer from one particle to another by contact. If, for example, a hot lump of copper is placed in contact with a cold lump of copper, as in Fig. 1.4, heat is said to be conducted from the hot to the cold, until the two will finish at an intermediate temperature which may be calculated by equating the total mass times the final temperature with the mass of the hot body times its temperature added to the mass of the cooler body times its temperature. If the two bodies are of different material, then the mass of each has to be multiplied by its specific heat capacity before this calculation can be made.

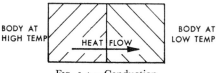

BODY AT HIGH TEMP

HEAT FLOW

BODY AT LOW TEMP

Fig. 1.4.—Conduction.

Conductivity is the property of bodies of being able to conduct heat, and the measure of conductivity is the number of units of energy transmitted per degree difference per unit thickness and per unit face area, in unit time.

Table 1.3 gives conductivities of various metals, building materials and insulators, and it will be seen that metals have a high conductivity, while materials like brick have a low conductivity. Certain bodies have such a low conductivity that they are known as *insulators*, and are used for covering warm surfaces to prevent heat losses from them. For convenience Table 1.3 also gives the density, specific heat capacity and coefficient of expansion of these substances. These terms will be referred to later.

Good conductors of heat are generally good conductors of electricity, though there is no rigid relationship between the two.

The conductivity of many materials varies considerably with temperature, and therefore figures should only be used within the range to which they apply. The insulation value is inversely proportional to the conductivity.

Porous materials are bad conductors when dry and good conductors when wet, a fact which is sometimes lost sight of when attempting to warm a newly-constructed building, where the heat losses may be far higher in the early months than they will be when it has properly dried out.

Convection is a transfer of heat which involves the movement of hot particles of a fluid medium from a hot body to a cold. A common illustration may be found in the ordinary radiator (so-called), see Fig. 1.5. This warms the air immediately in contact with it, which expands and so becomes lighter than the rest of the air in the room. It consequently rises, forming an upward current from the radiator, and travels to the colder portions of the room, to which it gives up its heat and eventually returns to the radiator for the process to be repeated.

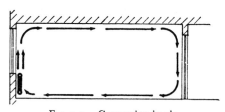

Fig. 1.5.—Convection in air.

TABLE 1.3

PROPERTIES OF MATERIALS

Material	Density (Specific Mass)	Specific Heat Capacity	Coeff. of Linear Expansion per 1° C $10^{-6} \times \ldots$	Thermal Conductivity (k)
	kg/m³	kJ/kg° C		W/m° C
Metals				
Aluminium (sheet) - - - - -	2 700	0·98	25·5	238
Brass (Cast) - - - - -	8 100	0·36	18·8	109
Copper (sheet) - - - -	8 800	0·39	17·5	385
Iron (Cast) - - - - -	7 400	0·51	10·2	47
Lead - - - - - -	11 400	0·14	29·0	35
Magnesium - - - - -	1 700	1·05	25·5	157
Mercury (0° C) - - - -	13 600	0·14	60·0	7
Mild Steel - - - -	7 800	0·48	11·3	48
Tin - - - - - -	7 300	0·23	21·4	64
Zinc (sheet) - - - -	7 200	0·39	26·1	112
Building Materials				
Asbestos cement (sheet) - - -	1 550	0·84	9·9	0·45
Asphalte - - - - -	2 250	1·68	—	1·2
Brick (exposed)- - - -	1 800	0·79	2·2	1·07
Concrete (exposed) - - -	2 400	0·84	9·9	2·55
Firebrick (at 400° C) - -	2 000	0·84	4·9	1·0
Glass (sheet) - - - -	2 500	0·84	8·4	1·05
Granite - - - - -	2 650	0·90	7·9	2·9
Limestone - - - -	2 200	0·86	6·3	1·5
Marble - - - - -	2 700	0·90	11·0	2·0
Plaster - - - - -	1 300	0·84	—	0·46
Plaster board - - - -	950	0·84	—	0·16
Slate - - - - -	2 700	0·75	19·6	1·9
Tiles (burnt clay) - - -	1 900	0·84	—	0·85
Timber				
Deal - - - - -	600	1·21	4 to 8 along grain	0·13
Oak - - - - -	750	1·88	20 to 80 across grain	0·16
Pitch pine - - - -	650	2·30	(when dry)	0·14
Insulating Materials				
Asbestos Millboard - - -	700	0·82	—	0·11
Lightweight Concrete - - -	600	0·84	1·4	0·18
Cork board - - - -	150	1·80	—	0·04
Diatomaceous Brick - - -	500	0·80	1·4	0·09
Fibreboard - - - -	380	—	—	0·05
Glass Fibre (quilt) - - -	80	0·82	—	0·04
Magnesia 85% - - -	200	—	—	0·06
Polystyrene (expanded) - - -	15	—	—	0·04
Vermiculite (loose) - - -	100	—	—	0·07
Wood wool (slab) - - -	600	—	—	0·11
Miscellaneous				
Water 4° C - - - -	1 000	4·205	0	0·6
15° C - - - -	998·5	4·186	65	0·6
100° C - - - -	958·4	4·214	250	0·67
Ice - - - - -	920	2·1	52	2·2
Air (at normal 20° C - - -	1·205	1·012	—	0·027
pressure) 100° C - - -	0·88	1·012	—	0·027

Another example is the heating of water in a boiler, as shown in Fig. 1.6. The water in contact with the hot surfaces over the fire becomes heated, expands, and produces an upward circulation, exactly as in the case of the air, eventually returning to the boiler for reheating. Convection, therefore, implies a medium capable of movement from the hot body to the cool body to be heated, and cannot occur in vacuum when no such medium exists.

Radiation is a phenomenon with which we are most familiar in its application to the problem of light.

In Newton's time the phenomenon of radiation was explained as a bombardment of infinitesimal particles from the source of heat, which were supposed to impinge on the cool body to be heated. At a later date radiation, whether of light or of heat, was supposed to be a wave action in an imaginary medium known as the ether, which was invented by mathematicians to account for this phenomenon. It is now known that all radiation is an electromagnetic process be it light, heat, X-rays or radio waves. Heat radiation occupies a band between the red end of the visible spectrum and the shortest of radio waves. Fig. 1.7 gives a pictorial representation of the order of the various forms of radiation.

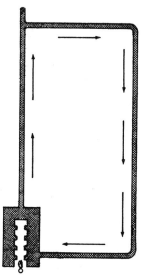

Fig. 1.6.—Convection in water.

Confining our attention to heat radiation, this may be regarded as a transference of energy which takes place in rays in such a way that the intensity varies inversely with the square of the distance, and is independent of any substantial medium such as air, i.e. it occurs just as readily across a vacuum as across a room filled with air, and does not depend on warming the medium through which it travels.

The so-called 'radiator' transfers its heat partly by convection and partly by radiation, the proportion depending mainly on the shape of the radiator. Those surfaces which face one another radiate very little to the rest of the room and depend principally on convection. Also, if anything is done to obstruct the free flow of air over the radiator, its proportion of convection is reduced. Flat surfaces such as a warmed wall or ceiling, however, may have a relatively high proportion of radiation as compared with convection.

In very rough figures, the proportion of radiation to the total heat emission in ordinary radiators may be about 20 per cent, in a wall panel about 50 per cent., and in a ceiling panel about 90 per cent. This shows that the term 'radiator' is a misnomer, since approximately 80 per cent. of its heat is transmitted by convection.

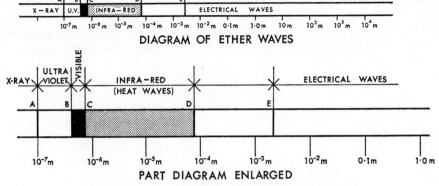

FIG. 1.7.—Types of Radiation. (Wave lengths are in metres.)

Fig. 1.8 shows the relatively concentrated radiant beam emitted from an electric bowl fire and the diffused radiation from a warmed ceiling panel.

The amount of radiation emitted from a surface depends on its texture and colour. Dead black is the best radiator and polished metal the worst. This effect is dealt with more fully later. (See p. 130.)

Surfaces which are good radiators of heat are found to be the best absorbers also: thus a black-felted roof is often found to be covered with hoar frost on a cold night, due to its good radiation into space, whereas

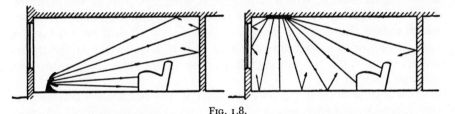

FIG. 1.8.

(a) Visible heat radiation from electric bowl-fire.　　　　(b) Invisible radiation from heated ceiling panel.

other surrounding objects may be apparently unaffected. Similarly, a black suit of clothes on a warm day is found to absorb much more heat from the sun than one of white material such as is worn in tropical countries. Again, asphalt on a flat roof does not become nearly so hot in summer if painted white.

Unit of Heat—As explained in the Introduction, it is fundamental in SI to treat heat as a form of energy. The unit is the *joule* (J), which is the energy of a mass of 1 kg accelerated to a speed of 1 metre per second in 1 second, i.e. a force of 1 newton (N) acting through 1 metre.

Heat flow is measured in joules per second, but 1 J/s = 1 watt and this latter unit or its multiples will be used wherever heat energy flow in unit time is the criterion.

Specific Heat Capacity—The old term *specific heat* related to the quantity of heat required to raise unit mass through unit temperature, water being taken as unity. In SI, the unit of specific heat capacity is kJ per kilogramme per ° C (kJ/kg° C). Thus heat flow to or from a substance is taken as

$$kJ = (kg) \times \left(\frac{kJ}{kg° C}\right) \times (° C)$$

If the problem involves heat flow in unit time (i.e. one second), then this is calculated as

$$kW = \frac{kJ}{s} = (kg) \times \left(\frac{kJ}{kg°C}\right) \times (° C) \times \left(\frac{1}{s}\right)$$

Table 1.3 gives specific heat capacities for materials in common use in buildings. For water the figure is approximately 4·2 kJ/kg° C.

Expansion—It will be found that Table 1.3 gives a column for the *Coefficient of Linear Expansion* of various materials.

All materials, with very few exceptions, expand on being warmed, and contract on being cooled. The Forth Bridge expands approximately 0·75 m as between warm and cold weather.

It is found that for most materials the expansion varies directly with the length and with the difference of temperature.

The coefficient of expansion for a material is defined as the proportion of its original length which it lengthens with 1° C rise of temperature. Thus the coefficient of expansion of mild steel is approximately 0·000011, which means that if a bar is say 1 m long at 15°, it will be 1·000011 m long at 16°, or, if raised through 100° C it will be 1·0011 at the higher temperature. A bar 100 m long will expand for the same temperature rise 0·11 m (110 mm).

It will be seen from this how necessary it is to make due allowance for the expansion and contraction of such things as long straight pipes, where the movement may be quite considerable. In practice, pipes beyond a certain length of straight have to be provided with expansion joints—a matter which will be dealt with in the appropriate place later. 'Invar' steel, a material much used in instruments, does not expand over a wide range of temperature.

The coefficients of expansion given in Table 1.3 are average figures within the range of normal temperatures—that is, between 15° and 100°—but a word of warning is necessary in applying these figures, as in some cases they vary considerably beyond this range. Water is an example of this, and near the freezing point it has a negative coefficient of expansion, i.e. it actually expands on lowering the temperature from about 4° C down to 0° C, the coefficient at 4° C being zero.

Superficial Expansion is the increase in area due to increase of temperature. The Coefficient of Superficial Expansion is taken as twice the linear coefficient. For an expansion of 'α' on the sides of a square of 1 unit initial length, increase in area $= (1 + \alpha)^2 - 1 = 2\alpha + \alpha^2$. α^2 is negligible, hence coef. sup. exp. $= 2\alpha$. *Cubic expansion* is the increase in volume due to increase of temperature. The Coefficient of Cubic Expansion is taken as three times the linear coefficient (by a similar argument to the above).

Specific Heat Capacity and Expansion of Gases—The perfect gas conforms to Boyle's and Charles' Laws, which state that the volume varies inversely with the pressure when the temperature is constant, and the volume varies directly with the absolute temperature when the pressure is constant. In other words, PV/T is a constant where P is the pressure, V is the volume and T is the absolute temperature.

Now the number of degrees Kelvin between freezing point and absolute zero is 273 (see Fig. 1.1). It therefore follows that, if a certain volume of perfect gas is contained in an envelope and its temperature is raised through 1° from 0° to 1° C or K without altering the pressure, the increase in volume will be $\frac{1}{273}$ of its original volume. This gives a coefficient of cubic expansion of $\frac{1}{273}$.

If, however, a volume of gas at the boiling point of water is raised through 1°, then its absolute temperature being 373°, its coefficient of expansion is $\frac{1}{373}$.

Most gases conform very closely to the properties of the perfect gas when at a temperature remote from their temperature of liquefaction. Ordinary air conforms very closely indeed, but CO_2, which can be liquefied at only moderate temperature reductions, does not conform quite so well.

Gases have two specific heat capacities according to whether the pressure or the volume is kept constant while the temperature is raised. If the pressure is kept constant, the increase in temperature is accompanied by an increase in volume, and, in the case of air, the specific heat capacity at 20° C is 1·012 kJ/kg° C. If the volume is kept constant, the increase in temperature is accompanied by an increase of pressure, and, again for air, the specific heat capacity is 0·72 kJ/kg° C at this temperature.

Power—Power and energy being synonymous terms within the present context, the old term *horsepower* is replaced by *watts* or *kilowatts*. (The hp was equivalent to 746 watts.)

It is found that an ordinary human being, by maintaining an internal temperature of approximately 37° C, emits approximately 120 watts when at rest at normal temperatures. This heat goes out partly as radiation and convection, partly as warm air containing water vapour from the lungs, and partly as evaporation from the skin. A man doing heavy physical work, such as carrying a 50 kg mass up a 20 m ladder in five minutes, is producing energy at a rate of 30 watts. In so doing his heat emission will rise to about 400 watts, which shows an efficiency of only $7\frac{1}{2}$ per cent! This is lower than that of the worst engine in common use, and should conduce to humility. A man can exert still more energy for short periods, and will then emit correspondingly greater quantities of heat.

Electrical Energy—It follows that electrical energy, whether in the form of power taken by a motor or heat due to passage through a resistance, is measured, as it always has been, in watts or its multiples. In the past, the *output* of a motor has been given in horsepower but it is now to be in watts as for the input, and some confusion may arise unless it is made quite clear as to which is referred to.

The Unit of Electricity is 1 kilowatt maintained for 1 hour, and may represent 10 amperes flowing in a circuit at 100 volts pressure for one hour or 100 amperes flowing in a 10 volt circuit for an hour, and so on.

Latent Heat—If water is heated from freezing point to boiling point (i.e. through 100° C), $4 \cdot 2 \times 100 = 420$ kJ are added per kg; and, if the supply of heat be continued, it is found that the temperature does not increase (assuming the vessel is open to atmosphere) but the water is gradually converted into steam at the same temperature. The experiment will show quite clearly that the amount of heat absorbed to effect this conversion is extremely large in comparison with the heat required to warm the water, and in fact it requires 2300 kJ to convert 1 kg of water at boiling point into 1 kg of steam at the same temperature.

It will be seen that this is approximately $5 \cdot 4$ times as great as the heat required to warm the water from freezing point to boiling, and this gives some idea of the importance of this phenomenon. This large quantity of heat required to convert a unit weight of any substance from the liquid to the vapour state is known as the *specific latent heat of evaporation*.

The reverse is equally true; when 1 kg of steam at 100° C condenses so as to produce 1 kg of water at the same temperature, 2300 kJ are liberated. Of all known substances, water has the greatest latent heat.

The boiling point of water varies with the pressure: at high pressures the boiling point is greatly raised and, conversely, it is reduced at pressures below atmospheric. It is often convenient to add the heat required to raise the water from freezing point to the temperature at which it is vaporized to the latent heat of vaporization, and this quantity is known as *specific enthalpy* at that temperature.

There is also a *latent heat of fusion* when ice is converted into water, or the reverse, and this has the value of 330 kJ/kg.

In all problems of humidification and de-humidifying, where water is either evaporated or condensed, the latent heat is the principal factor concerned.

All materials have a latent heat of fusion and one of vaporization, but these have lower values than in the case of water, and, of course, the temperatures at which the changes of state occur may be widely different.

Availability of Heat—The 'value' of heat depends on its temperature. Thus a certain quantity of heat at a high temperature may be useful for warming rooms or heating water. The same quantity of heat at a low temperature may be useless for these purposes. Heat is degraded in flowing from a high to a low level of temperature—increasing thereby in 'entropy.'

This is perhaps a subject which need not concern us here, but it is as well to remember that such a process exists in all heat-flow phenomena.

Table 1.4 summarizes a number of SI units and their equivalents.

TABLE 1.4

CONVERSION FACTORS: IMPERIAL UNITS TO SI

Unit	Imperial	SI	
		Exact	Approximate
Length	1 inch	25·4 mm	25 mm
	1 foot	0·3048 m	0·3 m
	3·28 feet	1 m	
	1 yard	0·9144 m	0·9 m
	1 mile	1·609 km	1·6 km
Area	1 sq. in	645·2 mm²	
	1 sq. ft	0·092 m²	
	10·77 sq. ft	1 m²	
	1 sq. yard	0·836 m²	
	1 acre	4046·9 m²	
Volume	1 cu. in	16·39 mm³	16 mm³
	1 cu. ft	28·32 litre	28 litre
	35·32 cu. ft	1 m³	
	1 pint	0·568 litre	0·6 litre
	1 gallon	4·546 litre	4·5 litre
Mass	1 pound	0·4536 kg	0·5 kg
	2·205 pounds	1 kg	
	1 ton	1·016 tonne	1 tonne
Density	1 lb/cu. ft	16·02 kg/m³	
Volume flow rate	1 gall/minute (g.p.m.)	0·07577 litre/s	0·075 litre/s
	1 cu. ft/minute (c.f.m.)	0·000472 m³/s	0·0005 m³/s
Velocity	1 foot/minute	0·0051 m/s	
	197 ft/minute	1·0 m/s	
	1 mile/hour	0·447 m/s	0·5 m/s
Temperature	1 Degree Fahrenheit	0·5556° C	
	$t = 32°$ F	$t = 0°$ C	
Heat	1 British Thermal Unit (Btu)	1·055 kJ	1 kJ
	1 Therm (100 000 Btu)	105·5 MJ	100 MJ
	1 Unit of Electricity (kWh)	3600 kJ	
Heat flow rate	1 Btu/hour	0·2931 W	0·3 W
	1 horsepower	745·7 W	750 W
	1 ton refrigeration (12 000 Btu/hour)	3·516 kW	3·5 kW
Density of heat flow rate	1 Btu/sq. ft hour	3·155 W/m²	3 W/m²

TABLE 1.4 (continued)

Unit	Imperial	SI Exact	SI Approximate
Transmittance (U value)	$\dfrac{1 \text{ Btu}}{\text{sq. ft hour } ^\circ F}$	5·678 W/m²° C	6 W/m²° C
Conductivity (k value)	$\dfrac{1 \text{ Btu inch}}{\text{sq. ft hour } ^\circ F}$	0·1442 W/m° C	
Resistivity (1/k)	$\dfrac{1 \text{ sq. ft hour } ^\circ F}{\text{Btu inch}}$	6·934 m° C/W	
Calorific value	1 Btu/lb 1 Btu/cu. ft	2·326 kJ/kg 37·26 J/litre or 37·26 kJ/m³	2·5 kJ/kg
Pressure	1 pound force per sq. in (lb/sq. in)	6895 N/m² or 68·95 m bar	7000 N/m² 70 m bar
	1 inch w.g. (at 4° C)	249·1 N/m² or 2·491 m bar	250 N/m² 2·5 m bar
	1 inch mercury (at 0° C)	33·86 m bar	34 m bar
	1 mm mercury	1·333 m bar	
	1 Atmosphere (standard)	101 325 N/m²	1 bar
Pressure drop	1 inch w.g./100 ft	8·176 N/m³	
Latent heat of steam (atmospheric pressure)	970 Btu/lb	2258 kJ/kg	2300 kJ/kg
Latent heat of fusion of ice	144 Btu/lb	330 kJ/kg	
Steam flow rate	1 lb/hr 8 lb/hr	0·126 g/s —	1 g/s
Heat content	1 Btu/lb	2·326 kJ/kg	
	1 Btu/gall	0·2326 kJ/litre	
	1 Btu/cu. ft	0·0372 kJ/litre	
Thermal diffusivity	1 ft²/hr	2·581 × 10⁻⁵ m²/s	
Moisture content	1 lb/lb 100 grains/lb	1 kg/kg 0·014 kg/kg	

CHAPTER 2

The Building and the Heating System

THE BUILDING

AS A GENERAL PRINCIPLE when approaching the problem of the heating of a building, it is desirable that we should consider the building and its heating system as one entity. The form and construction of the building will have an important effect on any question of how to heat it. The amount of heat required to maintain a given internal temperature can be greatly reduced by insulation. The present fashion to use vast areas of glass runs counter to this and imposes considerable loads on any heating system. The lightness or heaviness of the structure has a direct bearing on the type of heating system and on its control and running costs. A lightly constructed building requires a heating system which is quickly responsive to changes of temperature, whereas a heavy, massive building is probably better served by a system with a slow steady output.

The present vogue of building tall multi-storey blocks of offices and flats introduces fresh problems due to exposure to wind and sun as well as chimney effects within the building itself due to height. These again have an important bearing on the form and design of the heating installation.

In considering the problem of how to keep a building warm, we cannot overlook the fact that the people, electric lighting and other heat-producing items of equipment within the space contribute in some measure to the maintenance of the desired temperature. This is particularly so in the case of buildings with large expanses of glass, where the heat of the sun, even in winter, may often be sufficient to maintain the desired temperature without any additional aid. Whilst this may be so on one side of a building, however, it will not apply to all aspects at the same time, and therefore the system of heating has to be so devised that it can be shut off automatically in those areas where fortuitous heat gains of this kind are sufficient in themselves, whilst other areas continue to be supplied with warmth for the maintenance of the desired temperature.

An important question arises as to how far the designer of a heating installation for a given building should take into account fortuitous heat gains. If they are not allowed for, the system may be unduly large and unwieldy, and rarely, if ever, run at its full output. On the other hand, if the designer has made certain assumptions that heat gains will occur which in practice, due possibly to change of usage, do not actually arise, he might be faced with the building being inadequately heated.

The tendency in the past, when designing heating installations, has been to ignore entirely the effect of internal heat gains, in so far as the calculation of the heat necessary to maintain a given internal temperature is concerned; but, when it comes to a question of running cost and heat consumption per annum, the importance of such effects has been brought out by the study of actual consumptions of buildings heated electrically, and, by the same token, similar calculations for other sources of heat should make the same allowances.

HEAT LOSSES

The conventional basis for design of any heating system is the estimation of heat losses. It is assumed that a steady state exists as between inside and outside temperatures and that air temperature is the sole criterion.

In fact, steady-state conditions rarely exist and, as will be seen later, the radiant temperature of the enclosure if taken into account may call for higher or lower air temperatures for equal comfort. However, it will be best to deal with the conventional method first and to discuss these other effects subsequently. The method may be explained as follows.

Each room of a building is taken in turn and an estimate made of the amount of heat necessary to maintain a given steady temperature within the space, assuming a steady lower temperature outside. The calculation is in two parts: (a) that due to conduction through the materials of the walls, floor and ceiling, known as *transmittance*, and (b) the heat necessary to warm infiltration-air up to room temperature, this air escaping by some means or other to the outside—an effect which is known as *air-change*. This air-change is, in fact, ventilation, and without it a space would quickly become uninhabitable; but knowledge of what exact air-change rate to allow in any given case is debatable and hence this part of the calculation tends to remain empirical.

It might be thought that, with so many assumptions and 'rule of thumb' estimates, heat losses were little better than guesswork; but in practice they appear to give a satisfactory basis, partly no doubt due to the fact that all areas within the building are treated in a like manner and should therefore be consistent, and secondly due to the flywheel effect of the structure itself which, even with a lightly constructed building, still has floor slabs, partitions, furniture, etc. to absorb and emit heat and so avoid violent fluctuations.

Adjacent rooms maintained at the same temperature will have no heat transfer through partitions or other surfaces, and these are therefore ignored. All that concerns us is the leakage of heat from a room at one temperature to outside, or to other rooms at a lower temperature. If certain areas of the room are used for heating, such as the ceiling or the floor, then these, being warmer than the room, will not be taken into account in so far as the heat losses from the room are concerned; but they will have inherent losses upward or downward, dependent on the amount of insulation, and these will have to be taken account of separately from the heat loss calculations.

Transmission Losses—The transmission of heat through any material depends on the temperature difference between the two surfaces and on the conductivity of the material itself. We are concerned with building materials, and with the transmission of heat from air in a room to air outside. The air temperatures will not be the same as the surface temperatures of the material, due to a dead layer of air, which may be supposed to exist in contact with the surface, retarding the flow of heat. Fig. 2.1 illustrates this as it might be for a thin material such as a sheet of glass. The air inside the room being relatively still offers higher resistance to heat flow than that outside, where the effect of wind is to be considered.

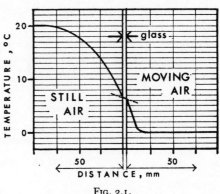

FIG. 2.1.

The effects of these boundary layers are defined as *surface resistances*, and values are put upon them as given later. In the case of the outside surface resistance, the value depends on the degree of exposure, such that a sheltered surface has a higher surface resistance than one with a severe exposure. In the extreme case, indeed, of say a very tall building, it might well be supposed that the force of the wind was such that the boundary layer completely disappears and the surface resistance is equal to zero. This would be the limiting case.

In order to calculate the transmission losses from a room, it is necessary to measure from drawings, or from the building, the area of each type of exposed surface, i.e. window, wall, roof, floor, etc. The measurements will be in square metres, and we then require to know the amount of heat which will be transmitted per square metre of surface, and this will depend on the material itself.

Conductivities of materials (k) are stated in watts for 1 metre thickness per square metre per ° C., i.e.

$$k = \frac{Wm}{m^2\,°C} = \frac{W}{m\,°C}$$

The resistivity of a material (r) is the reciprocal of the conductivity, thus

$$r = \frac{1}{k} = \frac{m\,°C}{W}$$

Conductivities of a few of the more common building materials are given in Table 1.3, but the *I.H.V.E. Guide* should be consulted for fuller information.

Note that the suffix '.. ivity' in these expressions signifies per unit thickness; the suffix '.. ance' signifies the value overall for the thickness stated.

U **Values**—The heat transmitted in watts per square metre per ° C. (W/m² ° C) for the purposes of heat loss calculations is termed *the transmittance coefficient U*, which is the reciprocal of the sum of all the resistances. Thus:

$$U = \frac{1}{R_{s1} + R_{s2} + r_1 L_1 + r_2 L_2 \text{ etc. } + R_a + R_h}$$

R_{s1} = surface resistance internal,
R_{s2} = surface resistance external,
r_1, r_2, etc. = resistivity of individual materials of which element of structure is composed, e.g. brick, plaster, etc.
L_1, L_2 = thickness of individual materials,
R_a = resistance of air gap (if any),
R_h = resistance of hollow block (if any) for which resistivity per unit thickness does not apply.

The values of the surface resistances R_{s1}, R_{s2}, vary with the type of surface, degree of roughness, and with the air movement over the surface. For the purpose of calculating U values they may be taken as follows:

Internal Resistance R_{s1}
    Walls - - - - - - 0·13 m² °C/W
    Ceilings and Floors
        upward heat flow - - - 0·11 m² °C/W
        downward heat flow - - - 0·15 m² °C/W

External Resistance R_{s2}
    'Normal Exposure'
    Walls - - - - - 0·06 m² °C/W
    Roof - - - - - 0·05 m² °C/W

The resistance of an air gap R_a in normal material such as brickwork, wall-board, etc., 20 mm wide or over, may be taken at the value of 0·18. If the air gap is lined with bright metallic surfaces, the value becomes 0·35.

If the transmittance coefficients multiplied by the surface areas of all the exposed surfaces in the room are added together, we then have the total quantity of heat which will be lost through the fabric per 1° C. difference of temperature. Table 2.1 gives a range of transmittance coefficients from the *Guide* by way of example of various forms of construction, but there are nowadays so many composite elements involved, that it is usually necessary to calculate the U factor from first principles.

For example, consider a curtain wall construction comprising:

Outside - - - 40 mm thick concrete panel
Air gap - - - 20 mm
Inside - - - 75 mm lightweight concrete block
Plaster - - - 20 mm

the resistances are summated thus

	k	$r = \dfrac{1}{k}$		thickness m		R
Concrete Panel	1·40	0·72	×	0·04	=	0·028
Air gap						0·18
Concrete Block	0·18	5·55	×	0·075	=	0·415
Plaster	0·46	2·18	×	0·02	=	0·044
Surface R_{s1}						0·13
R_{s2}						0·06
				Total R	=	0·857

$$U = \frac{1}{R} = \frac{1}{0 \cdot 857} = 1 \cdot 17 \ \text{W/m}^2 \, ^\circ \text{C}$$

It should be noted that the transmittance coefficients U quoted in Table 2.1 are for normal exposure. The *I.H.V.E. Guide* gives further values for sheltered and severe exposure, as follows:

'Sheltered'—including—the first two storeys above ground in the interior of towns;

'Normal' —the 3rd, 4th and 5th storeys of buildings in the interior of towns and most suburban and country premises;

'Severe' —the 6th and higher storeys of buildings in the interior of towns and all storeys of buildings on hill, coast or riverside sites.

U **Values and Moisture**—A recent review of basic data on which U values have been based has been made*, comparing calculated values with experimental data derived for BRS wall and roof laboratories. It has been shown that, in general, the conductivities of all masonry materials bear the same relationship with dry density and follow similar increases proportionately with increasing moisture content. The 1970 *Guide* is based on corrected practical values of moisture content thus:

Brickwork, protected from rain			1%	Moisture Content by volume	
Concrete,	,,	,,	,,	3%	,, ,,
Brickwork, exposed to rain			5%	,, ,,
Concrete,	,,	,,		5%	,, ,,

It is pointed out that driving rain and condensation may give higher moisture contents.

Thus, calculation of U values from conductivities should take account of moisture content where the material is exposed to weather. Inner skins of building elements should remain dry, as in the case of the inner material of a cavity wall, and the protected value is then used.

Table 2.1 from the *I.H.V.E. Guide*, 1970, is on this basis.

* A. G. Loudon: *BRS Current Papers*, Nov., 1968

U **Values for Floors**—Transmittance coefficients for solid floors on earth as given in Table 2.1 need some explanation. Where the building covers an extensive area the loss will be chiefly around the perimeter, as the earth temperature near the centre will have built up to nearly that of the room. In a small or narrow building the perimeter loss will be proportionately greater. The *I.H.V.E. Guide* gives the appropriate values for floors of varying extent of which the coefficients given in Table 2.1 are samples. They may be multiplied by the full temperature difference inside to outside.

TABLE 2.1

Transmittance Coefficients U
(W/m² ° C) for 'Normal' Exposure

(*including allowance for moisture as appropriate*)

WALLS

Brickwork	Thickness, mm*				
	105	220	260	335	375
Solid					
Unplastered	3·3	2·3	—	1·7	—
Plastered	3·0	2·1	—	1·7	—
Cavity (unventilated) brick outer skin					
Brick inner skin, plastered	—	—	1·5	—	1·2
Lightweight concrete inner skin, plastered	—	—	1·0	—	—
Lightweight concrete inner skin, 13 mm expanded polystyrene in cavity, plastered	—	—	0·7	—	—

Concrete	Thickness, mm		
	150	165	200
In situ			
Cast, unplastered	3·5	—	3·1
Cast with 50 mm woodwool on inside, plastered	1·1	—	1·1
Pre-cast			
Panel with 50 mm cavity and lined with 25 mm expanded polystyrene plus plasterboard finish	—	0·80	—

Framed Constructions

5 mm asbestos cement sheet
Bare on frame - - - - - - - - - - 5·3
On frame with cavity and aluminium foil backed plasterboard - - - - 1·8
Double skin with 25 mm glass fibre in between - - - - - 1·1
Tile hanging on timber battens and building paper, 50 mm glass fibre in cavity, plaster board finish - - - - - - - - - - - 0·65
Weather boarding on building paper and timber frame, 50 mm glass fibre in cavity, plasterboard finish - - - - - - - - - - - 0·62

Curtain Walling (typical examples)

With 5% bridging by metal mullions
Mullions projecting outside - - - - - - - - 1·2
Mullions projecting inside and outside - - - - - - 1·8
With 10% bridging by metal mullions
Mullions projecting outside - - - - - - - - 1·5
Mullions projecting inside and outside - - - - - - 2·8

GLAZING (*measured over wall opening*)
Single
Metal frames - - - - - - - - - - - 5·6
Timber frames - - - - - - - - - - 4·3
Double, air space 20 mm or over
Metal frames - - - - - - - - - - - 3·2
Timber frames - - - - - - - - - - - 2·5

* Note. 4½ inch brickwork taken as 105 mm, 9 inch, as 220 mm and 13½ inch as 335 mm.

TABLE 2.1 *continued*

Double, air space 3 mm
   Metal frames - - - - - - - - - - - - 4·8
   Timber frames - - - - - - - - - - - - 3·9
Roof Skylight
   Open - - - - - - - - - - - - - 6·6
   With laylight under, ventilated - - - - - - - - 3·8

DOORS

Solid timber, 25 mm thick - - - - - - - - - - - 2·6

ROOFS

Flat

Concrete 150 mm thick, covered asphalt or felt
   Uninsulated - - - - - - - - - - - - 3·4
   With 100 mm (average) lightweight concrete screed to falls, plaster soffite - - 1·8
Hollow tiles 150 mm thick, covered asphalt or felt
   Uninsulated - - - - - - - - - - - - 2·2
   With 100 mm (average) lightweight concrete screed to falls, plaster soffite - - 1·4
Woodwool slabs, 50 mm thick with asphalt or felt on 13 mm screed, on timber joists
   Aluminium backed plasterboard ceiling - - - - - - - - 0·9
   25 mm glass fibre above ceiling - - - - - - - - - 0·6
Hollow or cavity asbestos cement decking with asphalt or felt on 13 mm insulating board
   Cavity as void - - - - - - - - - - - 1·5
   Cavity with 25 mm glass fibre - - - - - - - - - 0·73

Pitched

Tiles on battens roofing felt and rafters, ceiling below joists
   Aluminium backed plaster board - - - - - - - - 1·5
   50 mm glass fibre over plaster board - - - - - - - - 0·5
Corrugated asbestos cement or plastic covered steel sheeting
   Uninsulated - - - - - - - - - - - - 6·7
   With cavity and aluminium backed plaster board lining - - - - - 2.0
Corrugated double-skin asbestos cement sheeting with 25 mm glass fibre insert - 1·1

	4 edges exposed	2 edges exposed
FLOORS		
Solid Floors in contact with earth		
Very long × 30 m broad - - - - - - -	0·16	0·09
Very long × 7·5 m -	0·48	0·28
60 m × 60 m broad -	0·15	0·06
60 m × 15 m broad -	0·32	0·18
30 m × 30 m broad -	0·26	0·15
30 m × 7·5 m broad -	0·55	0·32
15 m × 15 m broad -	0·45	0·26
15 m × 7·5 m broad -	0·62	0·36
7·5 m × 7·5 m broad -	0·76	0·45
3 m × 3 m broad -	1·47	1·07

Timber Floors

With ventilated airspace below - - - - - - - - 1·3
With 25 mm fibre board below floor boarding over ventilated airspace - - 1·08

	Heat Flow	
Intermediate Floors	Upwards	Downwards
Timber on joists, plaster ceiling - - - - - - -	1·7	
Concrete		
150 mm thick, 50 mm screed - - - - - - -	2·7	2·2
150 mm thick, 20 mm woodblocks - - - - -	2·0	1·7
Hollow tile		
150 mm thick, 50 mm screed - - - - - -	2·0	1·7
150 mm thick, 20 mm woodblock - - - - -	1·6	1·4

Heat Required for Air-Infiltration—What has been said so far relates to heat lost from a room through the fabric. There is an additional loss due to movement of air through the room, such air generally entering from outside through the lack of seal of the structure itself. If there is a flue in the room, this will frequently create a suction by which air is drawn into the room, sometimes at an excessive rate. Leakage through windows may be greatly reduced by weatherstripping or by the use of double windows. In single-storey sheeted factory construction, air leakage often occurs through joints in the sheeting, particularly at eaves and ridges. Large door openings may involve considerable air change. In tall buildings the chimney effect will produce strong inward air currents at doors at ground level.

The importance of air-infiltration is that it frequently accounts for as much as one-third or a half of the heat losses in total, and yet it is the most indeterminate. It is this question of the indeterminacy of air-infiltration losses that renders temperature guarantees for buildings difficult to sustain, since any test is more a test of the building itself than of the heating installation.

There are two methods of assessment for air infiltration. One is empirical and is based on the allowance of an assumed number of changes per hour of the content of the space in question. The first column of Table 2.2 lists the commonly accepted rates for various types of buildings.

Where leakage is confined to windows, as in a multi-storey office block or block of flats, another more precise method has been devised. To use this method it is necessary to know the building exposure, height, type of window and opening areas. The *I.H.V.E. Guide* gives in detail the graphs and factors applying to this technique, but it is too involved to summarise here. By way of example, however, taking an urban site, a wind speed of 10 m/s, a building of 30 m height with sliding windows weatherstripped, the infiltration rate is given as 0·75 litre/s m of opening joint; unweatherstripped, the figure is 1·5. Values range between 0·1 and 4·0 litre/s m. Having arrived at the air-flow rate in this manner, it may be converted into an hourly quantity for comparison with the air-change method previously referred to.

The quantity of heat per m³ required to warm the infiltration air is obtained from the specific heat capacity of air (at constant pressure) at 20° C = 1·01 kJ/kg ° C (from Table 1.3) and the mass per m³ = 1·2 kg. Thus the heat required to raise 1 m³ by 1° C = 1·01 × 1·2 = 1·212 kJ.

In the SI system, the hour is an unacceptable time unit and, of course, the use of air changes per second would lead to unwieldy values over the accepted range. Using the heat-requirement conversion, however, for one air change we have

$$m^3 \text{ (the volume of the space)} \times \frac{1 \cdot 212 \times 1\,000}{3\,600}$$

$$= 0 \cdot 33\ \text{J/s m}^3\ {}^\circ\text{C (or W/m}^3\ {}^\circ\text{C)}$$

TABLE 2.2

NATURAL AIR INFILTRATION FOR HEAT LOSSES

(Air Changes and Specific Ventilation Loss)

Room or Building	Air Change per hour	Ventilation Loss $(W/m^3 \, °C)$
Large spaces (e.g. Factories, Assembly Halls, Churches Canteens) Solid Construction		
up to 5000 m³	1	0·3
5000 to 25 000 m³	$\frac{1}{2}$	0·2
over 25 000 m³	$\frac{1}{4}$	0·1
Light Construction		
up to 5000	3	1·0
5000 to 25 000 m³	$1\frac{1}{2}$	0·5
over 25 000 m³	$\frac{3}{4}$	0·3
Living spaces, offices, libraries		
Windows exposed one side	1	0·3
Windows exposed two sides	$1\frac{1}{2}$	0·5
Windows exposed more sides (if windows are weather-stripped, halve the above rates.)	2	0·7
Circulating Spaces Generally	2-3	0·7-1·0
Lavatories	2	0·7
Laboratories	3	1·0
Hospitals, Schools etc See Department of Health and Social Security, Ministry of Education and Science publications setting standards		

NOTE: This Table refers to air-change rates for heat loss calculations. Where mechanical inlet and extract ventilation is provided, heat for additional air-change is extra.

Having regard to the arbitrary nature of the definition of air-change rate this may be rounded to 0·3 W/m³ ° C. The term *ventilation loss* has been adopted to define the SI 'air-change' units which have values as listed in the right-hand column of Table 2.2.

Whilst the heat quantity for air loss arrived at by either method will be that to be offset by heat emission from the appliances room by room, the sum total of all the rooms will not necessarily represent the total heat input required for the building. Windows on the windward side will be subject to inward leakage, whereas those to rooms on the leeward side will be the means by which warm air escapes. If, therefore, rooms are disposed equally on two sides of a corridor, the total air loss will be 50 per cent of the sum of the individual room air losses; this would not of course apply to a building one room wide with windows on both sides.

The precise method given in the *I.H.V.E. Guide* shows how the overall building-air loss can be derived from the room totals for various dispositions and proportions of glazing. It is noted that the building-air loss may be as little as one third of the sum of the individual room-air losses. If the air change or ventilation-loss method has been employed, some similar allowance can be made by assessment from the drawings. It will be clear that this procedure applies not only to the sizing of boiler plant, but also to the estimation of fuel consumption.

It is clear that in the interests of economy of installation and running costs for all time, money spent on weather-tightness of windows is a real economy, since uncontrolled ventilation goes on 24 hours a day, whereas ventilation necessary only during the occupied period is a matter of a few hours as a rule, and can usually be dealt with by some controlled means of ventilation, such as the opening of a window. During occupation, the heat of occupants alone is sufficient to allow of some additional ventilation; for instance, in the case of school classrooms it is recognized that the heat of the children is equivalent to three air-changes per hour, and so the heating system is designed accordingly.

If ventilation is necessary at a known rate, such as in workshops and factories coming under the Factories Act, or in places of public entertainment coming under local authority regulations, mechanical ventilation is involved (dealt with in later chapters); in which case, so far as the heating system is concerned, the designer has to make a choice as to whether he will install heating equipment to cope with the fabric losses only, or whether he will include this heat in the ventilation air, bringing it in somewhat warmer to allow for this loss. What has been said earlier relates to buildings and rooms where no such mechanical ventilation is involved.

TEMPERATURE DIFFERENCE

The sum of the transmission losses and the air-infiltration loss, both in $W/^\circ$ C difference, requires to be multiplied by the difference of temperature between that of the room inside and some assumed temperature outside.

The inside temperature depends on the purpose of the various rooms and the kind of building, and a list of conventional temperatures is given in Table 2.3.

With regard to the outside temperature, in the British Isles this is commonly taken at -1° C or 0° C, but such a basis, whilst it has sufficed for the types of buildings and central heating systems of the past, may not get by during periods of colder weather with lightly constructed buildings and heating systems with no inherent margin.

A committee set up in 1950 by the various Institutions, in its report 'Basic Design Temperatures for Space Heating'*, examined the frequency of cold spells, and the likely number of days per annum on average when the specified internal temperatures would not be met with designs based on -1° C. The argument brought out in this report is that a heavily constructed building has sufficient thermal time lag to tide over a period of a few days of exceptionally cold weather, whereas a lightly constructed building has not this thermal capacity, and so will suffer more quickly.

* P.W.B.S. No. 33, published by H.M.S.O., 1955.

TABLE 2.3

ROOM TEMPERATURES
(Generally accepted levels—rounded)

Space or Area	Temp. °C
Swimming Pool Halls	26
Art Classrooms, Bathrooms, Changing Rooms, Hotel Bedrooms	22
Banqueting Halls, Hotel Lounges, Operating Theatres, Residential Living Rooms, X-Ray Rooms	21
Banking Halls, Canteens, Laboratories, Libraries, Offices, Reading Rooms	20
Assembly Halls, Auditoria, Churches, Domestic Bedrooms, Exhibition Halls, Factories (sedentary work), Hospital Day Rooms and Wards, Restaurants, School Classrooms, Shops	18
Cloakrooms, Factories (light work), Gymnasia, Lobbies and Circulation Spaces, Sports Halls, Stores (working areas)	16
Factories (heavy work), Warehouses (storage areas)	14

Furthermore, central heating systems with boilers of some sort usually have some overload capacity which can be brought into use during periods of exceptional cold, whereas certain other systems have no such reserve, such as electrical systems or those relying on heat emitters of a fixed output capacity, such as unit heaters. Conclusions drawn by Jamieson[*] are approximated in Table 2.4.

TABLE 2.4

EXTERNAL DESIGN TEMPERATURES

Multi-storey buildings with solid
    intermediate floors and partitions - Type A
Single storey buildings - - - Type B

Type of Heating System	Thermal Time Lag of Building	
	Type A	Type B
Hot water with overload capacity 15–20% (25% if no tolerance on indoor temperature allowable) - - - - -	–2°C	–3°C
System without overload capacity (e.g. electric, steam unit heaters) - - -	–4°C	–6°C

* Journal I.H.V.E., 1954; Vol. 22, p. 465.

It will be noted that the temperature of $-2°$ C suggested is sufficiently close to the conventional $-1°$ C to warrant the latter continuing to be used as a convenient basis where some form of central heating with boiler is included.

It should perhaps be mentioned that, in calculating total heat losses, the temperature difference across inner walls or partitions or from one floor to the next (where the adjacent rooms are warmed to a different temperature or are unwarmed) are best taken off separately, and the loss per degree difference multiplied by some lesser assumed temperature difference. Transmittance coefficients should also be adjusted in such cases, as both surfaces are internal.

ALLOWANCE FOR HEIGHT

It appears reasonable to make allowance for the height of a room bearing in mind that warm air rises towards the ceiling. Thus, in a room designed to keep a comfortable temperature in the lower one and a half or two metres, a higher temperature must exist nearer the ceiling, which will inevitably cause greater losses through the upper parts of windows walls, and roof. This effect is greatest with a convection system, i.e. one which relies on the warming of the air in the room for the conveyance of heat. This would occur in the case of conventional radiators and convectors.

In the case of radiant-heated rooms the same tendency does not occur, and a much more uniform temperature exists from floor to ceiling. In the case of floor heating there is virtually no temperature gradient whatever. This effect is illustrated in Fig. 2.2 which gives the vertical temperature gradients which may be expected with radiators and warmed-floor and ceiling systems. Fig. 2.3 gives the same for warm-air systems and shows

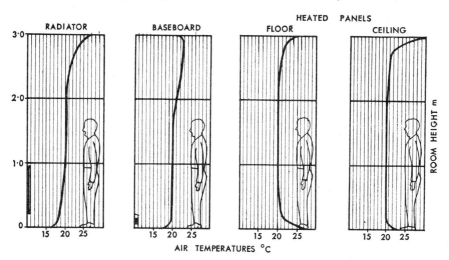

FIG. 2.2.—Vertical temperature gradients in a room with various types of heating, assuming rooms above and below heated by similar means. See also Fig. 2.3.

how a low-level inlet produces a much smaller temperature gradient than a high-level inlet due to better mixing with room air.

In the design of a heating installation, much can be done to reduce temperature gradient by the correct disposition of the heating surfaces under large cooling surfaces such as windows. In fact it might be said that a system producing a big temperature gradient is a bad system.

Table 2.5 gives height factors proposed in the *I.H.V.E. Guide* for various types of system.

TABLE 2.5

HEIGHT FACTORS

Type of Heating	Type and distribution of Heaters	% addition for height of heated space		
		5 m	5-10 m	over 10 m
Mainly Radiant	Warm floor	nil	nil	nil
	Warm ceiling	nil	0–5	*
	Medium and high temperature cross radiation from intermediate level	nil	0–5	5–10
	Ditto, downward radiation from high level	nil	nil	0–5
Mainly Convective	Natural warm air convection	nil	0–5	*
	Forced warm air crossflow from low level	0–5	5–15	15–30
	Ditto, downward from high level	0–5	5–10	10–20

* not applicable

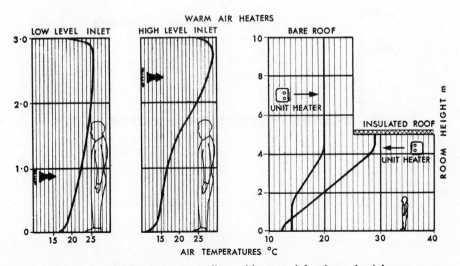

FIG. 2.3.—Vertical temperature gradients with warm-air heating and unit heaters.

HEAT LOSSES GENERALLY

Fig. 2.4 and the following heat loss calculation provide a simple example of the method normally adopted. The roof is partly exposed to the outside and partly has a warmed room over; so the latter has no loss.

The floor being solid, on earth, is taken at the appropriate coefficient for an area of this kind from Table 2.2.

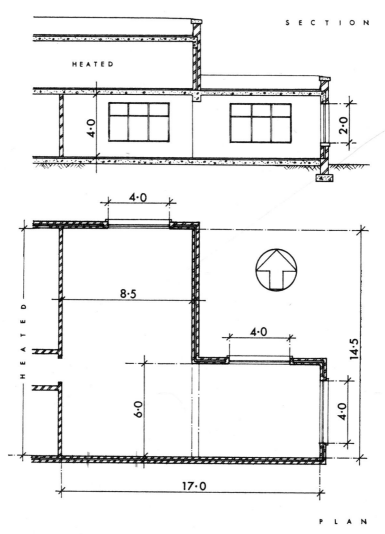

FIG. 2.4.—Heat Loss Example.

EXAMPLE: HEAT LOSS
(See Fig. 2.4, all dimensions in m)

Cube (allow $1\frac{1}{2}$ air changes)
$14\cdot5 \times 8\cdot5 \times 4\cdot0 = 493$
$6\cdot0 \times 8\cdot5 \times 4\cdot0 = 204$

$697 \times 0\cdot5 \ = 398\cdot5$

Windows
$16\cdot0 \times 2\cdot0 \qquad = 32 \times 5\cdot7 \ = 182\cdot4$

Wall
$48\cdot5 \times 4\cdot0 \qquad = 194$
less windows $\quad = 32$

$162 \times 1\cdot5 \ = 243\cdot0$

Floor
$14\cdot5 \times 8\cdot5 \qquad = 123$
$6\cdot0 \times 8\cdot5 \qquad = 51$

$174 \times 0\cdot62 = 108\cdot0$

Roof
$6\cdot0 \times 8\cdot5 \qquad = 51 \times 1\cdot8 \ = 91\cdot8$

$1023\cdot7 \ \text{W}/^\circ \text{C}$

Allow 0° C outside, 20° C inside $= 20^\circ$ C rise
$\therefore$ Loss $= 1023\cdot7 \times 20 = 20\ 474$ W (say $20\cdot5$ kW)
(This approximates to 30 W/m^3)

BUILDING INSULATION

In the past, insulation for a building was regarded as something which was applied and for which a good economic case had to be made out, which incidentally was not generally difficult to do. Building insulation has now come to be regarded as an essential element in construction and an inherent part of the building. It is indeed frequently possible, by the selection of the right materials, to achieve a high degree of insulation with little or no extra cost. Insulating materials are generally porous by nature, and hence weak structurally, so are more commonly used as inner skins. When used on flat roofs, they are prone to absorb condensation, and hence some vapour seal surrounding them, or means of venting, is needed.

The Thermal Insulation (Industrial Buildings) Act (1957) requires industrial buildings to be insulated in so far as the roofs are concerned, and certain maximum figures of U values according to temperatures are stated in the Schedule under the Act, in effect setting the maximum at $1\cdot7$ W/m^2 $^\circ$ C. The *Building Regulations*, 1965, applying in England and Wales, and the *Scottish Regulations*, 1963, stipulate standards of insulation. The latter are more specific and, for residential buildings, the maximum values are:

Roof including ceiling $\qquad$ $1\cdot13$ W/m^2 $^\circ$ C
Wall with no window $\qquad$ $1\cdot7$ $\qquad$,,
Wall with window average $\qquad$ $2\cdot38$ $\qquad$,,
Floor, if underside is ventilated $\quad$ $1\cdot13$ $\qquad$,,

A number of the more common types of insulating material are listed in Table 2.6, together with their conductivities.

TABLE 2.6

THERMAL INSULATING MATERIALS FOR BUILDING

Material	Bulk Density kg/m³	Conductivity W/m ° C
Asbestos insulating board	700	0·110
Asbestos, sprayed	240	0·075
Concrete, Lightweight block*	600	0·180
Corkboard	145	0·042
Fibreboard (insulating)	380	0·060
Glassfibre, quilt	80	0·040
Kapok, quilt	30	0·030
Mineral wool, loose mat	180	0·042
Polystyrene, expanded, board	15	0·037
Polyurethane board	30	0·023
Pumice, loose granules	350	0·070
Sawdust, loose	145	0·080
Thatch, straw	240	0·070
Urea formaldehyde foam	30	0·032
Vermiculite, granules	100	0·065
Wood-wool, slabs	600	0·110

* Protected, moisture content 3% by volume

With regard to glass, double-glazing reduces the heat loss from this element by half, and is desirable on its own account in cases of severe exposure. The cost of double-glazing is generally high, and it is often difficult to make a case on economic grounds. If windows are to be open for ventilation, the value of double-glazing as a noise barrier is much reduced. Hence double windows go with some means of mechanical ventilation.

Where translucence only is necessary, a sandwich form of insulating glazing may be used containing glass-fibre in the centre.

Figs. 2.5 and 2.6 show methods of insulating walls and roofs.

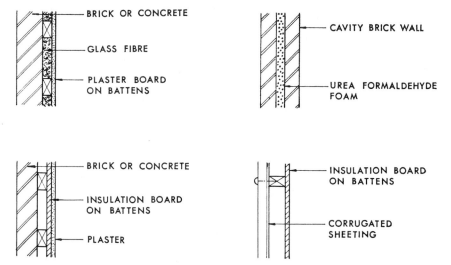

BRICK OR CONCRETE

GLASS FIBRE

PLASTER BOARD ON BATTENS

CAVITY BRICK WALL

UREA FORMALDEHYDE FOAM

BRICK OR CONCRETE

INSULATION BOARD ON BATTENS

PLASTER

INSULATION BOARD ON BATTENS

CORRUGATED SHEETING

FIG. 2.5.—Methods of applying insulation to walls.

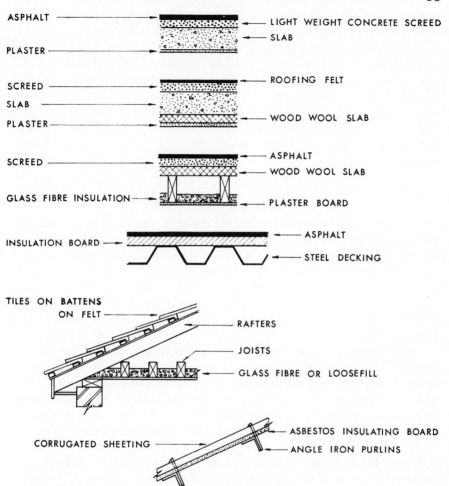

ASPHALT ——————————— LIGHT WEIGHT CONCRETE SCREED
                                          ——— SLAB
PLASTER ——————————

SCREED ——————————— ROOFING FELT
SLAB ——————————
PLASTER —————————— WOOD WOOL SLAB

SCREED —————————— ASPHALT
                                        ——— WOOD WOOL SLAB
GLASS FIBRE INSULATION ——— PLASTER BOARD

INSULATION BOARD ——— ASPHALT
                                        ——— STEEL DECKING

TILES ON BATTENS
    ON FELT ——————————
                                    —— RAFTERS
                                    —— JOISTS
                                    —— GLASS FIBRE OR LOOSEFILL

CORRUGATED SHEETING ——————— ASBESTOS INSULATING BOARD
                                        ——— ANGLE IRON PURLINS

FIG. 2.6.—Methods of applying insulation to roofs.

Another reason for the use of insulation in buildings is for the prevention of condensation. If the internal surface temperatures of walls and/or roofs drop below the dew point of the air within the space, condensation is bound to occur*. It is necessary then to calculate the thickness of insulation necessary to avoid this condition arising. Condensation on windows, where it is likely to occur through crowded occupancy or production of steam or vapour within the space, can be dealt with by double-glazing, or by heating the surface at the bottom of the glass.

Apart from the benefit of insulation as a means of reducing heat losses in winter, it will equally be effective in reducing heat gains in summer. Where a building is air-conditioned, as will be discussed later, double-

* But see the *I.H.V.E. Journal*, December, 1970.

glazing can generally be shown to be economically justified, owing to the much higher cost of cooling than of heating.

Where a concrete floor is suspended with a cavity below or indeed possibly an open space below, condensation may occur on the floor surface above unless insulation is provided of adequate thickness. It may well be found that this is best placed on the top of the slab rather than underneath, due to edge conduction. Edge conduction from exposed balconies can also be troublesome where the floor slab and the balcony are one homogeneous unit. This has been a fruitful cause of mould growth in multi-storey blocks of flats.

CONDITIONS OF COMFORT

It has been mentioned already that the ordinary thermometer measures air temperature. If placed in a room in which the air, walls, floor and ceiling are all at the same temperature, and if this temperature is one in which we say we are comfortable, then the thermometer may be regarded as giving a satisfactory index of comfort. That is to say if under the same conditions a lower temperature is registered, we shall no doubt say it is too cold, and likewise if a much higher temperature is registered that the room is too warm.

Such conditions of uniformity of temperature of air and surfaces, however, rarely pertain. In a warm room there is 'cold radiation' from the window, meaning in effect that heat is escaping from the body through the glass at a greater rate than to other surfaces, and, in order to counterbalance this, either the air temperature has to be raised or some form of radiant heating surface be provided. Under these conditions, where air temperature and surface temperatures of the enclosure are different, the ordinary thermometer fails as a true index of comfort. Furthermore, comfort is concerned with air movement. Even if air is warm, at a speed of much over 0·15 m/s most people would express discomfort.

Numerous attempts have been made to devise a scale of comfort of which the following deserve attention.

The Equivalent Temperature Scale. This combines the effect of air temperature, radiation and air movement, as measured by an instrument named the *Euphatheoscope*, a laboratory instrument. The late Mr A. F. Dufton, who developed this scale and instrument, defined equivalent temperature as that temperature of a uniform enclosure in which, in still air, a sizeable black body would lose heat at the same rate as in the environment, the surface of the body being one-third of the way between the temperature of the enclosure and 38° C (i.e. approximately blood heat). The instrument contained an electric heater, to maintain the surface temperature at a fixed point—the consumption of the heater being a measure of the Equivalent Temperature. Being a warm body it was sensitive to air movement.

Resultant Temperature is more easily measured than Equivalent Temperature. It is used as the most convenient index of warmth in the *Guide*. At 0·1 m/s the Resultant Temperature is the mean of air and mean-radiant temperature.

Environmental Temperature. This is not a comfort scale but is held to correlate more closely with measured heat losses at BRS wall and roof laboratories than other scales. For practical purposes, except where extremes of air temperature or radiant intensity occur, it can be regarded as equal to Equivalent or Resultant Temperatures. Environmental Temperature is a weighted mean between mean-radiant and air temperatures such that it is equal to $\frac{2}{3}$ mean-radiant temperature plus $\frac{1}{3}$ air temperature.

Globe Thermometer. This instrument comprises a thermometer inserted into a matt black sphere of copper or other thin material about 150 mm in diameter (Fig. 2.7). When suspended in a room, the black sphere will absorb heat radiation or emit radiation to cold surfaces, and so achieve some kind of balance, comparable with what is termed the *mean radiant temperature* of the enclosure: that is, the weighted sum of all the surfaces at their varying temperatures. In still air, or with air movements of a low order, up to say 0.15 m/s, the Globe thermometer temperature in conjunction with the air temperature may be taken as an indication of the equivalent temperature and will be closely akin to Resultant Temperature.

If strong draughts prevail, this thermometer tends towards air temperature. It is sluggish in response and, in normally heated rooms

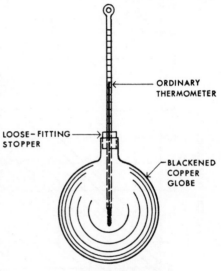

FIG. 2.7.—Globe Thermometer.

(such as offices) even with radiant ceiling heating, rarely reads more than 1° C above air temperature. In large spaces served with high temperature radiant surfaces, such as in factories, a Globe thermometer reading as much as 5° C in excess of air temperature may be found. In spaces heated by means of warm floors, with the thermometer 1 m above the floor, an excess temperature of about 2° C may be expected.

Effective Temperature. This is a scale which has been devised and developed in the United States, mainly applicable to air-conditioning, and combining the effects of air movement, air temperature and humidity, but having no reference to radiation. The scale is the outcome of a long series of subjective tests carried out on a number of people in a great variety of conditions, and has been revised from time to time. The results are expressed in a series of curves known as *Effective Temperature Lines*. This scale has little application where heating of a building alone is being considered, since, within the range of 15–20° C, variations in relative humidity between 40 per cent. and 70 per cent. have small effect on comfort. (See Fig. 2.8.)

A Corrected Effective Temperature Scale has been devised to include the effect of radiation, and, for this, four measurements are required: the Globe thermometer temperature, the wet bulb temperature, the air speed and the dry bulb temperature. By means of a nomogram, given these four measurements, the Corrected Effective Temperature can be obtained. Its reliability is restricted to the central part of the scale.

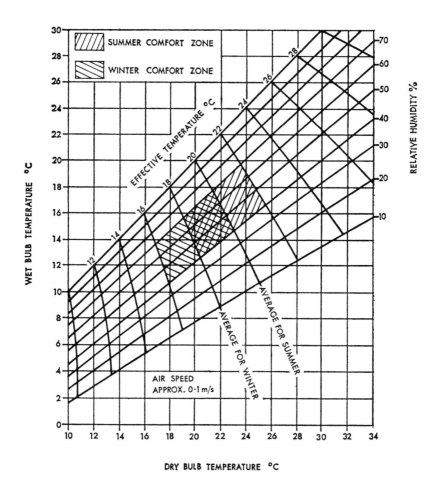

Fig. 2.8—'Effective temperature' chart. This chart is postulated from a variety of sources as applicable to Great Britain only. It gives a winter and summer range of conditions within which most people would feel comfortable, provided they remain long enough within the environment to become accustomed to the conditions.

Application—It will be clear that, whilst the various methods of measuring comfort may be of interest and concern as a matter of research or appraisal of one system against another, in so far as the design of heating systems is concerned, their application is difficult. In a space heated by means of radiant surfaces, the radiation falling on the various surfaces of the room and on the objects within the room, such as furniture, causes them to be warmed slightly so they in turn warm the air in contact by convection. The air temperature and radiant temperature thus tend to approach one another and eventually become nearly equal. With radiant heating, the air temperature as measured by an ordinary thermometer will be slightly cooler than the mean radiant temperature derived from the reading of a Globe thermometer and, conversely, if the heating is convective, the air temperature may be expected to be higher than the mean radiant temperature. The differences will be slight in the normal case.

Where the problem concerns heating by high-temperature radiant panels, such as in a factory, it is possible to estimate the mean radiant temperature, and a convenient method is given in the *I.H.V.E. Guide*. It is legitimate, however, to calculate on a normal air-temperature basis but to allow for a modest temperature difference, such as 3° to 5° C, on the assumption that a state of comfort will be achieved due to the radiant heating effect with air lower than the specified room temperature.

INCIDENTAL SOURCES OF HEAT

Reference has been made to fortuitous heat gains and their effect on the estimation of heat requirements. These heat gains may be evaluated from the following:

Heat From Occupants—Heat is emitted from the human body, partly as sensible heat and partly as latent heat in perspiration and vapour in the breath. The sensible heat alone affects the temperature of the room.

Table 2.7 gives average values at different states of activity in an atmosphere at 20° C and 60 per cent. relative humidity.

TABLE 2.7

HEAT FROM OCCUPANTS

Activity	Heat emission per occupant Watts		
	Total	Latent	Sensible
At rest - - - - -	115	25	90
Sedentary worker - - -	140	40	100
Walking - - - - -	160	50	110
Light manual work - - -	235	105	130
Heavy - - - - -	440	250	190

The sensible heat dissipated is reduced at higher temperatures, becoming zero at 37° C (blood temperature), and increases at low temperatures (e.g. at 0° C it is 70 per cent. greater than at 18°). The latent heat follows the opposite course, decreasing at lower temperatures and increasing at higher.

The heat of one person at rest is sufficient to heat 15 m³ of air per hour through 30° F. If the space per person allocated in a room approximates to this figure, and if the infiltration rate is the same, the air-change loss is covered by the heat from the occupants. If a building is designed so thermally perfect that fabric losses are insignificant, we can use the electric light to bring it up to temperature and thereafter rely on the occupants to keep it warm. This is the basis of the Wallasey School example. Limitations, however, arise when attempting to extend this principle to other types of building.

Heat from Machines—Energy expended by machines within a building results for the most part in the production of heat. In a lathe, for instance, in which metal is turned, the heat liberated is divided between the tools, the metal and the lubricant, which subsequently dissipate their heat in the room. The friction losses of shafting, belt drives, etc., are converted into heat. An electric motor driving machines transmits dynamic energy perhaps to a number of points within a building, where each machine in turn converts it into heat. The efficiency losses within the motor itself similarly come out in the form of heat. The exception is where work is done to produce potential energy, as in the pumping of water to a height, or in the lifting of goods in a hoist; in such cases the losses only are converted into heat.

Energy in the form of mechanical work being the same as in the form of heat, it follows that, for every kilowatt of power expended in a space, there is liberated 1 kilojoule of heat per second. If the power derives from an electric motor of 90 per cent. efficiency, 10 per cent. will come out as heat in the motor casing and air loss, and 90 per cent. as work which in turn will be degraded into heat—except where energy is conveyed outside of the space concerned.

Heat from Lighting—The energy consumed in the lighting of a space comes out as heat. Thus, if a space of 100 m² is lit to an intensity of 30 watts/m², the load will be 3000 watts, or 3 kW, and will be the heat energy involved.

Heat from Process Equipment—The liberation of heat within the enclosure from equipment such as steam presses, hot plates, drying ovens, gas or electric furnaces, etc., may be determined by summating the consumptions of all the items in watts or kilowatts, less any flue loss.

Heat from Sun—While the heat from the sun absorbed by a building is most important when considering cooling plants under summer conditions (and is treated later in Chapter 17), it is so unreliable in winter that no allowance can be made for it when designing a heating plant, as in the coldest weather it may be entirely absent.

Nevertheless, there are other days, even in winter, when its effect on the southerly aspect may be quite important, while rooms facing north receive no such benefit. Hence, there is much to be said for having rooms with south aspect separately zoned, so that they can have heat reduced without affecting rooms facing north.

Annual Total—A variety of interesting facts regarding the probable *annual* total of such adventitious heat input are set out in *Digest* 94 published by the Building Research Station. These relate to a semi-detached house of 100 m² area and are summarized in Table 2.8. The approximate total of these items represents something over a third of the theoretical *annual* heat requirement of such a dwelling.

TABLE 2.8

APPROXIMATE ANNUAL HEAT
FROM INCIDENTAL SOURCES

Source	Heat input per annum in kJ
Two adults, one child (body heat) - -	$2 \cdot 5 \times 10^6$
Radiation from Sun - -	$7 \cdot 0 \times 10^6$
Cooking, gas - - -	$6 \cdot 0 \times 10^6$
Cooking, electric - - -	$3 \cdot 5 \times 10^6$
Lighting, radio, TV and miscellaneous - -	$1 \cdot 0 \times 10^6$
Losses from water-heating appliances, circulating pipes etc. - - -	10 to 30×10^6

TEMPERATURE CONTROL

From what has already been said, it is clear that the ideal system is one in which each room has separate automatic control. It is then supplied with a heating element sufficient to warm it up to temperature, and then, as the heat from occupants, machines, light and sun begin to have effect, the heat input is cut down to prevent overheating and to save consumption. These factors probably operate at quite different times and in different degrees in all the rooms of a building.

A system which automatically allows for all such combinations is the ideal, other things being equal. Any departure from this ideal involves loss of comfort and waste of heat through the opening of windows in rooms which are otherwise too hot. In practice, a compromise is to be effected between the ideal, which may be expensive in first cost, and the primitive, which may be expensive in fuel. The more costly the fuel, the more important this item becomes.

The thermal characteristics of the building, its heating system and the system of temperature control are closely interlinked.

TIME LAG OF BUILDING—CONTINUOUS VERSUS INTERMITTENT HEATING

Continuous heating is necessary in buildings occupied for twenty-four hours during the day, like hospitals, police stations and three-shift factories.

Most buildings are, however, occupied for only a limited period during the day, such as schools (which are probably the lowest) at about 20 per cent. of the weekly hours. It is a matter of considerable importance when considering running costs to know whether, by operating the heating system intermittently, savings in fuel can be made whilst still maintaining satisfactory comfort conditions in the building.

The time lag of the building is perhaps best visualized by considering two cases. One, a shed of extremely light construction and having, in effect, no time lag whatever. A heating installation in such a building would, as soon as the heat is turned on, raise the internal temperature up to that desired, with a very small lag. The second kind of building— one with thick, massive walls, negligible windows and extremely heavy construction—involves a very great time lag. It might take a whole week to warm it up to the desired temperature. In the lightly constructed building, it is obvious that intermittent heating would achieve great economy, as heat would be supplied only during the hours of occupancy and no more. In the other case of the massive building, intermittent heating would be impracticable and thus no saving by such a means would be possible.

Between the two lie all the practical buildings of light, medium and heavy construction; and, according to the thermal time lag involved, so will intermittent heating show greater or lesser economy. Similarly,

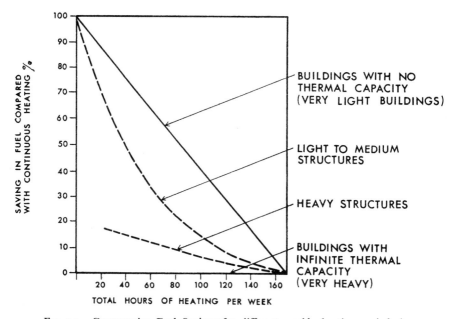

FIG. 2.9.—Comparative Fuel Savings for different weekly heating periods indicating effect of Thermal Capacity.
(With acknowledgment to H.V.R.A.).

according to the hours of occupancy per week, the possibilities of savings are greater or lesser in proportion to the hours when not occupied.

K. J. Colthorpe* summarizes these effects in a paper on the subject, by means of a graph, Fig. 2.9. This is based on a series of extensive tests on a number of buildings of various types in which savings compared with continuous heating were achieved, ranging from 5 per cent. to 50 per cent.

With an intermittently heated system, the building temperature will drop during the unoccupied period and, in order to bring this back to normal by the time the occupants enter in the morning, a preheating period is necessary. It might be considered that some excess capacity in the heating system would be desirable to bring this temperature up quickly:

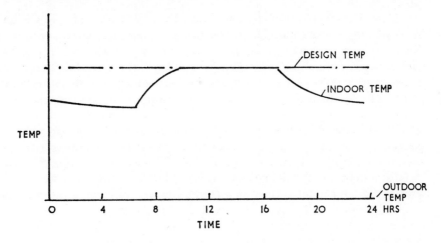

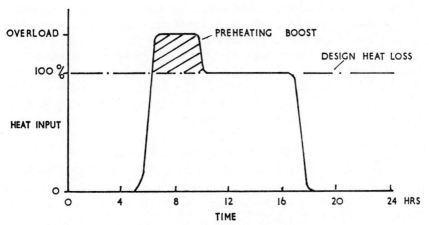

FIG. 2.10.—Intermittent Heating.

* H.V.R.A. Lab. Report No. 11, August 1962.

but what appears to suffice in practice is to start up the heat input at full rating, irrespective of the weather, and so drawing upon the spare capacity in the heating plant in the manner indicated in Fig. 2.10, which represents a normal cooling and heating curve for a building heated intermittently. The preheating time will be greater or less according to the outside temperature, and control systems have been devised to take account of this and adjust the preheating time automatically.

As the weather becomes colder, so the preheating time will become longer, until in extremely cold weather, or with outside temperatures below the basic design temperature, continuous operation of the heating system may be the only way of maintaining the desired internal temperature.

It has been shown that there is little advantage to be gained by installing a heating system of much greater capacity than that necessary to meet the hourly heat losses in an attempt to accelerate the warming up period, where such a system has its own time lag. This is due to the fact that as the system, boiler plant, etc., is increased in size, so its own time lag becomes longer and the advantages in a quicker heat-up are almost cancelled out. A 25 per cent. surplus is stated to be the maximum effective.

Where systems with a fixed input rate are involved, such as electric systems having fixed resistance elements, and gas systems using a set gas-rate, some additional capacity is necessary for a quick heat-up. Where an electric system is of the floor-warming type, operation so far as heat output is concerned is continuous even though the heat input off-peak is intermittent.

With regard to the type of system and its effect on fuel consumption under intermittent operation, it has been shown by Dick* and others that systems with small heat capacity and quick heating-up rate achieve greater economy than ones with a long time lag. This is an obvious conclusion, and points to the generalization that the greatest economy in buildings of light construction will be to run the system intermittently, the system itself to be of the shortest possible time lag. This accounts for the greater use of systems using warm air in some form or another in the present vogue of light construction. Older and more solid buildings achieve a satisfactory condition and reasonable economy with a steady supply of heat and with no fine degree of control. The light building with a quick heat-up system can obviously easily overrun its temperature, and hence a highly responsive control system is imperative.

PREHEAT AND PLANT SIZE

The 1970 *Guide* postulates a method of assessing plant size according to preheat time, plant response and type of structure. The basis is 'plant size ratio' which is the ratio between normal maximum output of plant and design load for 20° C rise.

* *Journal I.H.V.E.*, 1955; Vol. 23, p. 88.

Rounded up and in brief this shows:

	Preheat time	Plant size ratio required
for Heavy Structure		
Quick plant response	1 hr	3
	2 hr	$2\frac{1}{2}$
	$3\frac{1}{2}$ hr	2
Slow plant response	$4\frac{1}{2}$ hr	2
for Light Structure		
Quick plant response	1 hr	2
	3 hr	$1\frac{1}{2}$
Slow plant response	2 hr	$2\frac{1}{2}$
	3 hr	$1\frac{3}{4}$

The same table in the *Guide* gives the corresponding daily fuel consumptions taking 100 per cent. as continuous heating. For heavy structures the consumptions for the various preheat times range between 80 per cent. and 90 per cent., for light structures between 54 per cent. and 60 per cent. Exceptions are stated in respect of highly intermittent systems, radiant systems and storage systems.

Any approach such as this to the subject of intermittent heating must remain theoretical, and there is no point in attempting meticulous accuracy. It makes clear, however, that it is useless in a heavy structure to attempt a rapid heat-up as it is shown that for a one-hour preheat a plant capacity some three times normal is required, which would be uneconomic and often impracticable. This is why in church-heating a preheat time of twelve hours or more is required.

CHAPTER 3

Choice of Heating System

HAVING ARRIVED AT THE quantity of heat energy required for each room and thus for the building as a whole, as described in Chapter 2, it is now necessary to consider how this heat shall be supplied. It is proposed, therefore, to take a quick look at the whole range of available systems and to proceed thence to see how a choice can be made.

The number of heating systems is almost unlimited, if every combination of fuel, method of firing, transmission medium, and type of emitting element is considered. It is therefore quite useless to attempt to describe them at all clearly or systematically unless they are classified under the headings of the two main groups into which they all fall. This classification is indicated in Table 3.1, below and on the facing page.

TABLE 3.1

CLASSIFICATION OF HEATING SYSTEMS

DIRECT SYSTEMS

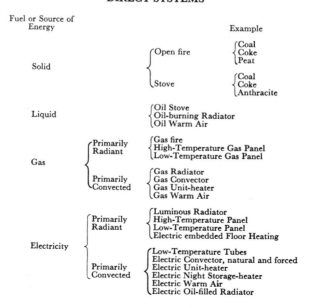

Fuel or Source of Energy		Example
Solid	Open fire	Coal / Coke / Peat
	Stove	Coal / Coke / Anthracite
Liquid		Oil Stove / Oil-burning Radiator / Oil Warm Air
Gas	Primarily Radiant	Gas fire / High-Temperature Gas Panel / Low-Temperature Gas Panel
	Primarily Convected	Gas Radiator / Gas Convector / Gas Unit-heater / Gas Warm Air
Electricity	Primarily Radiant	Luminous Radiator / High-Temperature Panel / Low-Temperature Panel / Electric embedded Floor Heating
	Primarily Convected	Low-Temperature Tubes / Electric Convector, natural and forced / Electric Unit-heater / Electric Night Storage-heater / Electric Warm Air / Electric Oil-filled Radiator

44

TABLE 3.1 (*Continued*)

CLASSIFICATION OF HEATING SYSTEMS

INDIRECT SYSTEMS

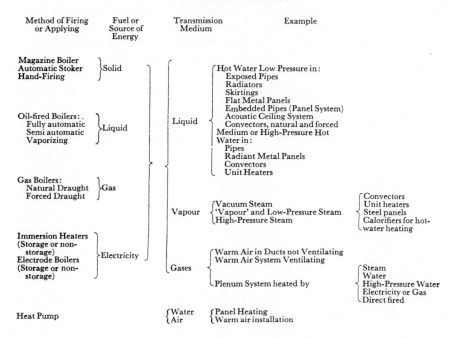

Method of Firing or Applying	Fuel or Source of Energy	Transmission Medium	Example

The two groups into which all systems may be divided are:

(*a*) '*Direct*' systems, in which the fuel or energy purchased is consumed in the room to be heated.

(*b*) '*Indirect*' systems, in which energy is transferred from some more or less centralized point outside the space to be heated to equipment within the space for liberation as heat.

The *direct systems* are separated into four main subdivisions, according to whether *the fuel (or energy source)* is solid, liquid, gaseous, or electrical. Each of these is again subdivided according to the nature of the emitting or radiating element. This group is simplified by the fact that there is no transmitting medium.

The *indirect systems* are subdivided according to the kind of *transmitting medium*, which may be liquid, vapour, or gaseous, and these in turn are subdivided according to their emitting element and other characteristics.

Within this large group, the fuels may be interchanged in any combination of transmitting medium and emitter, and a place is found for them in the Table.

We may here make the following comments on the different systems.

DIRECT SYSTEMS—SOLID FUEL

Open Fires—Open fires were, of course, the primitive source of heat, greater economy being obtained in mediaeval times by having the hearth in the centre of the room and allowing the gases to escape through a hole in the roof. Little heat was lost, but the system was not without its disadvantages.

Later, flues were invented, whereby economy was sacrificed for the greater advantage of smoke elimination.

The ordinary open fire is notoriously extravagent in fuel for the heat it gives out. Furthermore, it has a big open flue which draws in sometimes as much as eight or ten changes of cold air per hour into the room: five times as much as is necessary for good ventilation.

The modern version of the open fire is usually arranged for continuous burning, with means for controlling the air inlet, and with a restricted throat to the flue. It is designed to burn coal or smokeless fuel and efficiencies are of the order of 30 to 40 per cent. as compared with 15 to 20 per cent. for the old-fashioned type of grate.

Some forms of modern grates provide for warming convection air, and others for heating hot water.

Open fires of any sort are, of course, impracticable for buildings of any considerable size, partly owing to the labour and inefficiency already referred to, and partly owing to the great waste of space and cost of the multiplicity of flues required.

Stoves—Anthracite stoves in this country, and the large coal stoves so common on the Continent, are much more efficient, labour saving, and draught reducing than the open fire, and will keep alight all night. The fire is not visible in them, however, and for this reason their use in this country is generally confined to halls and corridors. Their efficiency may be as high as 50 per cent.

Again there are many modern forms of 'closable' stoves for burning coal, coke, anthracite, or one of the processed fuels. Sometimes these combine convection heating or water heating.

Coke stoves of the so-called slow-combustion type were a common method of heating churches but owing to the intense local heating which they produce, are often the cause of bad draughts. Their use is tending to die out although some have been converted to burn oil.

DIRECT SYSTEMS—LIQUID FUEL

Oil Stoves—The liquid fuel commonly used for room heaters is paraffin, burned either on a wick or on a vaporizing plate. The sulphur content of this fuel is as low as 0·04 per cent. so that the products of combustion are unobjectionable except in so far as they contain moisture.

Such heaters were commonly used in houses for background heating in cold weather, their merit being cheapness and portability. The efficiency can be as much as 100 per cent provided no flue is connected to the heater and provided the water vapour formed due to the combustion of the hydrogen in the fuel is condensed—as is frequently the case—causing moisture, however, to collect on the windows and cold walls.

It is this last factor—the condensation of the moisture in the products of combustion on windows and walls—together with the potential hazards of fire and asphyxiation which are now leading to a decrease in the popularity of such units.

Direct-Fired Warm Air System—A variation of the warm air system, used for industrial space heating, is one in which the air is heated by direct firing. The heater takes the form of a combustion chamber and passages through which the gases pass from an oil-fired combustion chamber over which the air is impelled by a fan. Oil is pumped from a central storage to a number of such units.

It should be mentioned, incidentally, that this type of heater is economical in first cost and it is also simple to run. An example is illustrated in Fig. 3.1.

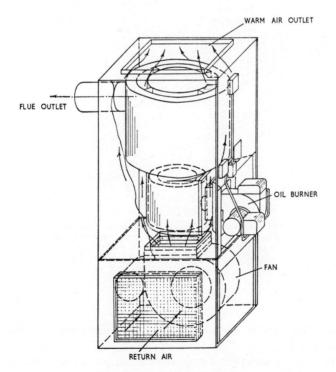

FIG. 3.1.—Oil-fired Warm Air Unit.

Direct-Fired Warm Air in Overhead Ducts—In this system, a unit similar to that just described is used to circulate hot air through ducts overhead which, in turn, radiate heat to the space below. This system is also applicable generally to industrial buildings, the radiant heating ducts being large and unsightly and, of course, necessarily visible.

DIRECT SYSTEMS—GAS FUEL (PRIMARILY RADIANT)

Gas Fires—Gas fires have been greatly improved over the last twenty years by increasing the radiation effect with special forms of radiant elements, and by combining convection surface for warming the air, see Fig. 3.2. Their efficiency is about 60 per cent. They have rapid response, but some people find the 'drying' effect of the radiation unpleasant for comfort.

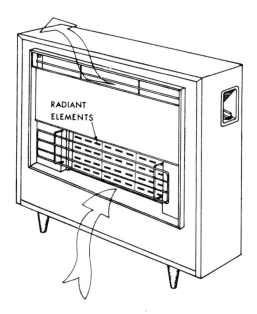

FIG. 3.2—Gas Fire. The heating elements provide radiant heat whilst convected heat is supplied by the cold air which is drawn in at low level and then warmed and delivered from the top outlet, as shown by the arrows.

A luminous type of gas radiant heater is illustrated in Fig. 3.3, on the facing page.

This heater is suitable for industrial application, and it may be thermostatically and remote controlled.

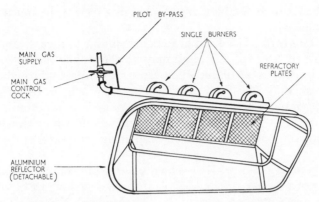

FIG. 3.3.—Gas Radiant Heater. (Harris Engineering Co., Ltd.).

DIRECT SYSTEMS—GAS FUEL (PRIMARILY CONVECTED)

Gas Convectors—Consist of a gas-heated element contained in a metal box with an opening at the bottom through which the air enters and another at the top through which it escapes after being warmed. The convection currents are then confined and a higher air temperature is the result. They are used chiefly in halls, shops, etc. and they have an efficiency of about 90 per cent. assuming that the vapour in the gases is not condensed.

Two types are shown: Fig. 3.4 shows an improved type in which the air for combustion is taken from outside and burnt gases returned to outside on a balanced flue principle; Fig. 3.5 is free-standing, drawing in

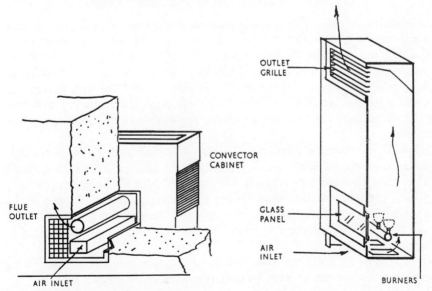

FIG. 3.4.—Cut-away view from outside wall of a Balanced Flue Convector (Drugasar).

FIG. 3.5.—Gas Convector.

E

air from the room at the base and discharging it with the burnt gas back into the room.

Gas Unit Heaters—Gas-fired heaters are similar to their steam counterparts described later (p. 206), and may be convenient if no steam or hot water for heating is available. The removal of fumes through a flue is normally essential. Gas-fired unit heaters are useful in factories, workshops, garages and other large spaces where noise is unimportant.

Gas Warm Air—The gas warm-air system is half way between a direct and an indirect system; it is much used in housing and is discussed in Chapter 6.

Flueless Heaters Generally—The discharge of products of combustion into the air of a room is only allowable if there is a good rate of ventilation to prevent concentration occurring. The Medical Advisory Committee reporting on this matter has, however, given it as their opinion that discomfort of overheating and high humidity would make itself felt long before any risk to health due to fumes arose.

DIRECT SYSTEMS—ELECTRIC HEATING

There are considerable savings to be made by using off-peak supply which, however, involves storage of heat by some means at night for use during the day. Systems of this type include *Electric Floor-Warming* and the *Block Storage Heater* either uncontrolled or with controlled output. These systems are dealt with in Chapter 12 (*Electric Heating*).

DIRECT SYSTEMS—RADIANT ELECTRIC SYSTEMS

Luminous Fires—These comprise a wire element forming a resistance which becomes incandescent on being switched on and rapidly achieves working temperature. A great variety of types and designs are available ranging in capacity from about $\frac{1}{2}$ kW to 3 kW. A particularly useful type has a polished reflector at the rear projecting a beam forward, and giving a concentrated band of radiation. The effect of any heater of this type is localized, and they are not generally applicable to the heating of large spaces.

High-Temperature Electric Panels—Consisting of a tile of ceramic material (often about 650 mm by 350 mm), enclosing a resistance element. They have been used for village halls, etc. (see Fig. 3.6). They are generally placed high up on the walls facing diagonally downwards, and reach a temperature of about 250° C with an emission of about 1·25 kW for the size stated. The radiation is about 80 per cent. These panels are much below red heat.

This system is naturally of limited application but it is an interesting example of what can be done by radiation alone.

Low-Temperature Electric Panels—These may consist of a flat plate of metal (Fig. 3.7), hardboard or other conducting material, to which is attached, or embodied, a heating element of nickel chromium, or other resistance wire or, in some special instances, a graphite coated mesh of synthetic fibre. The panels generally operate at about 70° C. One application is in the heating of churches, where they may be fixed in the pews, thus giving a quick effect of warmth to the congregation without fully heating the church.

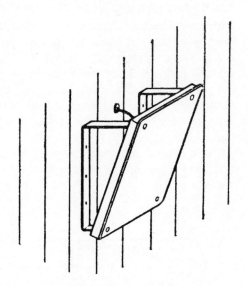

FIG. 3.6.—High-Temperature Electric Panel.

Another system consists of electric resistance elements embedded in a kind of wallpaper strengthened by fabric, which can be stuck to walls or ceilings with heat insulating material behind. The system generally runs at about 32° C and 180 watts per sq. metre with about 90 per cent radiation in still air when used on ceilings.

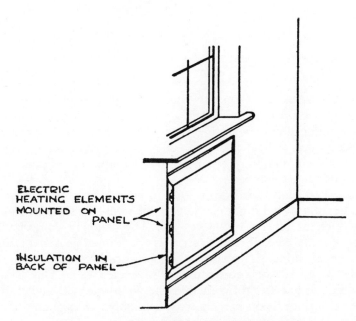

ELECTRIC
HEATING ELEMENTS
MOUNTED ON
PANEL

INSULATION IN
BACK OF PANEL

FIG. 3.7.—Low-Temperature Electric Heating Panel.

DIRECT SYSTEMS—ELECTRIC HEATING (PRIMARILY CONVECTED)

Tubular Electric Heaters—Are commonly of a round or oval shape (Fig. 3.8). The elements, enclosed in a thin steel tube for protection, are run at a temperature of 70° to 90° C. They are frequently fixed to skirtings under windows. They have also been applied to churches, placed under the pews, in which position they tend to keep the lower air warm without heating the whole building, and this naturally leads to economy for intermittent use. Downdraughts from the upper windows are prevented by more tubes on the sills at high level.

The tubes emit about 70 per cent convection and 30 per cent radiation. Tubes 40 mm to 50 mm diameter are rated at 180 to 240 watts per metre run.

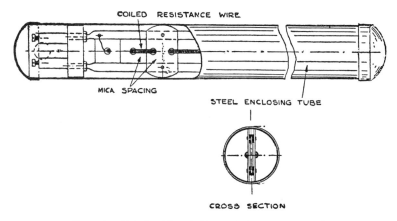

FIG. 3.8.—Tubular Electric Heater (diameter = 50 mm).

Electric Convectors—Generally consist of high-temperature elements in a casing with an opening below and a grating above through which an air current is induced. The warming effect of convectors is slower than with the radiant type of heater. A type of convector containing a fan gives a more rapid circulation of warm air in the room, one with a tangential fan being extremely compact.

Electric Oil-Filled Radiators—Are a convenient form of providing electric heating in buildings of the type otherwise furnished with central heating such as offices, residences etc. They usually comprise a steel radiator of neat design filled with oil so as not to be subject to freezing, and with an electric immersion element thermostatically controlled.

Electric Warm Air (commonly 'Electricaire')—This system is very similar to gas warm air in that the air is warmed and distributed by a fan to various rooms in a house. It is usually a storage unit taking off-peak current, and is dealt with in Chapter 12.

Electric Unit Heaters—These consist of a series of wire coil-heating elements, electric fan and adjustable louvres to direct the air (Fig. 3.9). They are usually fixed overhead and find a use in work-rooms, workshops etc. where no centralized system is possible.

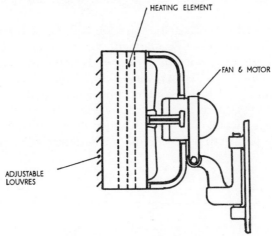

FIG. 3.9.—Electric Unit Heater.

INDIRECT SYSTEMS—LIQUID MEDIUM

Distribution of Heat—Hot water is the only liquid medium which need be considered under this heading. Water is cheap and has high heat capacity per unit volume. Its greatest merit is flexibility in control of temperature to meet variations in weather.

Its application is most common at low pressure: that is, with the system filled from a tank above the highest point of the system open to atmosphere. The water temperature is then controlled at some level below atmospheric boiling point, such as 80° C. Such a system is what is popularly called 'Central Heating'.

For large installations where larger distances have to be covered, the system is closed, so allowing pressure to be raised above boiling point. A medium-pressure system may operate up to about 130° C, and a high-pressure system up to 180° C or above. By this means the quantity of heat carried in a given size of pipe can be greatly increased. This subject is dealt with in detail in Chapter 11 on *High-Pressure Hot Water*.

Group heating or district heating exploits the possibilities of distribution of heat still further by serving a number of buildings or a whole town from one heat station. Chapter 23 explains the advantages of this. Plate II, facing p. 128, shows a pipe tunnel for distribution of services.

Whatever the method of distribution, the emission of heat requires apparatus to be fitted into the rooms to be heated. The following are representative types for use with low-pressure hot water.

Exposed Pipes—This is an old form of heating surface, and was much used in factories for economy in the past. The present tendency is to use small piping for conveying heat rather than as a heat-emitting surface. The downward radiation from overhead piping may be increased by the fitting of flat plates attached to the pipes. These are known as 'strip heaters'.

Hot-Water Radiators—Radiators are too well known to need detailed description. They are usually made of cast-iron or steel. The word 'radiator' is a misnomer, since the heat is only about 20 per cent radiant, the rest being convected. Many very slim designs of radiator are now available, having small water content and hence rapid response. Simple forms of radiator systems are shown in Fig. 3.10. For domestic use, radiators and other forms of heat emitters are often served from small-bore or 'mini-bore' piping with pump circulation as described on p. 175.

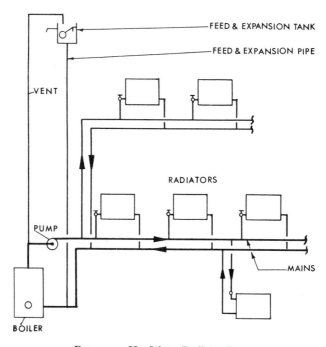

FIG. 3.10.—Hot-Water Radiator System

Flat Panels—Heating of a mainly radiant character is achieved by the use of metal panels comprising a large surface of flat plate heated in some way by the attachment (or otherwise) of waterways at the back. One of the advantages of this type of surface is its neat appearance in a room, since the surface may be treated to match the walls etc. and be inconspicuous.

Flat plate panels may be of steel or cast-iron (Rayrad), see Fig. 7.10, (page 137). Another version of the same type comprises a cast-iron waterway with a sheet steel front attached by screws. Yet another version is provided with ribs at the back to allow of convection currents assisting in the heating of the room.

Skirting Heating—The object of using skirtings for the heating of a room is to provide a well-distributed source of heat, unobtrusive in appearance. Developments by various makers make this type of system available in a convenient form. The units may be either flat-fronted to provide largely radiant heat, have apertures at the base and above to allow a certain amount of convection heat in addition or be of a type providing convection heating alone, i.e. a convector heater of small height.

Ceiling Heating—The use of the ceiling for heating originated in the 'panel' system in which pipes of 15 mm bore are embedded in the structural slab and plastered over. These were fed by hot water at low temperature (40 to 50° C), see Fig. 3.11.

A later variation was 'Panelite', in which the piping is encased in a sleeve, and water at a higher temperature (70 to 80° C) may be used.

The most common forms of ceiling heating in use today, however, make use of light metal trays perforated for acoustical effect and insulated above, being either clipped to a grid of heating pipes (Frenger, Fig. 7.15 page 142), or independently suspended below a series of pipe coils (Burgess, Sulzer). In another type, use is made of plaster panels perforated acoustically.

Characteristics of ceiling heating are cleanliness of decorations, uniformity of warmth, freedom in planning and economy in fuel consumption.

Floor Heating—Hot-water coils may be embedded in the floor screed either solid or of Panelite sheathed type. The floor surface being warmed to 24 to 27° C emits heat largely in radiant form. This system is commonly used in entrance halls, churches, banking halls and the like where occupancy is transient. Another valuable application is to the floors of buildings suspended over open-air car parks etc. where, whatever the air temperature maintained internally, discomfort can arise due to cold feet.

Hot-Water Convectors—A hot-water convector generally consists of a finned tube heater so placed as to induce an air current, as shown in Fig. 3.12. It is a neat and compact form of heating, but being entirely convective suffers from the disadvantages associated with the ordinary radiator as regards the carrying of dust and dirt by the air from the floor to the walls and breathing zone.

The absence of the radiant component makes it unsuitable for places requiring good through-ventilation.

An interesting feature of the convector is that the emission increases rapidly with the height of the flue, being nearly doubled when this is increased from 0·3 m to 1 m.

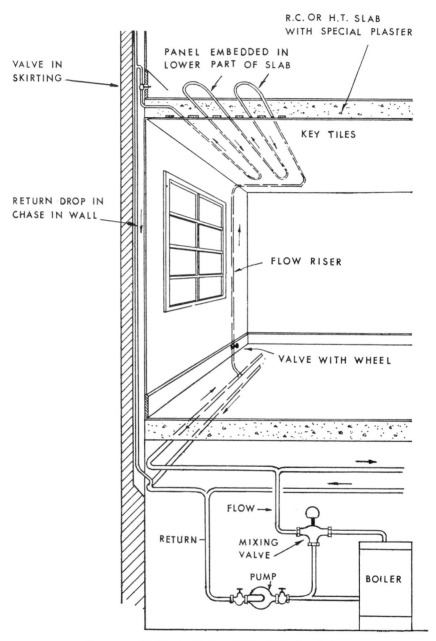

R.C. OR H.T. SLAB
WITH SPECIAL PLASTER

PANEL EMBEDDED IN
LOWER PART OF SLAB

VALVE IN
SKIRTING

KEY TILES

RETURN DROP IN
CHASE IN WALL

FLOW RISER

VALVE WITH WHEEL

FLOW →

RETURN

MIXING
VALVE

PUMP

BOILER

FIG. 3.11.—Hot-Water Embedded Panel Heating System.

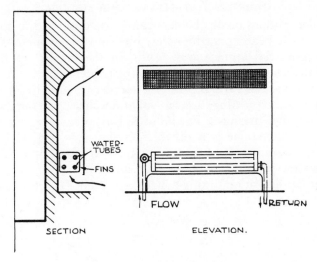

WATER-TUBES

FINS

FLOW   RETURN

SECTION   ELEVATION.

FIG. 3.12.—Hot-Water Convector.

Forced Convectors—Forced convectors are similar to natural convectors except that the air flow is produced by electrically operated fan or fans (see Fig. 3.13). The output is thus much increased in a given space, and they find particular application in schools, large halls, entrance spaces and the like. The fans are particularly quiet and are usually controlled thermostatically. A similar type of equipment has been developed for domestic use either in the form of units suitable for a single room, or for central installation to serve a whole dwelling when supplied with hot water from a centralised group or district-heating system.

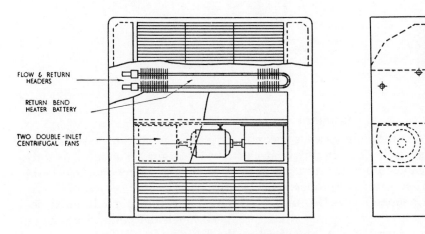

FLOW & RETURN HEADERS

RETURN BEND HEATER BATTERY

TWO DOUBLE-INLET CENTRIFUGAL FANS

FIG. 3.13.—A Forced Convector.

INDIRECT SYSTEMS—VAPOUR MEDIUM

In earlier editions of this book, steam was considered in some detail, but its use as a heating medium has died out in favour of hot water, except in industrial applications. Thus, in the context of the present chapter, it need have no more than a passing reference.

Steam was used at various pressures—under vacuum to preserve a low temperature in the emitting surfaces—at atmospheric pressure, and at low, medium and high pressures. For industrial purposes steam is commonly used in unit heaters, one type of which is shown in Fig. 3.14. Steam is also used in calorifiers (or heat exchangers) to heat hot water which is then used in low-pressure hot-water equipment.

Further consideration to this subject is given in Chapter 10 on *Heating by Steam*.

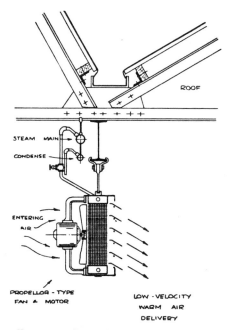

Fig. 3.14.—Suspended Type Unit Heater.

INDIRECT SYSTEMS—GASEOUS MEDIA

Air is the only gaseous medium now used in heating distribution, but due to its low specific heat capacity for a given volume it is only practicable for limited distances; otherwise the ducts sizes and heat losses therefrom become disproportionately large.

Flue Gases in Ducts (Roman System)—The culture of the Romans developed a type of heating which produced much the same effect as our radiant systems of to-day by warming the floors and walls from a furnace in the basement. Examples are brought to light from time to time in this country, and those at Bath are quite well known. Excavations at Verulamium and elsewhere show clearly that this form of central heating was an essental item of all the better-class houses of the time; yet the art was evidently completely lost when the Romans departed, and it has taken some sixteen hundred years or more to get back to their state of civilization in this respect.

Their system comprised a heating chamber below the ground from which the hot gases were conveyed in ducts under the floor to flues in the walls, emerging to the atmosphere at various points around the building. Sometimes proper ducts were formed, and in other cases the whole space under the floor, known as a 'Hypocaust', appears to have been used. The wall flues were invariably of hollow tiles, very similar to their modern counterparts. The floor consisted of a slab corresponding to our concrete slabs supported on 'pilæ', and over this a mosaic was laid. The heat was furnished by wood or charcoal. Probably the distribution of heat was not uniform, but there can be no doubt that the general effect was highly successful.

Warm Air in Floor Ducts—Part of the Anglican Cathedral at Liverpool is heated by the same means as used by the Romans, except that air, not flue gases, is circulated through ducts beneath the floor by means of a fan.

Warm Air Natural Circulation—In this system air is heated in a furnace in the basement, and rises through a grille in the floor over, by natural means, pervading the whole interior; return air is brought back to the furnace through other grilles. This system is still to be found in churches and may be fired by coke, oil or gas.

Gas Warm Air, Electricaire, Oil Warm Air—As mentioned under *Direct Systems*, these belong to a class of system which might equally be considered an indirect system as the heat is usually produced outside the rooms to be heated. Their use is preferably confined to domestic applications where the plan is such that ducts are short or are non-existent. In instances where long distribution ducts have been required, conditions of instability can occur due to differential wind pressures. In any event, the economies of the systems demand compactness.

Warm Air Forced Circulation (Plenum System)—The Plenum system of heating comprises a centrifugal fan, a heater warmed by steam, hot water, electricity or gas, and a system of ducts distributing the air to the points required, as in Fig. 3.15.

Air is drawn in at the fresh-air intake and is warmed at the heater to a temperature generally between 40° and 55° C. When this air is discharged into the room to be heated it cools to the room temperature, so giving up its heat to the walls, floor, ceiling, etc., just as the heated air from a convection radiator system cools down in its circulation through a room. There is, however, an important difference between these two systems in

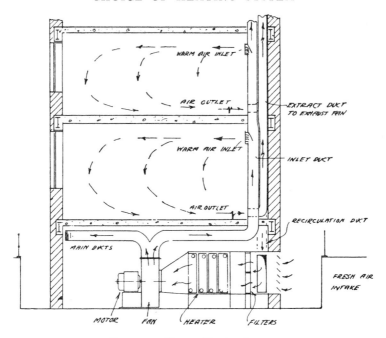

FIG. 3.15.—Plenum Heating System.

that the warmed air of the Plenum system is under a slight pressure, so that inward leakages of air through windows, etc., usual with radiator systems, are non-existent, and all air leakages are outward.

This air leakage means that a great loss of heat is continually taking place, since all air supplied by the fan has been warmed in winter from say 0° to 50° C, and has cooled only from say 50° to 18° C, so that one-third of the heat input is entirely wasted. It is therefore usual for such systems to incorporate a re-circulating arrangement by which air from the building is returned to the system for re-heating. This is usually effected by including a damper at the inlet, which can be set to give any proportion of fresh to re-circulated air. When 100 per cent. re-circulation takes place the pressure in the building is lost and infiltration will then occur, and so it is common practice to allow a maximum of 75 per cent. of air to be returned to the system.

Strong winds have the effect of counteracting the minute internal pressure generated by a Plenum system, so that on the windward side of a building air leakages inward through windows, etc., may still occur, with a corresponding drop in temperature.

Owing to the dirt brought in by the fresh air stream an air filter is generally interposed with the object of cleaning the air supply.

The system was in the past much used for the heating of factories but other systems have now largely replaced it.

Ventilating Systems—A ventilation system is not truly a heating system, and is mentioned here only to point out differences from the Plenum system.

When a building is ventilated mechanically it has a fan, heater, filter, and duct system similar to that described above, but the air is warmed only to room temperature or a few degrees below. The warming of the rooms is accomplished by direct radiation, or some other system entirely independent of ventilation.

It is necessary to remove the air from the room, and a system of exhaust ducts and fans is provided for this purpose.

The great advantage of such a combined system is that a feeling of freshness can be maintained in the rooms, and the objectionable features of a heated air supply are avoided.

FACTORS AFFECTING CHOICE

The choice can only be related to the type of building, since what may be suitable for a factory would be out of the question say for a block of flats.

Home Heating—Choice here may be influenced by personal preference, by sales pressure of various fuel interests, by close regard for economy or, where a public authority is concerned, by what is permitted by Ministry regulations.

Again, where the property exists with no more than a few open fire-places, the desire for some form of central heating may limit the choice to what is practicable.

Direct-heating systems using solid-fuel fires and stoves involve labour, dust and dirt and, with modern habits of families being out all day, are often inconvenient. Gas and electric fires are then preferred, but tend to be expensive in running cost. Gas Warm Air, and Electricaire using off-peak electricity, are then to be favoured. They can be controlled to give a set temperature at times when the house is occupied, with some minimal background warmth during the rest of the day to counteract, among other things, condensation.

Indirect systems for home heating can give whole-house comfort as well as hot water probably more consistently than any direct system. The capital cost may be higher but the running cost less, if burning an economically-priced fuel.

The question of fuel for central heating in the home, however, may not always be a matter of cost. Thus, the saving of space for an oil tank or coal store may give the advantage to gas. The saving of labour with gas or oil may rule out solid fuel. Maintenance of any fuel-burning appliance is small in comparison with fuel cost and, on the domestic scale, insignificant.

If central heating is decided upon, probably a small-bore or mini-bore system is to be recommended. The old gravity systems with large pipes are out-dated.

Some may prefer floor heating in a new house, avoiding convection-current marking of walls. In an old house, skirting heating may be less obtrusive than radiators, although the latter will be cheaper. If floor heating is preferred, electric off-peak may be a simple answer, but great care with insulation will be required to minimise consumption.

Control systems for home heating have been developed to a high degree of sophistication, being adjustable as to times and temperature in a variety of ways. Intending purchasers should not be misled by claims of one particular fuel or other interest, since what can be done by one can probably be done by all.

Thus, as will be seen, the choice for house heating is a bewildering one, there being in effect many answers to what, in all the circumstances, is the best. It is impossible to generalise except in the matter of cost of heat, but this again is dependent to some extent on the system and how it is run. Further reference to this is made later.

Flats—Multi-storey—Systems of heating in common use and from which a choice would no doubt be made are:

Electric floor-warming off-peak;

Electric warm air;

Gas warm air, subject to any embargo on gas in high buildings;

Oil warm air from common storage;

Central heating by low-pressure hot-water radiators;

Hot-water warm air from one fan convector per flat served from boiler on roof or in basement of each block;

Group or district heating by hot water to a number of blocks.

The housing authority or estate developer in making a choice will be influenced by the capital cost, fuel cost, maintenance costs and labour to run. There is also the question of how the heat is to be charged— whether by the public utility reading its own meters (the consumer paying direct), or whether the landlord is responsible for reading meters and collecting the money, or, again, whether the cost of heat is included in the rent.

All these matters have to be considered as well as the class of tenant and what kind of expenditure can be afforded before a recommendation can be made.

Commercial—Offices being usually in blocks are most economically heated by a central system in some form. Choice is then confined to the kind of emitting surface, radiators, convectors, ceiling heating etc., and to the kind of fuel.

For reasons already explained, an office block with larger expanses of glass and light construction will be subject to rapid swings of temperature;

hence a system which is quickly responsive to change of output is needed. Thus, embedded-coil systems, or electric-floor warming, are not to be recommended although they may find a place in buildings of heavy construction.

Small blocks, where a central plant may be unsuitable for a variety of reasons, will probably best be served by one of the direct systems such as electric block-storage heaters, gas fires or convectors.

A study of probable temperature rise in modern office blocks reveals a tendency to overheating in summer, due to solar heat gains in addition to heat from office machinery and personnel. It is then necessary to consider air-conditioning instead of heating alone, which matter is developed in Chapter 17.

Public Buildings and Schools, Universities, Halls of Residence, Swimming Pools, Hospitals, Hotels etc.—This general class of substantial buildings will in most cases be in the hands of a consulting engineer, or public authority engineer, who can be expected to advise which system and fuel should be used. Matters to be reported on will cover, among other things, architectural appearance, the need for ventilation, capital cost, effect on redecoration costs, fuel, plant space, controls, labour costs, maintenance costs, noise and amenities generally.

Industrial—Here the choice will be largely governed by economics. Which system will produce adequate conditions with minimum upkeep, fuel consumption and labour?

If steam is produced for process work, the choice may fall in the direction of using steam for heating also—unit heaters being then one answer. If no steam is required, hot water at medium pressure or high pressure is to be preferred. For heat emission, unit heaters may again be selected on account of low cost, but they involve maintenance which does not apply with radiant panel or strip heating.

Direct oil or gas fired air heaters meet certain cases where a boiler plant is not desired or where extensions cannot be dealt with from existing boilers.

Choice of Fuel—The basic cost of heat energy produced from a given quantity of fuel at 100 per cent. efficiency depends on the cost per unit quantity and the heat energy content therein. Simplicity is achieved if basic costs are reduced to a common unit, namely the gigajoule (GJ)*.

Electricity

The unit of electricity is the kWh $= 3 \cdot 6$ MJ or $0 \cdot 0036$ GJ

If the price per unit, in pence (p)† is

1	p,	the cost of heat	$=$	280 p/GJ
0·5	p,	,, ,, ,, ,,	$=$	140 p/GJ
0·3	p,	,, ,, ,, ,,	$=$	84 p/GJ

* 1 GJ = 10 therms of 100 000 British Thermal Units (approx.)

† 1p = 2·4d

Gas

Assuming that gas is charged in MJ (and at the time of going to press this question is unresolved), then:

if the price per MJ, in pence (p) is

0·1 p,	the cost of heat	=	100 p/GJ
0·05 p,	,, ,, ,, ,,	=	50 p/GJ
0·03 p,	,, ,, ,, ,,	=	33 p/GJ

Oil

The heat content of 1 litre will depend upon the grade of oil and its density. Taking Class D (gas oil) the heat content = 0·045 GJ/litre.

Thus, if the price of oil per litre, in pence (p) is

2 p,	the cost of heat	=	45 p/GJ
1·5 p,	,, ,, ,, ,,	=	33 p/GJ
1 p,	,, ,, ,, ,,	=	23 p/GJ

Solid Fuel

The heat content of solid fuel will vary with the type and source. As an average, with a calorific value of 0·03 GJ/kg, a figure of 30 GJ/tonne may be assumed.

Thus, if the price of solid fuel per tonne is

£15,	the cost of heat	=	50 p/GJ
£10,	,, ,, ,, ,,	=	33 p/GJ
£5,	,, ,, ,, ,,	=	17 p/GJ

Of course, these figures do not represent the cost of the heat arriving in the room, except with direct electric heating. In an indirect central system there is a loss from piping—mains loss—and a loss in combustion in the boiler due to hot gases leaving by the flue. The first loss may be 10 per cent., the second loss 25 per cent. with gas and oil, and somewhat more with solid fuel. If, for simplicity, we take the overall loss as 30 per cent.—a system efficiency of 70 per cent.—the costs above become as shown in Table 3.2.

Other factors affect fuel consumption, such as controllability, running hours and no-load losses. For instance an off-peak electric system may have losses during the night when otherwise heating would be off.

Each case merits separate evaluation. The costs arrived at in Table 3.2 may, however, be useful as a guide and can be adjusted *pro rata* up or down as necessary according to the actual fuel charges and estimated efficiencies in use.

TABLE 3.2

Cost of Heat from Various Fuels

Type of fuel and unit cost	p/GJ
Electricity (direct heating, 100% efficiency)	
at 1·0p per unit - - - - -	280
at 0·5p per unit - - - - -	140
at 0·3p per unit - - - - -	84
Gas (70% efficiency)	
at 0·1p per MJ - - - - -	143
at 0·05p per MJ - - - - -	71
at 0·03p per MJ - - - - -	48
Oil (70% efficiency)	
at 2·0p per litre - - - - -	60
at 1·5p per litre - - - - -	44
at 1·0p per litre - - - - -	30
Solid fuel (70% efficiency)	
at £15 per tonne - - - - -	71
at £10 per tonne - - - - -	47
at £5 per tonne - - - - -	24

F

CHAPTER 4

Boilers and Combustion

IF THE CHOICE OF system has been to adopt some form of central heating using hot water as the medium of transmission, the next step is to determine the size of boiler plant, where it shall be located, which fuel is to be used and what flue is required.

Boiler Power— The sizing of the boiler plant is to be such as to meet the heat losses from the building under basic design-temperature conditions. As mentioned already, the total of the maximum heat losses may exceed the actual peak demand, due to the fact that infiltration air entering rooms on one side of the building leaves by rooms on the other side, so that a correction should be made to avoid taking the same air-change twice. In addition, where heating is continuous, it has been proposed that diversity factors apply as follows:

Single space	1·0
Single building or zone controlled centrally	0·9
Single building individual room control	0·8
Group of buildings with similar use	0·8

The boiler output must, however, be in excess of the corrected total in order that the design temperature can be achieved in a reasonable time. If heating is continuous, the excess will be a minimum; the more intermittent the usage, the greater excess capacity needed.

There is a further case for some margin of capacity, where a boiler is thermostatically controlled, in order to provide what may be termed 'acceleration'—the ability of the boiler to surmount the load under peak conditions and still be under control. A still further purpose of a margin is to deal with the declining efficiency of a boiler due to fouling of the heating surfaces by soot or ashy deposit, especially where solid fuel is used.

The reduction on air change may be set against the increase required for a margin for the reasons stated, and to some extent these cancel out. In the past, these refinements of calculation have generally been ignored and a margin of around 25 per cent added to the total of the heat losses and mains losses to cover a multitude of sins. Where time is short for design and estimating, this rough and ready method may suffice, but it would be worth while in due course checking with the data now available.

By way of example, using the approximate method, let us
assume that the heat loss total for the building is 500 kW
Also let us assume the heat losses from main piping in
trenches etc. is 10 per cent of this figure 50 kW

_____

550 kW

Allow margin of 25 per cent 137 kW

_____

687 kW

_____

Reference to makers' catalogues may show that the nearest boiler size
up is 750 kW. If this is chosen, the margin would be 36 per cent and might
be considered excessive. The next size down might be 650 kW, still giving
a margin of 18 per cent which would be more reasonable.

In the old days of hand-stoked boilers a big boiler was considered an
advantage as it meant longer periods of run on one charge of fuel. Nowadays, with automatically-fired boilers burning oil or gas, it is often a disadvantage to overdo the boiler power as overall efficiency may be lowered.
If the requirement has been accurately calculated, a margin of 10 to
15 per cent may be adequate. Any margin is likely to be a boon during
exceptionally cold spells.

Boiler Ratings and Margins—Boilers are to be rated in kW* and fall
broadly into the following ranges:

Small 10–50 kW, mainly domestic
Medium 50–500 kW
Large 500–2500 kW

In any sizeable installation over say 300 kW there is a case for providing
more than one boiler. If there are two boilers, one will suffice in mild
weather working near its full output, which has advantages in efficiency
and avoidance of corrosion. The second is brought in during cold weather
and can also act as a standby during breakdown or cleaning. In the past,
it has been customary with two boilers to arrange each to take two-thirds
of the net heat load; thus, together, they have $33\frac{1}{3}$ per cent margin, though
this might well be reduced due to the diversity which undoubtedly exists
the bigger the installation.

It will be obvious that for very large installations three, four or more
boilers may be used, giving greater flexibility than with fewer larger units.
Several boilers with a small margin on each then give almost a complete
one-boiler standby.

Selection of the size of individual units for a multi-boiler installation is
a matter for compromise. Limitation of the number of spare parts to be

* for convenience of those more familiar with the Btu:

	10 000 Btu/hr = 2·93 kW or rounded up 3 kW
thus	100 000 Btu/hr = nearly 30 kW
	1 000 000 ,, = ,, 300 kW
	10 000 000 ,, = ,, 3000 kW

stocked suggests the use of equally sized units (e.g. three boilers at 166 kW each for a total requirement of 500 kW). Conversely, to obtain maximum efficiency whilst meeting a load which varies from day to day with the external temperature, there is a case for sizing the units unequally in order to provide 'steps': e.g., for the same total requirement, one boiler at 100 kW and two at 200 kW each would allow an output at 100, 200, 300, 400 and 500 kW as required (five 'steps' as against the three provided by equal-sized boilers).

Hot-Water Supply—Although this is dealt with in a later chapter, it is customary where a central system is installed to couple with it the supply of hot water. The same boiler or boilers can serve both heating and hot water by means of a heat exchanger or calorifier, so that, during the heating season, the hot-water load is added to that for heating. Bearing in mind that hot-water demand is spasmodic, it is possible to supply it by slight 'robbing' of the heating for short periods.

In summer, when heating is 'off' in a single-boiler installation, it becomes a question of proportion as to whether the hot-water supply load is sufficient to warrant the running of the boiler; if not, then a small boiler for summer use is required. In a multi-boiler installation it is likely that one boiler may be of a smaller size than the others for the same purpose.

BOILER TYPES

Cast Iron—By far the most common type of heating boiler is the cast-iron sectional, which has been in use for nearly a century. Originally designed for burning coke on a set of fire bars, it has been developed and refined

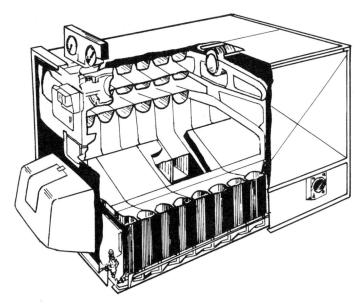

FIG. 4.1—Cast-iron Sectional Boiler (Crane Ltd.)

for application of oil firing or gas firing at high efficiency. Sizes range from the smallest up to 3000 kW.

The boiler is made up of sections usually connected together by means of three nipples, one at the top and two at the bottom, all pulled together and held water-tight by steel tie-bars externally. Some of the larger types are in two halves, having four nipples. Back and front sections differ from the intermediate ones as they make provision for firing, cleaning and outlet for products of combustion. Sections may be added for extensions or replaced on failure.

With the decline in availability of coke and of labour for stoking, this method of firing may soon become a thing of the past. Fig. 4.1 shows a type designed for oil or gas firing in which the flue passages are arranged for higher gas velocities and hence greater output for a given size. This type has a waterway bottom instead of being open, and hence the base of brick or concrete on which it stands is subject only to the relatively low temperature of the water, otherwise an insulating base is necessary to prevent undue temperature build-up in the floor structure.

The normal cast-iron sectional boiler is designed for working water pressures up to about 40 m head of water, which covers most normal heights of building; but certain makes for larger sizes, using a special grade of cast iron, are suitable for heads up to 100 m of water. Test pressures are double the above.

Cast iron is less prone to attack from sulphurous corrosion products arising from combustion of fuel, especially oil, than is steel.

Steel—Wrought iron was at one time a favourite material for the manufacture of boilers, being very ductile and resistant to corrosion. The art

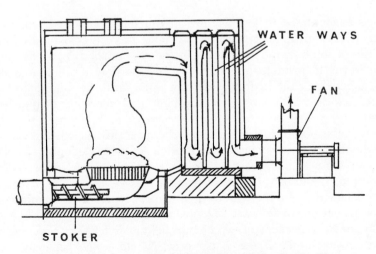

FIG. 4.2—Sectional Diagram of Welded-Steel Boiler for Hot Water with Underfeed Coal Stoker (Centrax).

of 'puddling' of iron being largely a manual process, however, wrought iron has disappeared from the market and steel has taken its place. Steel is homogeneous in structure and, unlike wrought iron, is very liable to attack by sulphurous corrosion products.

For long life, a steel boiler requires the system which it serves to be designed to operate at a temperature outside the region where severe attack is liable to occur. Given these conditions, steel is probably more versatile than cast iron—which feature has been taken advantage of in the wide range of designs. One example is illustrated in Fig. 4.2. In this type of boiler the heating surfaces are formed by welding together pressed-steel plates; other types use fire tubes for the convection surface.

Sizes in a variety of forms range from about 30 to 3000 kW, above which, tubular and shell types are used, as referred to in Chapter 11, these being applicable to higher pressures.

Concrete—Experiments have been made in France using concrete as a material for sectional-type boilers. A special aggregate is used and each section is cast in halves to a pattern, including flue-ways for the passage of the combustion gases. Within the concrete, an extended coil of steel pipe is embedded, and this has terminal ends brought outside the half section at top and bottom. A great advantage would seem to be that no refractory lining is required to the combustion chamber for firing with oil or gas and, since the actual concrete reaches red heat, combustion is complete and virtually no cleaning of the flue-ways is needed.

COMBUSTION OF FUEL

It is desirable that the principles of combustion be understood before considering methods of firing; but the subject is involved, and the treatment herein should be regarded as an outline as affecting heating boilers only.

Fuels encountered in practice are mixtures or compounds chiefly of carbon, hydrogen and oxygen. In addition, there are generally small quantities of sulphur, nitrogen and—in the case of solid fuels—ash. The carbon burns to form carbon dioxide (and possibly small amounts of carbon monoxide), the hydrogen to form water, and the sulphur to form sulphur dioxide which later may combine with more oxygen to form trioxide.

The combustion of fuel is a physico-chemical process which is accompanied by the liberation of heat. Combustion reactions can only take place at a high temperature known as the *ignition temperature*, which varies between 400° C and 600° C according to the fuel.

The elements combine with oxygen in proportion to their molecular weights, which are as follows: $O_2 = 32$, $C = 12$, $H_2 = 2$, $S_2 = 64$, $N_2 = 28$. Air contains 23·15 per cent. of oxygen by weight and 76·85 per cent of nitrogen, etc. (by volume: $O_2 = 20·8$ per cent, N_2 etc. $= 79·2$ per cent). The basic combustion reactions are given in Table 4.1.

TABLE 4.1

COMBUSTION REACTIONS

Reaction	Products	Requirement for combustion kg/kg of combustible		Heat Liberated MJ/kg of Combustible
		Oxygen	Air	
$C + O_2 = CO_2$ $12 + 32 = 44$		2·7	11·6	34
$2C + O_2 = 2CO$ $24 + 32 = 56$		1·3	5·8	10
$2H_2 + O_2 = 2H_2O$ $4 + 32 = 36$		8·0	34·6	145
$S_2 + 2O_2 = 2SO_2$ $64 + 64 = 128$		1·0	4·3	9
Methane $CH_4 + 2O_2 = CO_2 + 2H_2O$ $16 + 64 = 44 + 36$		4·0	17·3	56
Ethylene $C_2H_4 + 3O_2 = 2CO_2 + 2H_2O$ $28 + 96 = 88 + 36$		3·4	14·8	49

The Calorific Value of a fuel is the quantity of heat released on the complete combustion of unit weight. There are two such values always given—*gross* or higher, and *net* or lower. The *gross calorific value* includes the heat given up in the reaction to supply the latent heat of vaporization to any water forming part of the products of combustion. The *net calorific value* is the gross value minus the latent heat referred to. The greater the amount of hydrogen in the fuel the greater the difference between the two values.

The calorific value of a fuel may be calculated approximately from a knowledge of its analysis and the heat due to the reaction of oxygen with each element, but it is necessary to check any such computation by experimental determination.

Solid Fuel—Solid fuels contain carbon as *fixed* or uncombined carbon, or as *volatiles* or hydrocarbons. The former burn direct to form CO and CO_2 giving a short blue flame, the volatiles give a yellowish flame which is liable to cause smoke if cooled too quickly on water-backed boiler surfaces or by cold air entering above the fire. As the volatiles are consumed, the residue is largely coke.

These fuels also contain hydrogen not combined with carbon, referred to as *inherent moisture. Free moisture* is that on the surface or in the interstices often augmented by rain in open trucks. Whilst the moisture content in both forms constitutes a loss in available heat, it appears to have some beneficial effect on combustion with some coals such that arrangements are sometimes made to spray the coal before firing, or to admit steam below the fire bars.

Coal contains	40 to 50% fixed carbon
	30 to 40% volatiles
	10% moisture
	1 to 1·9% sulphur
	the remainder ash
Anthracite contains	about 90% fixed carbon
	5% volatiles
	2% moisture
	1% sulphur
	the remainder ash
Coke contains	about 84% fixed carbon
	3% volatiles
	7% moisture
	$1\frac{1}{2}$% sulphur
	the remainder ash

Typical gross calorific values are:

Coal	about 26 to 33 MJ/kg net
Anthracite	,, 34 MJ/kg net
Coke	,, 29 ,, ,,

The *proximate analysis* gives the percentage mass of fixed carbon, volatiles, moisture and ash, as above. From this the probable characteristics of combustion can be forecast. The *ultimate analysis* gives the percentage mass of the various elements or compounds in the sample. From this the theoretical air for combustion may be estimated and the combustion efficiency.

Assume a sample of coal has the following analysis:

		Air for combustion
Carbon	80%	$0·8 \times 11·6 = 9·28$
Hydrogen	4%*	
Oxygen	1·1%	$\left(0·04 - \dfrac{0·011}{8}\right) \times 34·6 = 1·33$
Sulphur	1·0% (ignored)	—
Ash, moisture etc	13·9%	
	100·0%	10·61 kg/kg fuel

$$\text{of this } O_2 = \ 2·46$$
$$N_2 = \ 8·15$$

The hydrogen combining with oxygen to form steam is condensed out before the CO_2 is sampled and therefore does not enter into the assessment of CO_2 percentage. The maximum CO_2 content possible from combustion of this sample with no excess air is:

---

* The oxygen burns one-eighth of its equivalent in hydrogen:

i.e. $4 - \dfrac{1·1}{8} = 3·875$ per cent hydrogen is left requiring air for combustion.

$$
\begin{array}{lcl}
\text{C} & = & 0\cdot80 \\
\text{O}_2\ 2\cdot7 \times 0\cdot8 & = & 2\cdot16 \\
\hline
\text{CO}_2 & = & 2\cdot96 \\
\end{array}
$$

by volume:

$$
\begin{array}{lcl l}
 & & & \% \\
8\cdot15 \quad \text{N}_2 \div \text{mol wt } 28 & = & 0\cdot290 & 81 \\
2\cdot96\ \text{CO}_2 \div\ \text{,,}\ \text{,,}\ 44 & = & 0\cdot068 & 19 \\
\hline
 & & 0\cdot358 & 100 \\
\end{array}
$$

The maximum CO_2 from combustion of this sample of coal will thus be 19 per cent, assuming no CO or other unburnt products. If the air supply is in excess, it can be shown that the CO_2 percentage with 50 per cent excess is $11\cdot5$ and 100 per cent is $8\cdot5$. The method will be clear from the following section on *Oil Fuel*.

The theoretical calorific value will be:

$$
\begin{array}{lll}
\text{C } 0\cdot8 \times 34 & 27\cdot2 \\
\text{H } 0\cdot04 \times 145 & 5\cdot8 \\
\text{S} \quad\text{—} & \text{—} \\
\hline
 & 33\cdot0 \text{ MJ/kg} \\
\end{array}
$$

Oil Fuel—A sample of oil may have the following analysis by mass:

Carbon 86%   Hydrogen 11%   Sulphur 3%

The theoretical air for combustion would then be (using Table 4.1):

$$
\begin{array}{lccl}
 & \begin{array}{c}\textit{kg air/kg} \\ \textit{combustible}\end{array} & & \textit{air, kg} \\
\text{C—CO}_2 & 11\cdot6 \times 0\cdot86 & = & 10 \\
\text{H—H}_2\text{O} & 34\cdot6 \times 0\cdot11 & = & 3\cdot79 \\
\text{S—SO}_2 & 4\cdot3 \times 0\cdot03 & = & 0\cdot13 \\
\hline
\text{Total air kg/kg fuel} & & & 13\cdot92 \\
\end{array}
$$

$$
\begin{array}{lcl}
\text{Proportion of O}_2\ 23\cdot15\% \text{ mass} & = & 3\cdot21 \\
\text{Density of air at } 20^\circ \text{ C} & = & 1\cdot2 \text{ kg/m}^3 \\
\text{Thus volume per kg fuel} & = & \dfrac{13\cdot92}{1\cdot2} = 11\cdot6 \text{ m}^3 \\
\end{array}
$$

The theoretical calorific value would be:

$$
\begin{array}{llll}
\text{C} & 0\cdot86 \times 34 & = & 29 \text{ MJ/kg} \\
\text{H} & 0\cdot11 \times 145 & = & 16 \quad\text{,,} \\
\text{S} & 0\cdot03 \times 9 & = & \text{—} \\
\hline
 & & & 45 \quad\text{,,} \\
\end{array}
$$

The actual calorific value would be obtained experimentally, as mentioned earlier, typical values being given for various grades of oil in Chapter 5 (page 91).

Assume sample as before contains 86% carbon:

Air for combustion			$=$	13·92 kg/kg
of this O_2			$=$	3·21 ,,
thus N_2			$=$	10·71 ,,
C		$=$	0·86	
O_2 2·7 × 0·86		$=$	2·3	
				3·16

(The balance of O_2 (3·2–2·7) is in the hydrogen and sulphur reactions)

				%
by volume 10·71 N_2 ÷ mol wt 28	$=$	0·38		84·5
3·16 CO_2 ÷ ,, ,, 44	$=$	0·07		15·5
		0·45		100

The maximum CO_2 from combustion of this sample of oil will thus be 15·5 per cent, assuming there is no CO or free O_2 in the products. If the air supply is in excess:

	25%	50%	75%
CO_2 as above	3·16	3·16	3·16
N_2 10·7 × 1·25 etc.	13·4	16·1	18·7
O_2 3·2 × 1·25 etc.—3·2	0·8	1·6	2·4
by volume			
CO_2 ÷ 44	0·07	0·07	0·07
N_2 ÷ 28	0·48	0·57	0·64
O_2 ÷ 16	0·05	0·10	0·15
	0·60	0·74	0·86
in per cent			
CO_2	12	9·5	8·2
N_2	80	77·0	74·5
O_2	8	13·5	17·3
	100	100·0	100·0

Gaseous Fuels—A typical natural gas (such as North Sea) contains by volume, 90 per cent methane, 7 per cent hydrocarbons and 3 per cent nitrogen. Table 4.1 shows that the heat liberated by the combustion of methane is 56 MJ/kg. Thus, using a specific mass of 0·72 kg/m³ for this constituent,

$$\text{volumetric calorific value} = 56 \times 0·72$$
$$= 40 \text{ MJ/m}^3$$

The measured calorific value of natural gas varies little from this calculated figure, the other constituents being unimportant. The air for combustion may likewise be taken from Table 4.1 as 17.3 kg/kg gas. Thus, the specific mass of gas being 62 per cent of that for air,

$$\text{air volume for combustion} = 17 \cdot 3 \div 1 \cdot 2$$
$$= 14 \cdot 3 \ m^3/m^3 \ gas$$

Combustion of Gas—Taking natural gas as pure methane,

C   H₄
12   4   =   16   C   =   75%   H   =   25% by mass

Air for combustion 1 kg gas   = 17·3 kg
  N_2 17·3 × 0·77   = 13·4 kg
  O_2 2·7 × 0·75   =   2·0
  C   =   0·75   = 2·75 kg CO_2

by volume 13·4 N_2/28   =   0·48   89%
          2·75 CO_2/44   =   0·06   11%

                                0·54   100%

The maximum CO_2 in this case is therefore 11 per cent.

Towns Gas contains by volume typically:

Carbon Dioxide	3·2%
Carbon Monoxide	15·0 ,,
Methane	21·8 ,,
Hydrogen	46·9 ,,
Hydrocarbons	3·0 ,,
Nitrogen	9·0 ,,
Oxygen	1·1 ,,

It has a typical calorific value of 18·6 MJ/m³ (i.e. approximately half that of natural gas) and a specific mass of 45 per cent to 50 per cent of that for air. The air required for combustion is approximately 8·8 m³/m³ gas.

Flue Gas Analysis—Analysis of flue gases resulting from combustion of fuel is the way used to assess excess air and thus 'efficiency' of burning.

The standard apparatus is the *Orsat*, by which a sample of the gas is first dried to remove water vapour, and is then subjected to adsorption by three liquids in turn. The first is caustic soda which adsorbs CO_2. The second is an alkaline solution of pyro-gallol which adsorbs O_2. The third is acid cuprous chloride which reacts with CO. Measuring burettes enable the proportion of gas adsorbed in each case to be measured and hence the volumetric content of the gases present.

Maximum CO_2 contents may be set down approximately thus:

Pure carbon	21·0%
Coke	20·0 ,,
Coal 80% carbon	19·0 ,,
Oil 86% carbon	15·5 ,,
Natural gas (as methane)	11·0 ,,

The presence of CO, as mentioned earlier, indicates incomplete combustion and, if present in any quantity, steps should be taken to reduce it to zero.

More convenient instruments than the Orsat for day to day use are now available for determining CO_2 content when setting up and adjusting automatic boilers. For permanent indication and recording in large boilerhouses, various types of automatic instruments are used, some depending on relative electrical or thermal conductivity of gases, some on relative mass, and some on chemical reaction as in the Orsat.

Flue Gas Temperature—Gases at boiler exit are measured by pyrometer or high temperature thermometer, which may be nitrogen filled mercury in glass, mercury in steel, or electrical type.

Chimney Loss—The chimney loss is directly proportional to the mass gas flow, the temperature of gases leaving the boiler and the specific heat capacity of the gases. The loss also includes the latent and sensible heat of the steam content due to combustion of hydrogen in the fuel, the steam being in a superheated condition. The mass gas flow varies according to the excess air quantity, which in turn can be assessed from the CO_2

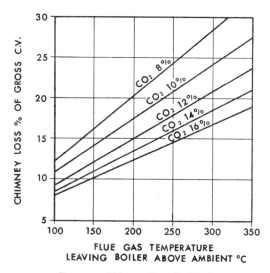

FIG. 4.3—Chimney Loss for Coal.

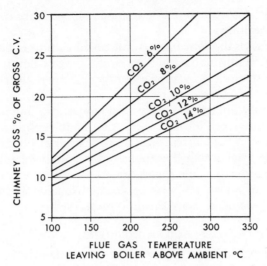

FIG. 4.4—Chimney Loss for Oil.

analysis. The formulae for chimney loss are involved, and hence it is usual to refer to graphs prepared for particular fuels such as are to be found in the *I.H.V.E. Guide*; one for coal is as Fig. 4.3, and one for oil is as Fig. 4.4.

Unless some excess air is admitted, combustion will be incomplete, producing CO and soot. Fuels containing hydrocarbons (coal and oil) will produce smoke.

If the flue gas temperature is too low, there will be condensation, with serious corrosion hazard in the boiler. There is a water dew-point at which any steam in the gases from the combustion of hydrogen is deposited. This is generally in the region of 50° C. In the case of sulphur-bearing fuels (coal, oil) there is an 'acid dew-point' in the region 100° to 130° C, where the sulphur being converted from SO_2 to SO_3 or SO_4 combines with water, forming sulphurous or sulphuric acid. There is a great volume of research on this topic which may be studied by those interested.

In the practical low-pressure hot-water heating boiler, the endeavour is to keep the temperature of the water-backed surfaces up, so that they are above the water dew-point but below the acid dew-point—generally about 80° to 85° C. Boilers serving high temperature water systems are best operated at 140° C or over.

Flue gases, on leaving the boiler, may condense in the flue system (which is referred to later), but, so far as the boiler is concerned, the practical lower limit of exit temperature is generally about 200° C.

SMOKE INDICATION

The *Ringelmann* charts are a series of four grids of increasing darkness marked on cards 100 mm square, against which the smoke from a chimney may be compared. The Clean Air Act prohibits smoke darker than No. 2

on the Ringelmann scale, but where smokeless zones are in force smoke of any shade is prohibited.

Another scale is the *Bacharach*. A controlled volume of flue gas at a constant rate is passed through a filter paper and the discoloration is compared with standard shades 0 to 9 smoke numbers. This is clearly a laboratory instrument.

Smoke-density indicators consist of a light source on one side of a flue and a light sensitive cell on the other. Any shade causing a diminution of the light received by the latter is caused to register electrically on a dial in the boilerhouse from which the smoke number (Ringelmann) is read. In addition, a recorder may be incorporated for use as evidence in case of dispute. Also, an alarm bell is incorporated to give audible warning if smoke is being produced above the limit prescribed.

BOILER EFFICIENCY

The efficiency of a boiler expressed as a percentage is

$$\frac{\text{Heat output} \times 100}{\text{Heat input}}$$

Heat Input—With steam boilers, the water pumped into the boiler and evaporated is conveniently measured by a water meter. From the inlet-water temperature and outlet-steam pressure, the total heat per unit quantity of water evaporated may be determined, and hence the total heat output. With hot-water boilers, measurement of water flow through the boiler is more troublesome and costly due to the large volumes involved and is not commonly included except in large installations. Where such measurement is made, the quantity, multiplied by the temperature difference inlet to outlet, gives the total heat output per unit of time.

Heat input is measured according to the fuel: solid fuel by weighing and deducting the weight of ash, liquid fuel by meter or tank gauge, and gaseous fuel by gas meter. Calorific values of fuel and ash must be determined. Detailed methods of carrying out tests are covered by B.S. 845. Duration of tests is generally six hours with one control hour before and after. From such tests, including combustion analysis, a heat balance may be struck. The following is an example with a certain class of coal:

Overall thermal efficiency	78%
Loss due to sensible heat in chimney gases	15 ,,
Loss due to unburnt CO	1 ,,
Combustible matter in ash and clinker	2 ,,
	96 ,,
Radiation and other losses	4 ,,
	100%

Boiler test efficiencies with modern automatically-fired and controlled plant range between 75 and 85 per cent. Hand-fired boilers are not capable of sustained high efficiencies: no more than 65 per cent may be expected. Seasonal efficiencies may be 10 per cent less than these values.

AIR SUPPLY TO BOILERS

It has been shown above how the theoretical air supply for combustion of fuel is determined. Table 4·2 gives a summary.

TABLE 4.2

THEORETICAL AIR FOR COMBUSTION PER KG OF VARIOUS FUELS

Fuel	Theoretical air required	
	in kg	in m³ at 15° C
Oil	13·9	11·5
Bituminous Coal	10·6	8·8
Anthracite Coal	9·6	8·0
Coke	9·1	7·6
Natural Gas	17·3	14·3
Towns Gas	10·6	8·8

In the design of a boilerhouse, provision must be made for the theoretical air to enter plus excess air, which, due to lack of adjustment, may be as high as 100 per cent. Thus the theoretical quantities may well be doubled. The entering velocity through grilles or openings of various sorts may be sized on a velocity of 2 m/s. For example, for combustion of 100 kg of oil per hour, the free area of inlet-air openings should not be less than:

$$\frac{100 \times 11·5 \times 2}{2 \times 3600} = 0·32 \text{ m}^2$$

CHIMNEY SIZING

The calculation of the size and height of a chimney can be a very involved matter, but for the purposes of heating boilers it may be simplified by reducing the number of variables. The following notes show such a method.

1. *Products of combustion. Quantity*

Assume excess air   -   -   -   -   75%
Average temperature in chimney
above ambient   -   -   -   - 200–260° C
Boiler efficiency   -   -   -   - 75%

It can be shown that the products of combustion per MJ of boiler output are, by volume for coal, coke or oil, approximately,

at 200° C   -   -   -   -   - 1·3 m³/MJ
at 260° C   -   -   -   -   - 1·4 m³/MJ

These temperatures cover the range of most heating boilers but volumes at other temperatures may be derived in proportion to degrees Kelvin. Flues for gas appliances are discussed in Chapter 6.

2. *Time*

Bringing a time scale into consideration, the above (for one second) become the products deriving from fuel burnt at a rate to give an output of 1 MJ/sec:

$$1 \text{ MJ/sec} = 1 \text{ MW } (1000 \text{ kW})$$

3. *Velocity of flue gases in chimney*

Assume the range of velocity with natural draught from 5 to 8 m/s; with mechanical draught, from 10 to 12 m/s.

4. *Area*

The cross-sectional area may then be derived according to boiler duty, fuel and type of draught. Thus, if the boiler duty is 100 kW, oil-fired, with natural draught (5 m/s) chimney temperature average 260° C,

$$\frac{1 \cdot 4}{10 \times 5} = 0 \cdot 028 \text{ m}^2$$
$$= 190 \text{ mm diam.}$$

Fig 4.5 gives areas and diameters direct, having selected a velocity.

5. *Efflux Velocity*

In order that the plume of gas should rise clear of the chimney top and not flow down the outside of the chimney (down-wash), the diameter at the top should be reduced so as to maintain as high a velocity as practicable. For small boilers with natural draught a velocity of 6 m/s is advised. Larger boilers with mechanical draught should achieve 7·5 to 15 m/s.

The velocity head corresponding to these rates may be read from Fig. 19.14 (page 435).

6. *Draught required*

At boiler exit consult makers' data.

Oil-fired boilers vary from	7 to 50 N/m²
Solid fuel fired, if burning rate is 5 kg/m² grate area	70 N/m²
Flue connection boiler to chimney (depending on number of bends and other losses) average	15 to 30 N/m²

Note: The following may be calculated from the data on duct sizing in Chapter 19:

Efflux velocity head for 6 m/s	22 N/m²
The total for an oil-fired boiler may then be from about	40 to 100 N/m²
,, ,, ,, a solid fuel ,, ,, ,, ,, ,, ,,	100 to 120 N/m²

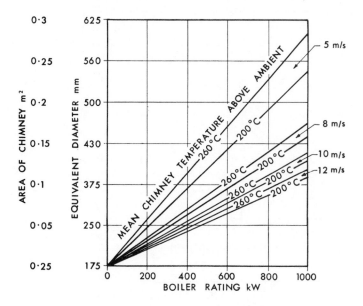

FIG. 4.5—Chimney Areas for Stated Velocities.

7. Draught produced by Chimney

The theoretical draught of a chimney at the two mean temperatures named is

per metre of height at 260° C   4·8 N/m²

  ,,   ,,   ,,   ,, at 200° C   4·0 N/m²

Fig. 4.6 is drawn on this basis.

Thus a chimney of 30 m height will produce theoretically a draught of 144 N/m² at 260° C.

8. Draught loss in Chimney

The pressure loss per metre height may be taken from Fig. 4.7. Values for other velocities may be interpolated.

Fig. 4.8 gives the pressure loss for chimneys of given heights having determined the loss per unit length from Fig. 4.7.

9. Rectangular Equivalents

The sides of a square flue of brick or concrete etc. may be taken as the same as the diameter of a circular flue; the resistance will be taken as rough. If the flue is rectangular, the area can be judged from that of the inscribed figure with circular ends, it being generally a rule that a ratio of sides of 3:1 should not be exceeded.

G   K.H.A.C

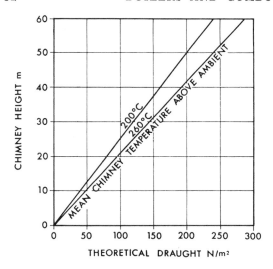

FIG. 4.6—Theoretical Draught for a Chimney of a Given Height.

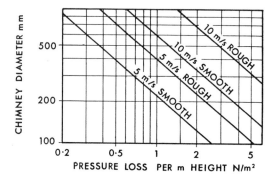

FIG. 4.7—Pressure Loss per metre Height of a Chimney.

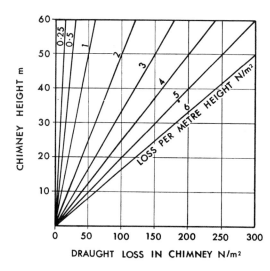

FIG. 4.8—Draught Loss for a Chimney of a Given Height.

10. *Procedure*

(1) Using Fig. 4.5 select a velocity according to whether draught is natural or mechanical and find chimney area and diameter.

(2) From Fig. 4.7 determine draught loss per metre of height for this diameter.

(3) Make an assumption as to chimney height and hence from Fig. 4.8 note loss in chimney. Add for loss through boiler and flue connection and efflux velocity head.

(4) Using Fig. 4.6, the available draught may be found for the same assumed chimney height. If this is equal to or exceeds the sum of the losses, the assumption as to height may stand. If the available draught is insufficient, either the height may be increased or the velocity reduced, or both. If the calculated draught is much in excess of requirements, a smaller chimney or less height may be the solution, or if neither is possible, or desirable, a damper may be used.

11. *Clean Air Act*

The chimney height obtained, as above, is that necessary for combustion. It is necessary to consider it next in relation to mandatory requirements of the Clean Air Act, 1968. For this purpose reference should be made to the second edition of *Memorandum on Chimney Heights*, 1967.

The *Memorandum* is concerned to limit the SO_2 contamination near the chimney by making the chimney higher as the rate of SO_2 increases. The type of locality is taken into account as well as the relationship of building height to chimney height.

The *Memorandum* excludes installations releasing less than $1 \cdot 5$ kg of SO_2 per hour (about 200 kW with fuel oil and 400 kW with solid fuel). Further reference to this matter occurs in Chapter 5 on *Oil Firing*.

The Local Authority needs to be consulted in every case of a new boiler installation. The *Memorandum* is intended as a guide and calls for intelligent interpretation.

BOILERS WITH MECHANICAL DRAUGHT

Any boiler may be fitted with an induced-draught fan for exhausting the products of combustion and discharging them up the chimney, as in Fig. 4.9. Many boilers are obtainable with such a fan fitted as part of the standard unit. In others, the fan supplying combustion air has sufficient power to expel the products under pressure.

The advantages of mechanical draught are: first, that the boiler can be designed for higher velocities over the heating surface, so giving a higher rating for a given size; second, that draught produced by stack height is unimportant, and the chimney may thus be short (subject to the Clean Air Act), or, if the boiler is on the roof, non-existent.

Induced-draught fans usually take in some diluting air and makers' data should be consulted for the volume to be handled by the chimney,

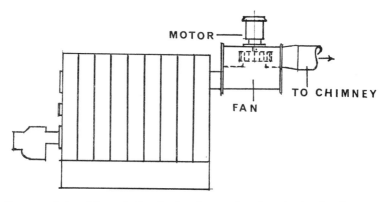

FIG. 4.9—Induced-Draught Fan fitted to Beeston Cast-Iron Sectional Boiler, F. 4.

the area of which will then be determined by the velocity decided upon, such as 10 to 12 m/s. The resistance will be calculated as before—it is usual, where the fan is part of the boiler, for about 60 N/m^2 surplus pressure to be available for chimney loss and efflux velocity. A purge period before and after firing is incorporated in the control system to clear the boiler of stray products.

CHIMNEY CONSTRUCTION

Domestic Chimneys—*Building Regulations*, 1965, requires that all chimneys be lined with some impervious material such as tile. For advice on details of construction, reference may be made to the Building Research Station's *Digest No. 60*

Larger Installations—The importance of keeping the products of combustion in the chimney warm is explained in the next chapter. With this in mind some form of insulation is a necessity and various forms of construction are illustrated in Fig. 4.10. They are:

(A) steel, enclosed in an aluminium outer shell having about 6 mm air gap. The air gap constitutes the insulation. These chimneys may be self-supporting;

(B) pre-cast concrete in sections lined with insulating moler concrete;

(C) brick-lined with fire-clay tile;

(D) brick-lined with moler brick;

. (E) brick or concrete outer stack, and independent lining of fire brick or moler brick with air gap between.

The relative heat losses from these constructions may be calculated as for U values—the inner surface resistance being assumed as zero. Cost, permanence and appearance if free-standing, will always be determining factors.

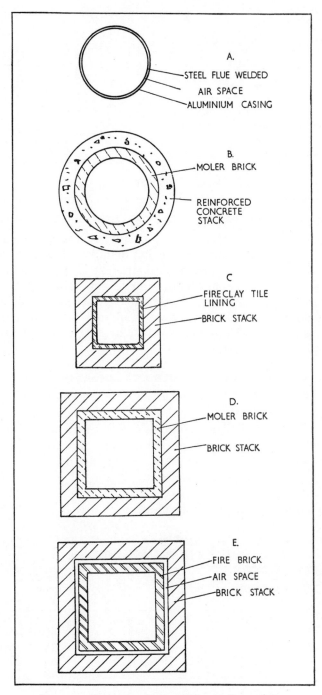

Fig. 4.10—Various methods of Chimney Construction.

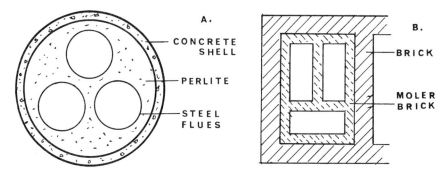

FIG. 4.11—Multiple Flues

Multiple Flues—It will be apparent that, where more than one boiler occurs and there is a common chimney, it would be impossible to maintain the design efflux velocity with anything less than the full number of boilers in use. Furthermore, where mechanical draught is used and the flue is under pressure, back draught might occur to any boilers idle. Hence, it is now advocated that each boiler should have its own individual flue connection and chimney, and that they should not be combined into one large stack as has been the practice in the past. The vertical flues may be grouped into one stack, as in Fig. 4.11.

BOILERHOUSES: SIZE AND LOCATION

Size—Various attempts have been made to give guidance to architects for planning a boilerhouse in advance of design being worked out. There are so many designs of boilers and kinds of arrangement that any advice can only be in very general terms. Some boilers are long and narrow, others short and tall. Access for installation may decide whether sectional must be used. A cylinder for hot-water supply must often be housed in the same space.

Table 4.3 may be used as a rough guide.

TABLE 4·3

APPROXIMATE BOILERHOUSE SIZES

Boiler Rating kW	Length m	Width m	Minimum Height m
30	2·5	3·5	2·3
75	3	4	2·3
100	3	4·5	2·5
150	4	5	2·5
200	5	5	2·5
300	6	5	2·8
500	7	6	3
750	7	7	3
1000	7	8	3

Location—It was always usual to house the boiler in the basement, but construction below ground is expensive and a flue must be accommodated up to the roof, which is costly and takes up space. Hence, where conditions permit, there is often a case for putting the boilerhouse on the roof. With mechanical draught the flue may be quite short. Ventilation and access are easy.

Roof boilerhouses are only applicable with oil or gas, the former being pumped-up from storage at ground level. Considerations of weight, noise and vibration will affect the structure and acoustical treatment, but are usually capable of being simply resolved.

Alternatively, boiler plant may be at ground level—possibly detached from the building with an independent stack. A group-heating scheme would most likely require this sort of planning.

METHODS OF FEEDING BOILERS WITH SOLID FUEL

Solid fuel is fed into boilers in one of three ways:

(*a*) by hand;
(*b*) by gravity from a magazine;
(*c*) by automatic stoker.

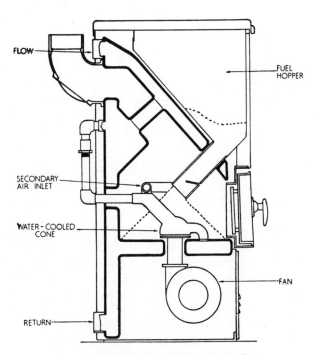

FIG. 4.12.—The 'Earleymil' Magazine Boiler.

Hand Feeding—The vagaries of hand feeding have already been mentioned briefly. The high cost and the difficulty of obtaining labour for the hand-firing of boilers, apart from other considerations, is causing automatic firing methods to be generally adopted.

Magazine Boilers—The fuel is fed into a magazine and descends by gravity into the burning zone where it is consumed, the ash ending up as clinker to be removed from the base. Fig. 4.12 illustrates a magazine boiler for burning anthracite grains.

Automatic Stokers—One of the most usual forms of automatic stoker for heating boilers operates on the worm-feed principle. Fuel is fed into a hopper, at the bottom of which is situated a worm or screw, rotated at a slow speed through reduction gearing from a motor drive. The worm is enclosed in a tube beyond the hopper, and serves to convey the fuel into the firepot, which is built into the firebrick inside the boiler, as in Fig. 4.13. The fuel is bituminous coal of small size.

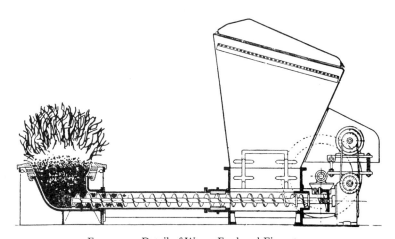

Fig. 4.13.—Detail of Worm Feed and Firepot.

A second tube delivers air into the firepot from a fan driven from the same motor that operates the worm. This air is discharged through a series of slots or openings in the firepot (tuyeres), so disposed as not to be closed up by ash or coal.

Thus, forced draught is provided, and a very high combustion rate is possible, so high, in fact, that a grate is unnecessary. The fuel is burnt as it passes over the edge of the firepot, and all the ash is reduced to clinker in the process. This, however, does not impede combustion, as the fresh coal brought in by the worm pushes the waste material to one side where it remains for periodical removal.

Safeguards are generally provided to prevent jamming of the worm from damaging the mechanism, also to prevent the fire going out if the machine is shut off for lengthy periods by its thermostat. The former is

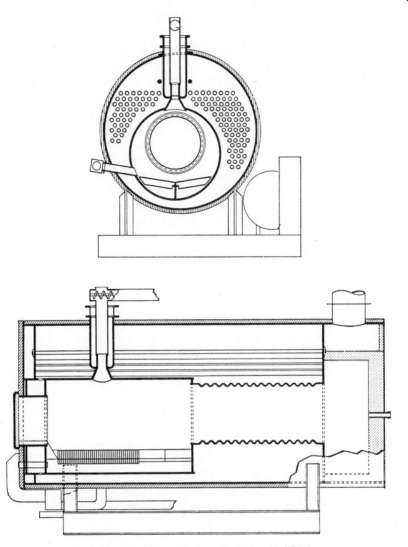

FIG. 4.14—The Vekos Powermaster Boiler (G.W.B.).

accomplished by a shearing pin, or slipping clutch. The latter, by arranging the motor to start up every hour or so for a few minutes, whether required by the thermostat or not.

Thermostatic control is applied either as stated above, by the stop-and-start method, or by a variation in the rate of the fuel feed. With the latter method it is necessary to vary the air supply at the same time, or considerable excess will result when operating at low outputs.

Another method of feeding is direct from bunker. In this, the worm extends into the main fuel storage and the hopper is eliminated.

For smokeless combustion, makers provide a method of supplying preheated secondary air.

The advantages of automatic stokers are that labour is reduced and higher efficiency is obtainable than with hand-stoking. Provision for grit arresting may be required under the Clean Air Act. Plate III (facing page 128) illustrates an example of under-feed stokers fed by gravity chutes from coal hoppers overhead. The boilers serve a block of flats.

Vekos Boiler—Another type of automatically-fired boiler for burning coal is the Vekos, shown in section in Fig. 4.14. The coal is delivered pneumatically or by worm to the top of the boiler and is then 'sprayed' into the combustion chamber. Ash and unburnt fine fuel is removed by a grit collector, which is integral to the flue system, and is conveyed back to the combustion chamber for re-burning. Ultimately, ash is burnt into clinker and this may be removed daily through access at the boiler front. It is common practice to incorporate a clinker-crusher system with the boiler plant so that the waste matter may be raked directly into a hopper at the boiler front and then crushed and conveyed pneumatically to a storage cylinder, from which it may be removed with minimal effect upon menities.

CHAPTER 5

Oil Firing

AN AUTOMATIC, LABOUR-FREE heat service is demanded today. One method of achieving this is by oil firing. At the same time, the economics of the matter have for a long time been in favour of oil, though the competition of natural gas may bring about some change in the comparison. Table 3.2 (page 65) gives the cost of heat produced by oil and other fuels at an assumed efficiency of combustion, the basic rates being the sort of brackets within which prices range at the date of this edition.

A strict comparison requires cost of maintenance, capital charges on storage space, flue and other factors to be taken into account, which can best be done for specific cases.

Types of Fuel Oil—Four types of oil are in common use for heating purposes, having characteristics as given in Table 5.1. A fifth, kerosene, is referred to later under *vaporizing burners*. Of the four, the heavier the grade the cheaper the price, but the greater the amount of preheating and attention required.

Class D oil requires no preheating, is clean in burning in fully-automatic burners, and is used in domestic boilers. It is increasingly being used in larger plants up to about 1000 kW on account of its lower sulphur content, ease of maintenance and absence of any preheating complications

TABLE 5.1

CHARACTERISTICS OF TYPICAL FUEL OILS

Description	Class D Gas Oil	Class E Light	Class F Medium	Class G Heavy
Specific Gravity at 15·6° C - - - - -	0·835	0·93	0·95	0·97
Flash Point (closed) min. - - - - -	66° C	66° C	66° C	66° C
Viscosity				
Redwood No. 1 at 38° C secs. - - - -	34	250	1000	3500
Kinematic, centistokes, at 38° C secs. - -	4	62	247	864
Pour Point, °C - - - - - - -	− 18	− 7	21	21
Calorific Value MJ/kg, gross - - - -	45·5	43·4	42·9	42·5
net - - - -	42·7	41·0	40·5	40·0
Storage Temp., min. °C - - - - -	−	7	20	32
Handling Temp., min. °C - - - - -	−	7	27	38
Sulphur Content, max. % by weight - - -	0·75	3·2	3·5	3·5
Ash Content, % by weight - - - - -	0·01	0·05	0·12	0·2
Mean Specific Heat Capacity, kJ/kg 0–100° C -	1·93	1·82	1·78	1·78

and cost. Price differentials when balanced against these advantages are often found to be worth while.

Class E oil requires a small amount of heating in the storage tank if exposed outside, but none where inside a building. It requires preheating at the burner for atomization to a temperature of about 60 to 70° C. It is used in boilers up to about 2000 kW in fully-automatic burners.

Class F and G oils require continuous heating of the storage even if within a building, and heating before entering the burner to temperatures up to 105° C and 125° C respectively. They are used in large boiler plants where trained operating personnel is available.

With reference to the characteristics of these oils shown in Table 5.1, the following should be noted:

The Flash Point, Pensky Marten (closed), is the minimum temperature at which a flash can be obtained on the apparatus when the oil in it is heated. While the flash point limits the amount of low boiling-point materials which can be incorporated in the oil, from a combustion point of view it has little or no significance.

The Viscosity, which is given above as Redwood number 1 at 38° C, is generally determined nowadays in a U tube viscometer; it is also usual to quote the viscosity in terms of Kinematic Viscosity (centistokes).

The Pour Point generally follows the viscosity, and is the temperature at which the oil ceases to run freely. For practical reasons this point must be below the normal temperature of the storage vessel.

The Calorific Values (gross and net) have already been explained in Chapter 4.

The Sulphur Content determines the quantity of sulphur trioxide which is produced on combustion. It is the chief cause of pollution from chimneys and corrosion in boilers, as discussed already.

In normal boiler plant, where high superheat temperatures are not employed, the *ash* in the fuel is of little importance.

Oil Consumption—The quantity of oil to be burnt for a given heat output depends on the calorific value of the fuel and the efficiency of the boiler. For instance, the boiler for the example on page 67 was 650 kW. Taking Class E oil, the calorific value is 43·4 MJ/kg gross. An efficiency of 75 per cent is assumed. The consumption will be:

$$\frac{650}{43\cdot4 \times 1000 \times 75} = 0\cdot02 \text{ kg/s} = 72 \text{ kg/hr}$$

$$\text{or} \quad \frac{72}{0\cdot93} = \qquad 78 \text{ litre/h}$$

The preheat required for this oil quantity for a temperature rise of 50° C is

$$0\cdot02 \times 1\cdot82 \times 50 = 1\cdot82 \text{ kW}$$

ATOMIZATION AND TYPES OF BURNERS

Atomization—The whole problem of burning oil efficiently and without smoke is one of atomization, that is to say the intimate mixture on a molecular scale of carbon in the fuel and oxygen in the air. All kinds of attempts to solve this problem have been made over the years, but for heating boilers they have now been resolved into relatively few. Systems have included:

Vaporization—The oil is heated as in a blow lamp or Primus stove. Vapour is formed which is ignited. The heat of the flame continues the process. It is suitable only for very light oils such as kerosene.

Pressure jet—The earliest system simply pumped oil at high pressure through a fine jet giving it a swirling motion. Natural draught air entered through a front register giving a swirl in the opposite direction. Used mainly in marine practice, it was also used for large industrial plants in the early days. Skill was required in adjustment of the burner and the air to achieve reasonable efficiency and to avoid smoke. Heavy crude oils were used, requiring considerable preheating.

Rotating cup—The atomization in this system is caused by a spinning cup. Little or no pressure is required to deliver the oil to the cup, which has a serrated edge. The cup in some of the methods is driven by the primary-air blast, in others by mechanical means from the motor-driven fan and the oil pump.

Compressed air—The principle of atomization in this case depends on admission of a jet of compressed air around the nozzle, delivering oil under pressure.

Steam—This is the same as the last except that steam is used in lieu of the air blast. Its use is confined to steam boilers.

Emulsifying—The emulsifying burner relies on the principle of pre-mixing air and oil before delivery to the burner. A reduction in preheat requirements results, but the system has not been developed extensively.

Combustion Air—Natural draught as a means of introducing air for combustion is used in some of the small vaporizing burners. Apart from this, mechanical draught is employed, either supplying the primary air to the burner—the remainder entering by natural draught as secondary air—or supplying 100 per cent of the combustion air. The tendency now is to adopt the latter method, in which case the combustion chamber of the boiler is sealed and pressurized. Alternatively, the burner delivers the oil and air mixture into the combustion chamber and relies on natural draught or mechanical-induced draught to remove the products of combustion.

It will be seen that there are various possible combinations of atomizing methods with systems of air introduction, and hence as many variations in design of the burner. Of these the following are described.

Vaporizing-Type Burner—This is a burner for small boilers up to about 20 kW and uses kerosene marketed by the oil companies for this purpose and known as Class C. Its properties are: Calorific Value 43·6 MJ/kg net, Flash Point 38° C, Sulphur 0·2 per cent. One form of vaporizing burner is illustrated in Fig. 5.1, being the pot type. Oil enters the

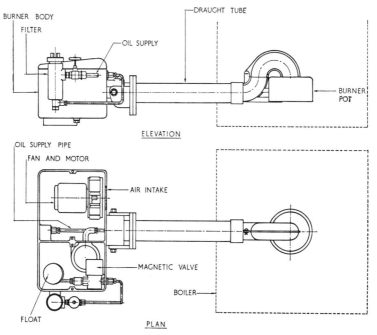

FIG. 5.1.—Vaporizing Oil Burner (Nu Way).

chamber under control of a ball float and is thence fed to the burner pot by gravity. Ignited material is placed in the pot to commence the flame and, thereafter, the heat from the oil flame is used to vaporize the liquid. Air is supplied by means of a small fan delivering into the pot where it is mixed with the vaporized oil and burns in the combustion chamber of the boiler immediately above. The flame is controlled by a thermostat actuating the magnetic valve, but a small flame is always maintained sufficient to preserve ignition. In other words the flame does not go out.

Other forms of vaporizing burner are fully-automatic and therefore include means for ignition and safety devices for flame failure. One such is the *wall-flame* which is available up to about 50 kW. The vaporizing burner is relatively cheap, quiet in operation and suitable for small domestic applications.

Gun-Type Burner—This is by far the most common type of oil burner, ranging in size from a boiler output of 15 to 2500 kW. It contains an electrically-driven fan delivering 100 per cent combustion air, see

Fig. 5.2. Oil is delivered under pressure from an oil pump coupled to the spindle driving the fan. The oil is sprayed through a fine calibrated jet, the size of which controls the output of the burner and is fixed. Different jets can be used according to the duty of the boiler. Air from the fan, delivered *via* swirling vanes in the nozzle, is controlled as to quantity by adjustable slots or dampers so as to give the desired oil/air mixture. This

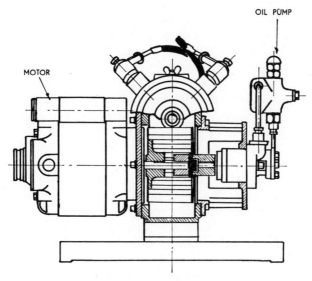

Sectional Front Elevation

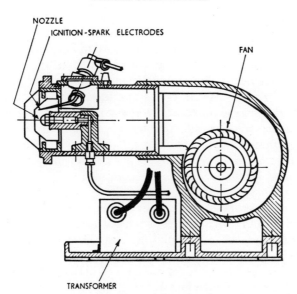

Sectional Side Elevation

FIG. 5.2.—Gun Type Fully-Automatic Oil Burner.

adjustment is pre-set when the burner is tested by CO_2 measurement to give say $10\frac{1}{2}$ per cent by flue gas analysis.

Ignition is by electric spark from two electrodes located near the nozzle tip, the high-tension supply at about 8·5 kV being from a transformer which is energized when the burner starts up. The spark is maintained for a sufficient period to cause ignition, and is then cut off by a time delay in the control system.

The burner is under the control of a thermostat in the case of a water boiler, or pressure-stat in the case of a steam boiler, this being the 'control stat'. In addition, a high-limit thermostat is incorporated to cut off the burner if an unduly high temperature results—this being of a manual reset type.

In the smaller size up to about 300 kW, control is usually on–off. Above this, either high–low or modulating control is incorporated to reduce the shock on the boiler of a sudden big flame. In the high–low method the burner starts up on one-third load, and after a delay changes to full load, being cut down to one-third as the temperature is reached and finally to off. In the modulating system, the burner starts up on one-third load and changes to a modulated condition between that and full load, according to demand. On the downward run, if temperature is satisfied at one-third load, the burner becomes on–off below that point.

The flame shape of this burner may be varied to suit the geometry of the fire-box by selecting the angle of divergence of the jet and by adjusting the angle of the directional vanes in the nozzle.

When Class D oil is used, no preheating is necessary. For Class E oil, an electric preheater is incorporated as part of the burner unit. The heater contains a thermostat set to maintain the required temperature for atomization of 60 to 70° C. The control system is interlocked so that the burner cannot start until this temperature is reached.

Rotary-Cup Burner—The main elements of a burner of this type are illustrated in Fig. 5.3. The burner is mounted on the front of the boiler. Oil is delivered from the pump at low pressure through a hollow spindle to which is attached the fan and spinning atomizing cup. The periphery of the cup is surrounded by an annular air nozzle with swirling vanes. The air from this nozzle may represent 20 per cent of the air for combustion. The remaining (secondary) air for combustion is introduced at low pressure by a separate fan, being preheated to some extent by its passage over the primary-air and secondary-air quarls. (See Plate I, *frontispiece*.)

Ignition is usually by gas such as bottled gas, ignited by electric spark. Control of output is on a modulating basis from one-third to full load, being on–off below one-third, or on high–low.

Control of output is by adjustment of oil delivery linked with damper control of primary and secondary air.

This type of burner is applicable to larger outputs from 150 kW to 5000 kW, with oils of Class E, F or G.

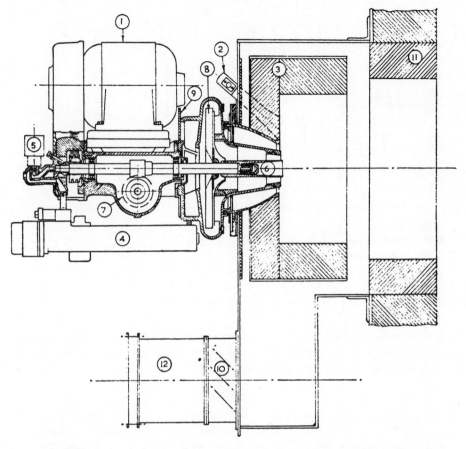

(1) Motor   (5) Solenoid valve   (9) Primary air control valve
(2) Photocell   (6) Atomizer   (10) Secondary air control damper
(3) Primary air quarl   (7) Pump Drive   (11) Secondary air quarl
(4) Heater   (8) Fan   (12) Forced draught fan

FIG. 5.3.—Rotary-Cup Burner (Hamworthy).

Oil Preheating—For the larger installations using Class F and G oils, it is the usual practice to circulate hot oil from a heating and pumping unit mounted in the boilerhouse separately from the burner, as in Fig. 5.4. Heating is by means of a hot-water or steam-heat exchanger with electric immersion heaters for start-up. The circulation is taken right up to the burner so as to avoid any cold oil which might cause smoke on start-up. Oil lines are kept warm by heated piping or electric tracing. Un-used oil from the ring is returned to the pump suction. Pumps are usually in duplicate, one being a stand-by. Where the type of burner does not have a self-contained pump, a slightly different arrangement applies. The burner

H   K.H.A.C.

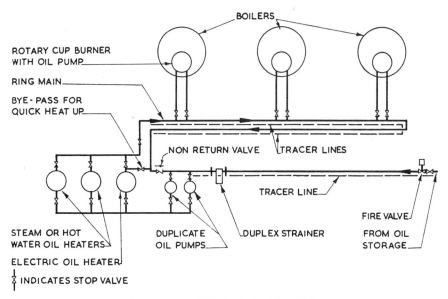

FIG. 5.4.—Hot Oil-Circulating Ring Main.

connections are then in parallel across the flow-and-return mains and, at the end of the ring, a pressure relief valve is inserted thus setting up a pressure at the burner for atomization.

CONTROLS

The ideal always aimed at is to make an installation completely automatic and thus dispense with the labour of attendance altogether. Fully-automatic types of burner have been referred to and it will be obvious that they inherently include means for self-igniting and self-extinguishing, the rate of firing or the period of operation being controlled by temperature, i.e. thermostatically in the case of a hot-water boiler or air-heater, or by pressure in the case of a steam boiler. One thermostat is provided for control purposes and a second high-limit stat to cut off the burner in the event of over-run of temperature. The latter is for hand re-set.

In self-contained unit type burners, in which one motor drives fan and oil pump, there is little likelihood of oil being delivered into the boiler without the fan running and supplying the necessary air; hence safeguards necessary with the old tailor-made systems to guard against this possibility are not required. The worst that can happen, however, is that the flame is not established due to failure of ignition or, having been established, is subsequently extinguished from some cause such as water in the oil or maladjustment of the burner. To guard against this hazard the flame failure cut-off device has been developed and will be found to be incorporated in one form or another in most of the self-contained burners referred to earlier. The most reliable and rapid device of this nature depends on a light-sensitive element sometimes called an 'electric eye' so positioned as to receive light from the burner flame. When starting up, this device is shorted out for a period and, if by the end of this period the flame is not established, the burner will cut off automatically. In some types of control, after a delay period to allow any gases from the first attempt to have cleared from the boiler, a second attempt at ignition is made. If the second attempt fails, the burner is locked out permanently and it is then necessary for an attendant to discover the fault and reset the control.

In the normal course, the flame having been established, the light-sensitive cell is actuated and maintains the burner in operation. Should the flame fail, however, under this normal running condition, the burner is caused to shut down immediately.

Complete automation of a heating system, whether the source of heat is oil, gas or electricity, brings with it control of operation according to time, involving clock control. In the case of oil firing, the burner may be switched off completely at night or may change over to another thermostat with lower setting. This is sometimes further extended on the clock switch to remain at the lower level over a week-end when applied to offices, shops etc. Where a boiler or boilers are supplying heating only, it is tempting to allow the control to vary the temperature of the water delivered from the boiler, but this may result in unduly low water temperatures at times of mild weather with danger of condensation forming within the boiler causing corrosion. Excepting in small installations it is therefore preferable to run the boiler at constant temperature and to rely on a mixing valve to give the variable temperature for the heating system, as referred to on page 181. Where boilers are supplying, in addition to heating, hot-water supply or hot water for air-heater batteries in a ventilation system, this method will be adopted in any event for obvious reasons.

PACKAGED-BOILER UNITS

As applying to hot-water boilers, the term 'packaged unit' implies a boiler specially designed for oil- (or gas-) firing, having all the necessary equipment fitted to it of the type best suited to the characteristics of the boiler.

These units generally operate with high combustion rates, the flue passages being designed for high velocity and, hence, for maximum heat transfer for a given heating surface.

The burner will have been selected to give the best flame shape for the combustion chamber and may be arranged to pressurize the chamber so that flue pull is not relied upon. Fig. 5.5 illustrates a unit of this type. Alternatively, an induced-draught fan may be included as part of the equipment.

The unit will be complete with control gear for fully-automatic operation, including safety controls. Certain types include a boiler-circulating pump.

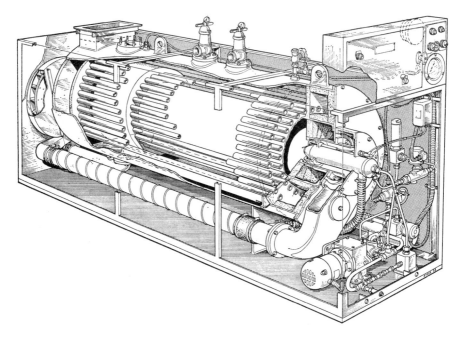

FIG. 5.5.—Oil-fired Packaged Boiler: Paxman 'Autonomic'. *With acknowledgements to Ruston and Hornsby.*

The chief merit of this kind of approach is that all parts are suitably matched and pre-set, so that the task of setting-up and commissioning is reduced to a minimum.

REGULATIONS

Owing to the fact that the burning of oil contains the elements of a fire hazard, various bodies have drawn up regulations governing the installation of oil-firing equipment, oil tanks, etc. British Standard No. 799 specifies the requirements for atomizing burners and associated equipment in Part 1, 1963. Part 2, 1964, covers vaporizing burners and associated equipment.

A British Standard *Code of Practice* has also been produced, No. CP. 3002, Part 1, being in respect of oil-firing installations burning Class D fuel oil. Part 2 covers vaporizing burner installations and Part 3 installations using preheated fuels, Classes E, F and G.

The various local authorities, fire authorities and insurance companies have their own regulations and these should be ascertained when any new installation is under consideration. Local authorities also have responsibilities under the Clean Air Act which may affect the grade of oil selected and the dimensions of the chimney, referred to later.

FUEL OIL STORAGE

It is usual to provide a storage based on two to three weeks' running on full load. Except in special cases, the full load would not apply over twenty-four hours, and it is generally reasonable to assume the equivalent of about twelve hours per day. In the case of small installations, such as in domestic premises, certain oil companies now deliver oil metered through a pump, according to requirements, in which case a tank of about 1300 litres may serve. Otherwise, it is usually considered desirable to have a tank large enough to take at least a 2500 litre load, which, allowing for some reserve, gives a minimum tank capacity of say 3500 litres.

In larger installations, it is frequently necessary to provide two or more tanks. This has the advantage that one tank can be filled while the others are in use, thus allowing sludge and water, if any, to settle.

Table 5.2 gives the consumption of various sized plants based on three weeks' run at full load (twelve hours per day), from which the appropriate storage may be estimated.

TABLE 5.2

OIL STORAGE CAPACITY

21 days × 12 hours = 252 hrs. Boiler efficiency 75%. For 1 kW output heat

$$\text{input to boiler} = \frac{252 \times 3600}{0.75} \text{ kJ/s}$$

Oil: take 41 MJ/kg $\dfrac{252 \times 3600}{41\,000 \times 0.75} = 29.3$ kg

Take sp.gr. 0.9 = say 35 litres per kW

Thus:

Boiler Rating kW	Storage for 3 Weeks Litres
20	700
40	1400
60	2100
80	2800
100	3500
150	5250
200	7000
300	10 500
400	14 000
500	17 500
750	26 250
1000	35 000

Note: The thousands column gives storage in m³. (m³ × sp. gr. = tonnes).

Storage Vessels in Buildings*—Oil-storage vessels are usually of steel construction and may be either rectangular or cylindrical. They may be brought into the building in one piece, or welded *in situ*. The latter is necessary in confined situations. Table 5.3 gives capacities of tanks, cylindrical and rectangular.

TABLE 5.3

OIL-TANK CAPACITIES

Cylindrical

Diameter m	Length m	Gross Capacity Litres	Net Capacity* Litres
1	2	1570	1300
	2·5	1960	1600
1·5	2	3530	3200
	2·5	4420	4000
	3	5300	4800
2	3	9430	8800
	3·5	11 000	10 200
	4	12 570	11 800
2·5	3·5	17 180	12 900
	4	19 640	14 800
	4·5	22 090	16 600
3	4	28 280	27 500
	5	35 350	34 200
	6	42 420	40 200

Rectangular

Length m	Width m	Depth m	Gross Capacity Litres	Net Capacity* Litres
1	1	1	1000	700
1·5	1	1	1500	1100
2	1	1	2000	1500
3	1·5	1	4500	3300
3	1·5	1·5	6750	5600
3	2	1·5	9000	7500
4	2	2	16 000	14 000
4	3	2	24 000	21 000
4	4	2	32 000	28 000
6	4	2	48 000	42 000

* Allowance 150 mm up to outlet and 100 mm ullage (at top), rounded to nearest 100 litres, lower.

Tanks deeper than 1 m often have iron access ladders inside. To allow proper inspection outside of the tanks, they should have a walking-way of 0·5 m clear all round. Cylindrical tanks are supported on steel cradles, rectangular tanks on steel joists, in either case bearing on sleeper walls about 0·3 m high with bituminous felt or lead packing on top.

* See B.S. 799 and C.O.P. 3002 for detailed recommendations.

Pipe-work and tanks used in connection with oil should not be galvanized.

Tank Rooms, separated entirely from the boiler house by a brick or concrete construction, are called for in C.O.P. 3002. An oil-storage room should have the lower portion oil-tight to hold the full capacity of the tanks, should they leak; this calls for an access door at a higher level with steps outside and inside. The door must be fire-resisting; also, separate inlet and outlet ducts direct from outside are required for the ventilation of the tank room.

Fig. 5.6 shows this arrangement, together with other details which will be referred to later.

Tank Fittings—Various fittings are required in connection with storage tanks, the principal ones being given below.

A manhole must be provided to each tank, a common size being 0·5 m diameter. The joint with the tank top should be made gas-tight.

The filling terminal should be standard 2 in, $2\frac{1}{2}$ in or 3 in (50, 65, or 80 mm, according to length and grade of oil) male gas thread for hose coupling with gunmetal cap. From this, the pipe connects to the tank with a steady fall. Where more than one tank exists, a 3-way cock or set of valves will be required, or one filling pipe per tank.

Vent pipe. A vent pipe from each tank is required. It should be 80 mm in diameter, be carried up to the roof of the building or to some point where smell will not be troublesome, and be fitted with a wire guard. It should be separate from the filling pipe.

In the event of the overfilling of a tank from a road wagon which is delivering with compressed air or by pump, a combined vent pipe arrangement allows oil to flow into the next tank, through the pipe, without obstruction.

In the event of overfilling, oil may rise in the vent pipe to a considerable height if the building is tall, so placing an undue pressure on the tank. An alarm device is desirable, therefore, to give warning of this condition, and, in addition, an oil seal or one of the proprietary unloading devices should be branched from the vent pipe.

It is unnecessary to point out that vent pipes should have a steady rise to the top, as any dip which might become filled with oil would obstruct the free passage of vapour.

The outlet connection should have a valve next to the tank and should be 100 to 150 mm above the tank bottom. The size is determined by the number and size of burners connected, each of which should again be separately valved.

If a number of boilers is being served from the oil-storage tank, some designers favour the use of a daily service tank. This enables the boiler plant to be started up before steam or electric current is available for heating the fuel in the main storage tank. The capacity of the service tank is limited by C.O.P. 3002 to 900 litres per boiler.

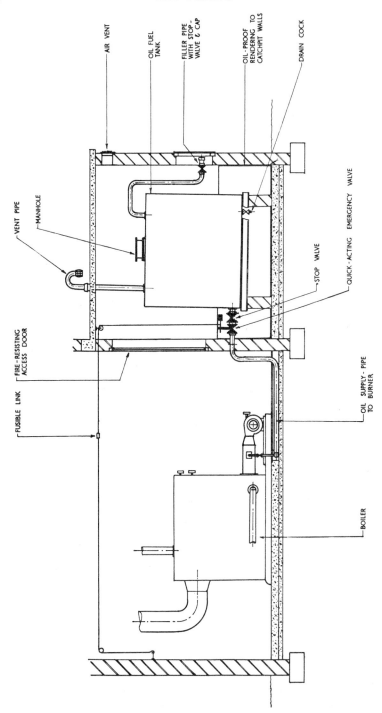

AIR VENT

OIL FUEL TANK

FILLER PIPE WITH STOP-VALVE & CAP

OIL-PROOF RENDERING TO CATCHPIT WALLS

DRAIN COCK

VENT PIPE

MANHOLE

STOP VALVE

QUICK-ACTING EMERGENCY VALVE

FUSIBLE LINK

FIRE-RESISTING ACCESS DOOR

OIL SUPPLY-PIPE TO BURNER

BOILER

FIG. 5.6.—Oil Tank Layout.

The daily service tank also serves to limit the amount of oil which could escape in the case of pipe-fracture. If an electric pump is provided for filling the daily service tank, it is necessary to arrange for hand-starting and automatic stopping.

Sludge Outlet. Water, being denser than oil, collects at the bottom of the tank, generally in the form of mud, since it mixes with the sludge and solid matter which gradually settles out from the oil. After a few years of partial emptying and refilling of the tanks the accumulation increases, and, before there is any chance of its reaching the outlet, it must be removed. A sludge outlet is therefore provided at the lowest point and is fitted with a valve. The end may terminate in the tank room, provided it is easily accessible. Oil companies generally make arrangements for the collection and removal of sludge, as it cannot, of course, be put down the drains. A small oil sump with hand pump in the tank chamber is an advantage.

Level Indicator. Gauge glass indicators, being fragile, are unsatisfactory. Direct reading dial gauges for mounting in the side of the tank are available, or there are various forms of remote reading gauges operating by sealed pressure system (see Fig. 5.7). There are also pneumatically and electrically operated types.

Filters. To protect the burners and pumping equipment, it is desirable to fit a coarse filter, either immediately on the outlet or at the burner installation, the size being such that it causes no excessive resistance to the flow of the oil. The object of this filter is to stop any extraneous matter, which may have gained access to the storage tank, from reaching the transfer or burner pump, and therefore protect them from mechanical damage.

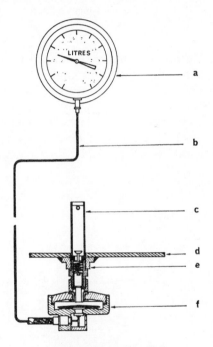

(a) Remote reading dial
(b) Capillary
(c) Anti-sludge unit
(d) Tank bottom
(e) Tank boss
(f) Transmitter box

FIG. 5.7.—Oil Tank Contents Gauge.

Heaters. In the case of heavy oils it is necessary to provide means of heating in the main storage tanks, either by steam or hot-water coils, or by electric immersion heaters. Electric heaters are usually confined to the proximity of the outlets the object being to heat the oil at the point of exit rather than the whole tank. This device is termed an *out-flow*

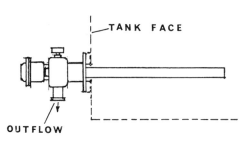

FIG. 5.8.—Electrical Out-flow Oil Heater
(Landon Kingsway Ltd.).

heater, one form of which is illustrated in Fig. 5.8, being so designed that oil cannot be withdrawn below the level of the heating element.

In addition, steam or hot-water tracer lines are run in contact with the outlet-pipe to the pumping and heating unit, or, alternatively, electric heating cable is wound round the pipe. The heating lines and pipe are then lagged in one envelope.

Underground Storage Tanks—Where it is possible, tanks may be placed out of doors underground. This economizes building space and is a very safe and practical arrangement. If buried without an enclosing pit, the tank should be properly protected with bituminous paint on the outside and buried in concrete. It is generally necessary to anchor the tank to a block of concrete to overcome buoyancy when empty. A better arrangement (as in Fig. 5.9) is to construct a pit, so that inspection may be made all round.

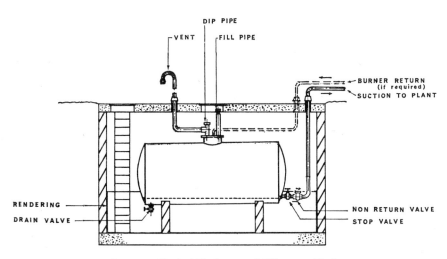

FIG. 5.9.—Typical Underground Oil-storage Tank.

Roof-top Boilerhouse—The advantages of a roof-top boilerhouse have already been referred to, particularly for multi-storey construction. The arrangement for oil supply from main tankage at ground, or below ground, is shown in Fig. 5.10.

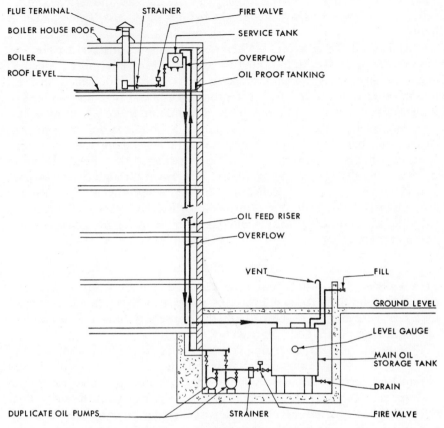

FLUE TERMINAL
BOILER HOUSE ROOF
BOILER
ROOF LEVEL
STRAINER
FIRE VALVE
SERVICE TANK
OVERFLOW
OIL PROOF TANKING
OIL FEED RISER
OVERFLOW
VENT
FILL
GROUND LEVEL
LEVEL GAUGE
MAIN OIL
STORAGE TANK
DRAIN
DUPLICATE OIL PUMPS
STRAINER
FIRE VALVE

FIG. 5.10.—Roof-Top Boilerhouse.

Piped Oil Supplies—An interesting development in the use of oil for heating is the provision of a central oil-storage to serve a group of flats or residences each of which has its own oil-burning appliance. The oil is continuously pumped from a main storage on the estate and fed to each consumer by meter. Piping is laid underground.

FURNACE LININGS

The firebrick of oil-fired combustion chambers calls for a material which will withstand temperatures of 1400 to 1600° C without fusing or premature disintegration. The brick, therefore, requires to be burnt to a temperature in excess of this, and several standard brands are available for the purpose.

One important point in the building-in of this brickwork, particularly where a sectional boiler is concerned, is to watch that the material is not carried solid up to the metal without an air-gap of half an inch or so being left around it. The great temperature in the combustion chamber causes considerable expansion of the bricks, and the air-gap allows freedom for this to take place without any strain being put on the boiler plates or joints.

Under-Hearth Insulation—Where a boiler does not have a waterway bottom, it is necessary to insulate beneath the firebrick forming the hearth of the furnace, so as to avoid cracking the concrete floor which, in turn, might cause damage to any waterproof membrane that may exist. For boilers of any size up to about 600 kW, this insulation may comprise, say, 6 in. of Moler insulating brick. For larger boilers it is desirable to construct a sub-base of honeycomb form to allow air-cooling and, with some types of boiler and burner, this must be so arranged that the air drawn in by the forced-draught fan for combustion is drawn through these passages, so serving as a means of preheating the air. Alternatively, the sub-base may be cooled by circulating return water through a pipe coil embedded therein. It is desirable that the concrete below a boiler should not exceed a temperature of 70° C.

SAFETY PRECAUTIONS

Fire Valve—The outlet from a tank should include a fire valve. This generally takes the form of a lever-type valve heavily weighted, kept open by a taut wire stretched across the boiler house and having a fusible link of low melting-point alloy over each boiler front.

Should a fire occur, the rising flames or hot gases will melt the link and the valve will shut immediately, thus cutting off the supply of oil and minimizing further damage. A hand release is also provided for testing and this may incorporate a solenoid release in circuit with the sump float referred to later. Alternatively, an electrical system may be used in which a solenoid valve is kept in the open position only so long as current passes. The circuit is broken by a fusible stat or stats in series over the burners. Such devices require hand reset.

Further Safety Precautions—Further devices for protection against oil fires, particularly on large plants, are often called for. These include the following:

(*a*) Boiler dampers to be removed or locked open with an indicating plate showing position of vane.

(*b*) Alternatively, the burner valve and damper must be provided with an interlock operated by a common key which cannot be withdrawn from the damper lever until this is in the open position.

(*c*) A foam chemical extinguisher of portable or permanent type to be installed, serving both the boiler room and oil-tank chamber.

(*d*) Alternatively to (*c*), foam pipes to boiler house and tank chamber to which the fire brigade can connect their apparatus in the street.

(*e*) A sump in the boiler house provided with a ball float and electrical contact so as to cut off the oil supply to the burner(s) should there be an oil leak (see Fig. 5.11).

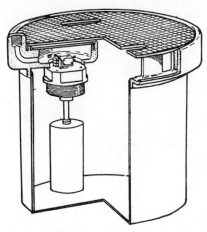

FIG. 5.11.—Oil Sump and Float Switch (Landon Kingsway Ltd.).

CHIMNEYS FOR OIL-FIRED BOILERS

The corrosive effect of sulphurous gases on the heating surfaces of a boiler has been alluded to previously (page 77). If the products of combustion on entering the chimney are allowed to cool to the region of the acid dew point, a new hazard is set up with oil-firing, namely acid smuts.

It appears that the minute unburnt particles of carbon always present in flue gases are liable to form nuclei on which condensation takes place, and they then agglomerate into visible black oily specks. In a chimney they are particularly liable to collect on any roughnesses, or at points of change of velocity. In so doing, on starting up, the sudden shock may cause them to be discharged from the top—often giving rise to complaints from surrounding property.

It has come to be recognized that, in order to avoid this trouble, the following principles should be adhered to:

(*a*) to keep the velocity up the stack as high as possible with the draught available;

(*b*) to keep the gases hot by insulation of the chimney as referred to in the previous chapter;

(*c*) to avoid cooling the gases by a draught stabilizer or air leaks;

(*d*) to design flue connections with easy bends and gradual changes of velocity;

(*e*) to keep the chimney smooth inside;

(*f*) on multi-boiler plants, to use one flue per boiler; or, if a common flue is unavoidable, the dampers on idle boilers must be gas-tight.

Acid smut formation is less of a problem with a low sulphur oil such as Class D, and draught stabilizers are often part of standard boilers on a domestic scale.

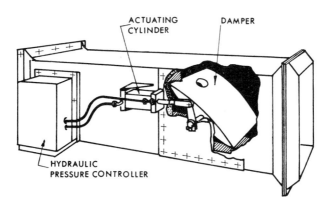

FIG. 5.12.—Automatic Draught-Control Damper (Ivo).

For the control of draught on plants burning the heavier oils, in lieu of a draught stabilizer admitting cold air, some form of damper control is necessary which may be automatic, of a type such as Fig. 5.12.

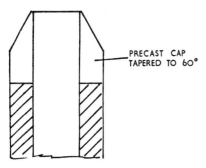

FIG. 5.13.—Form of Chimney Top.

Treatment of flue gases by means of a proprietary additive to the fuel or in the hot-gas outlet may be considered where smut nuisance is particularly troublesome.

The top of a chimney for an oil-burning installation for good dispersal is preferably of the form shown in Fig. 5.13. It will be noted that the cap is tapered at 60° to give a thin edge which has the effect of reducing the tendency to down-wash.

CHAPTER 6

Gas Firing

A COMPLETE TRANSFORMATION IN the energy resources of the Nation has come about due to the discovery of natural gas in the North Sea. The potential supply contracted for over the next thirty years is given as 100 million cubic metres per day, equal to double the consumption of the whole of Great Britain at present, or four times in terms of heat capacity. A high-pressure trunk main system has been constructed for distribution of gas to the various Area Boards, this being also capable of receiving gas from Algeria brought in tankers. Lower tariffs are being offered which should make natural gas competitive with other fuels.

For heating purposes, therefore, it may be assumed that gas will in the future take a much bigger share of the load for new construction, and there will no doubt be conversions from other fuels.

Gas has the advantage of requiring no space for storage on the premises, but storage at or near source is necessary in order to balance-out variations of supply and demand. This storage is in liquified form (ergogenic) in underground chambers of immense size, some 40 m in diameter and 40 m deep. Methane is liquid at atmospheric pressure below $-163°$ C, and in this system the ground is frozen so preventing escape. Another system makes use of certain geological formations, the gas being stored in the pores of greensand or the like.

The projected storage capacity is stated to be equal to one day's supply. However, it appears there may be an inducement made to large consumers to interrupt the supply at periods of heavy demand by offering a considerably lower charge on the basis of an 'interruptible supply'. This then involves the consumer in having some alternative fuel available, such as liquid petroleum gas or oil. If the latter, dual fuel-burners are called for probably using class D oil.

Natural Gas and Towns Gas compared—

Analysis (average values) by volume

	Natural Gas	
Methane	92%
Hydrocarbons	5%
Inert gas	3%
	100%

Towns Gas (by volume)

Hydrogen	48%
Carbon monoxide	5%
Methane	34%
Carbon dioxide	13%
	100%

Properties (average values)—

	Natural Gas	Towns Gas
Maximum burning velocity	0·34 m/s	1·0 m/s
Maximum flame temperature in air	1930° C	1960° C
Wobbe Number	54	29
Calorific value MJ/m³	41	20
Density kg/m³	0·7	0·6
Relative density air = 1	0·55	0·47
Sulphur compounds mg sulphur/m³	0 to 20	122 to 392
Theoretical air required for combustion m³/m³ gas	9·8	4·5
Toxicity	non-toxic	toxic

The principal differences are in burning velocity, and calorific value. The slow-burning rate of natural gas is coupled with a tendency for the flame to lift, for which reasons the neat-gas or flat-flame burner no longer functions. The flame must be aerated and the burner so designed as to remain stable. One method of achieving this is by means of a small keeping-flame near the base of the burner. By way of compensation for this difficulty is the fact that the natural-gas burner has little inclination to light-back.

The calorific value of natural gas is double that of towns gas; hence for a given duty only half the volume is required. This means less ability to entrain air for aeration, yet twice the quantity is needed. In consequence, it is necessary to increase pressure to produce a higher velocity whilst reducing jet size.

As change-over proceeds from towns gas to natural gas, so is pressure raised in the mains, new governors being fitted as well as filters. The raising of the pressure means that existing mains can carry twice the volume of gas and, at the same time, with the higher calorific value, four times in terms of heat. Thus, although there are advantages, there are also problems involved in the use of natural gas not encountered with towns gas.

BURNERS

There are two broad classifications of burners suitable for natural gas, though equally suitable for towns gas: one is *natural-draught*, the second is *forced-draught*.

The natural-draught burner is usually quiet in operation, is limited to about 150 kW and is cheaper. Its disadvantage is that it is dependent on

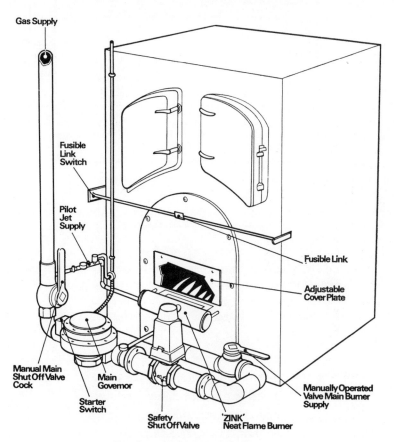

FIG. 6.1.—Natural-Draught Burner.
(*By courtesy of the Gas Council*)

flue conditions which may affect burner performance and efficiency. The presence of a permanent pilot also adds to consumption. One type of natural-draught burner, as applied to a sectional boiler, is shown in Fig. 6.1.

The forced-draught burner, as in Fig. 6.2, is supplied as a factory-tested unit complete with all safety and other controls, and can be expected to achieve the maximum efficiency of combustion with or

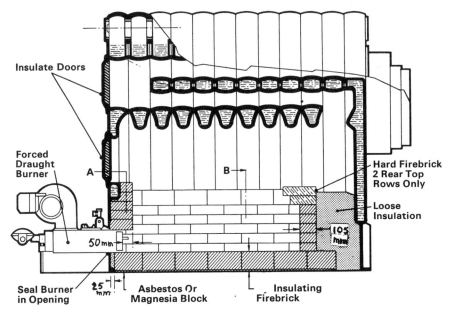

FIG. 6.2.—Forced-Draught Burner fitted to Sectional Boiler.
(*By courtesy of the Gas Council*)

without a flue. It is noisier than the natural-draught burner, but probably similar in this respect to any oil burner. Its capacity is virtually
unlimited.

The forced-draught or 'automatic' burner-control system is electrical
and comprises:

> a pre-purge period;
>
> pilot-flame ignition by spark;
>
> a period for pilot-flame proving;
>
> main flame establishment and control;
>
> on shut-down, a post-firing period.

In addition there is a flame-failure control which will shut off the
boiler in one to two seconds. Types at present in use include either a
flame-failure probe or an ultra-violet scanner. The luminosity of a gas
flame is not adequate to operate a light-sensitive cell as in an oil burner.
The arrangement of main and pilot governors, safety shut-off valves,
flame probe and other features for one particular make will be apparent
from Fig. 6.3.

Standards for automatic gas burners are the subject of close specification by the Gas Council, having special regard to safety.

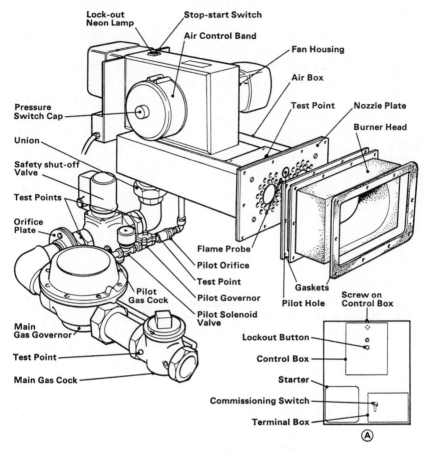

FIG. 6.3.—Automatic Burner Assembly.
(*By courtesy of the Gas Council*)

Boiler/Burner Units—Burners such as those described may be applied to new or existing boilers with either natural or towns gas; but where a new installation is concerned it is obviously preferable to adopt a packaged or prefabricated boiler/burner unit, similar to that shown in Fig. 5.5 (page 100), but with a gas burner instead of an oil one. In this way, any teething troubles associated with flame shape, combustion-chamber volume, application of controls and so on will have been overcome, and this unit should require no more than connecting-up to function as intended. Sizes range from the smallest for domestic use to about 750 kW in sectional form, and 6000 kW or over in steel-shell or tubular form.

Gas-boiler Controls—In the atmospheric burner there is a permanent pilot which is only open when a flame is applied to ignite it, the flame itself then keeping its supply open. The main flame is brought on under control of the control-thermostat which is either of direct-acting or relay type.

There are several designs of the direct-acting type. One of the most reliable is the rod type, shown in Fig. 6.4. This consists of a metal tube which expands with heat, to the inside of which is attached at one end a non-expanding metal rod such as Invar steel. Movement of the rod due to variations in temperature of the outside tube is transmitted to a mushroom type valve, which is either raised or lowered on to a seating, thus increasing or decreasing the flow of gas to the burner. The thermostat is usually used direct in the water or return. A bye-pass screw is provided to keep the gas alight on the burner when the thermostat shuts down the gas supply. Other direct-acting thermostats utilize the expansion of a capsule containing a volatile liquid, such as ether, for opening or closing a gas valve.

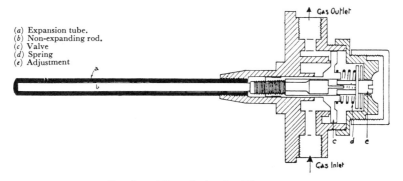

(a) Expansion tube.
(b) Non-expanding rod.
(c) Valve
(d) Spring
(e) Adjustment

FIG. 6.4.—Direct-Acting Gas Thermostat.

In the case of a relay type thermostat, a diagrammatic sketch of which is shown in Fig. 6.5, a modified rod type thermostat is used, either in the room, or in the water flow or return. This varies the quantity of gas passing a small orifice and consequently the pressure, which passes over the top of a weighted diaphragm placed in the gas stream to the burner. As the pressure increases on top of the diaphragm, becoming more nearly equal that on the bottom, the valve closes, until finally, when the thermostat is closed, the pressure on top equals the pressure underneath, and the weight of the diaphragm lowers it and shuts off the gas supply. An adjusting screw is provided to allow a certain amount of gas to pass to keep the burners alight. The small amount of gas passing the thermostat when it is open is piped to the burner, where it burns.

Thermostatic control with gas gives a perfectly modulated action. Thus, the heat requirement at any particular time is in direct proportion to the amount of gas passing the thermostatic valve. This is a great advantage, and gas can be adapted to thermostat control more easily than any other fuel.

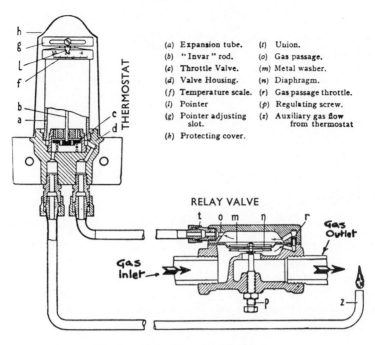

(*a*) Expansion tube.	(*t*) Union.
(*b*) "Invar" rod.	(*o*) Gas passage.
(*c*) Throttle Valve.	(*m*) Metal washer.
(*d*) Valve Housing.	(*n*) Diaphragm.
(*f*) Temperature scale.	(*r*) Gas passage throttle.
(*l*) Pointer	(*p*) Regulating screw.
(*g*) Pointer adjusting slot.	(*z*) Auxiliary gas flow from thermostat
(*h*) Protecting cover.	

FIG. 6.5.—Relay Type Gas Thermostat.
('Perfecta'.)

AIR FOR COMBUSTION

With Natural draught—

After diverter, assume 100% excess air.
Volume of air required:

	Towns Gas	Natural Gas
Air m³/m³ gas (theoretical × 2)	9	19
CV MJ/m³	20	41
Air per MW m³/s input (i.e. approximately the same)	0·45	0·46
Air per MW of boiler rating at 75% efficiency	=0·6 m³/s	

With Forced draught—

Assume 30% excess air.

Air m³/m³ gas	6	13
Air per MW m³/s input	0·3	0·3
Air per MW boiler rating at 75% eff.	=0·4 m³/s	

Inlet area—

Taking air-inlet velocity at 2 m/s.

Free area required:

Natural draught   0·3 m²/MW boiler rating
Forced draught   0·2 m²/MW boiler rating

According to BS *Code of Practice* 332, *Part* 3, the required free area for admission of air for combustion for a gas appliance should be the equivalent of

at low level   0·001 m² per 3 kW of boiler output;

and for ventilation

at high level   0·001 m² per 6 kW of boiler output.

This applies up to boiler ratings of 600 kW, and it will be found to agree as to the low-level inlet with the foregoing for natural draught. For larger duties, the area should be calculated on the lines indicated. The area given for ventilation as distinct from combustion, at double the latter, is no doubt safe enough for removal of heat, but can only be empirical.

CHIMNEYS AND FLUES FOR GAS BOILERS

Natural Draught—The requirements for gas boilers differ from those previously discussed for oil and solid fuel, due to the air admitted by the draught diverter acting as a dilutant to the combustion gases. The CO_2 content of the latter may be 9 per cent with a gas temperature of 240° C, but the normal for the 'secondary' flue is about 4 per cent and a gas temperature of about 120° C. Curves for sizing and data as to resistance factors are given in the *I.H.V.E. Guide*, from which, by way of example, Table 6.1 is taken.

TABLE 6.1

SAMPLE GAS-BOILER CHIMNEY SIZES

Boiler Rating kW	Chimney diameter (m)				
	height 6 m	10 m	15 m	20 m	30 m
30	0·13	0·13	—	—	—
50	0·15	0·15	0·15	0·15	—
100	0·25	0·2	0·2	0·2	0·2
250	0·35	0·3	0·3	0·3	0·25
500	0·5	0·45	0·45	0·4	0·4
1000	—	0·6	0·55	0·5	0·5

These sizes assume flue to chimney 2 m long, 1 bend, 1 draught diverter, 1 terminal.

Sizes are to nearest 0·05 m.

Fan-Diluted Draught—In this system, one form of which is shown in Fig. 6.6, the combustion products, together with a quantity of diluting air, are exhausted to the atmosphere by a fan. By this means it is possible to dispense with a chimney entirely by discharging through the wall of the boilerhouse. The diluting-air quantity is such as to bring the CO_2 content of the mixture down to 1 per cent, which involves a fan to handle approximately 100 m^3 per m^3 of natural gas burnt, or half this quantity with towns gas. Included in the system is a fan-failure device to shut off the burners in the event of draught failure.

The duct sizing with this system should be on the basis of a gas velocity of 0·3 to 0·5 m/s.

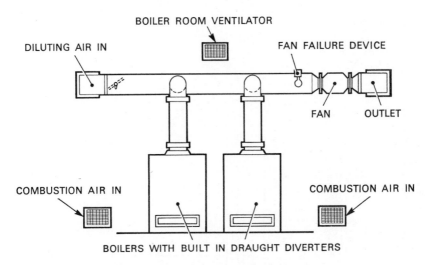

FIG. 6.6.—Fan-Diluted Draught.
(*By courtesy of the Gas Council*)

Materials for chimneys and flues—Condensation is likely to occur in gas and chimneys, particularly on start-up; hence they should be lined with an impervious and acid-resisting material such as glazed stoneware or asbestos cement. It is desirable to keep the chimney warm to assist draught, either by enclosing it in brickwork or concrete, or by insulation. Connecting flues between boiler and chimney are usually of asbestos cement. In the fan-diluted system they may be of galvanized steel with sealed joints and painted edges.

Explosion Doors—In the event of the malfunctioning of safety controls, the remote contingency of an explosion must be faced and means provided for its relief. If the boiler has an open skirt, this will serve to relieve pressure. Where, however, the combustion chamber is sealed, as when a boiler has been converted from another fuel or where a forced-draught burner is fitted, other provision must be made, such as by relief doors on the boiler or panels in lieu of doors which will shatter under pressure.

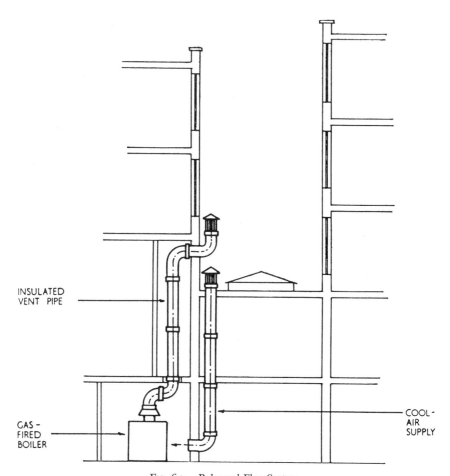

INSULATED
VENT PIPE

COOL-
AIR
SUPPLY

GAS-
FIRED
BOILER

FIG. 6.7.—Balanced Flue System.

Terminals—Care must be taken to see that flues discharge the products of combustion freely to the atmosphere. In buildings with return walls, areas, etc., peculiar atmospheric pressure conditions occur and it is advisable to extend the flue to a height of at least 18 in above the eaves of the roof. The possibilities of downdraught are then much reduced.

Every flue should be fitted with a terminal of approved design to prevent birds nesting, etc. A good terminal will ensure the maximum efficiency being obtained from a flue.

Bafflers, or more correctly 'draught diverters', should be fitted to the flues of all natural-draught gas boilers and heaters if these are not incorporated in the design of the heater.

Balance Flues—In difficult situations where the flue terminal cannot be carried above roof level, but must terminate in some position such as in an internal light well, differences of pressure are found to cause back draughts. These may be overcome by the provision of a balance flue, as in Fig. 6.7, and comprises a return flue similar to the rising flue except that it is carried down to near floor level in the boiler chamber. The sizing of this flue must be generous, as it has to convey not only the air for combustion, but also that drawn in by the baffler; and the resistance should, in any event, be kept as low as possible.

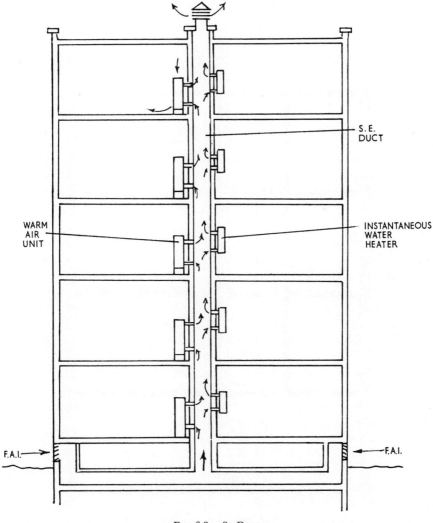

Fig. 6.8.—Se Duct.

Another flue system is the SE, as shown in Fig. 6.8. This is suitable for multi-storey buildings and, as will be seen, air enters at the base and the products of combustion from the various appliances discharge into the same duct. Thus the gas combustion space is virtually sealed off from the occupied space.

An alternative to the SE duct, for use where a bottom inlet is not possible, is the U-duct shown in Fig. 6.9. Fresh air taken in at the roof is conveyed down a shaft running parallel with the rising shaft, which acts as a shared flue as in the SE duct, the products of combustion being exhausted at the roof from a terminal adjacent to the intake, but at a slightly higher level. All gas burning appliances take their combustion air from, and return the products of combustion to, the rising duct.

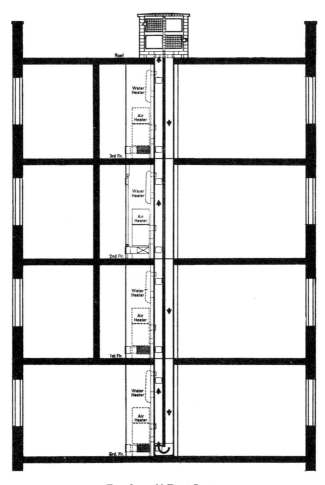

FIG. 6.9.—U-Duct System.
(*By courtesy of The Gas Council*)

Roof-top Boilerhouse—The flue problem is avoided altogether in a boiler installation if the boilers are on the roof. Fig. 6.10 shows such an arrangement with short outlets from the boilers carried through the roof of the penthouse to the atmosphere. Gas boilers on the roof are obviously simpler to deal with than oil—there is no delivery pumping and negligible fire risk. Thus, in many examples of multi-storey buildings, the use of gas in this location for boilers has much to commend it.

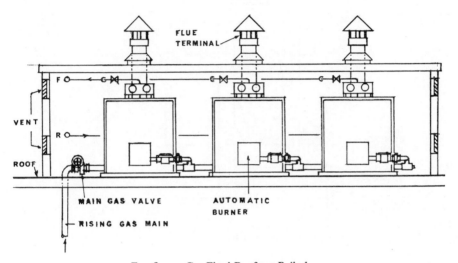

FIG. 6.10.—Gas-Fired Roof-top Boilerhouse.

GAS WARM AIR

This system has been developed for residential use, i.e. for houses and flats.

The air-heating unit is shown diagrammatically in Fig. 6.11 and comprises a combustion chamber formed of some non-corrodible metal, a fan for blowing air over the combustion chamber, and a casing into which air is drawn from the rooms and from which it is returned after warming.

Units range from 8 to about 24 kW, and are suitable for heating a small 2-bedroom flat up to a 4-bedroom house.

As generally planned in a flat, the unit has short duct connections to discharge the warmed air near floor into two or three rooms, the return being at high level from the hall, or from one of the rooms. When applied to a house, delivery ductwork carries the warm air *via* the under-floor space or roof space to floor grilles in the various rooms to be heated, the return being via the hall. So as to allow free movement of return air, some gap below the door or a transfer grille is necessary. It is important that the ducts are airtight, that all ducts are well insulated, and that gas pressure is according to maker's requirements.

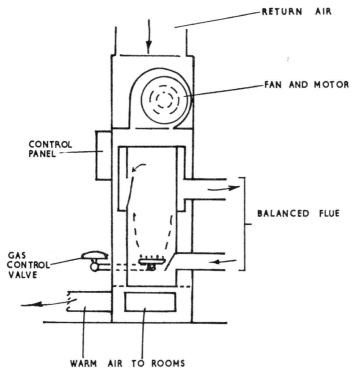

FIG. 6.11.—Gas Warm Air Unit.

The control of gas warm air units is generally by stopping and starting the fan and a magnetic gas valve so arranged that the gas can only be 'on' when the fan is running. A thermostat is provided in one of the rooms, together with a clock switch and often a low limit stat in order to maintain some background warmth such as 10° C even when 'off'.

The gas warm air system is economical in first cost, readily responsive to demands for warmth and has become a popular feature in multiple housing schemes.

Industrial Gas Warm Air—For industrial applications the type of air-heater shown in Fig. 3.1, page 47 (but with gas burner in lieu of oil) is applicable. The absence of storage is an advantage and, like the oil version, the unit is mobile if change of lay-out is required.

GAS-SUPPLY PIPES

At the point of entry there will be

> main gas cock,
> filter,
> governor,
> meter.

From this point onwards the supply piping is to be sized to suit the demands of the apparatus. The *I.H.V.E. Guide* gives Tables for flow of gas against pressure drop for pipes, from which Table 6.2 is extracted for steel pipes (see page 126).

It will be noted that, retaining the kilowatt as the basis of calculation, a flow of gas of 1 litre/sec is equivalent to 41 kW with natural gas and 20 kW with towns gas. Thus, the input rating of any boiler or other appliance in kW may be converted into gas flow by dividing these equivalents. Alternatively, makers' lists usually give gas rates volumetrically, from which litre/sec may be derived.

In terms of pressure loss, the pipe loss from point of entry to terminal may be 100 N/m²*.

Then, taking as an example,

Input rating of boiler, 800 kW

Natural gas $\div$ 41 $\qquad$ = 20 litre/s

Pressure drop assume available $\qquad$ 100 N/m²

Run from intake to boiler,
including single resistances $\qquad$ 30 m

per m run $\dfrac{100}{30}$ $\qquad$ = 3·3

From Table 6.2, size $\qquad$ = $2\frac{1}{2}$ in

The pressure available for distribution within the building is bound up with governor settings and burner pressures which, with natural-gas conversions, are matters on which the technical departments of the Area Gas Boards will advise.

FUTURE DEVELOPMENTS

This chapter has so far been concerned with the application of gas heating only to single buildings, dwellings, et cetera. Now, however, the availability of cheap natural gas opens up developments on a very much wider scale for schemes such as group heating and district heating— matters which are referred to in Chapter 23.

Furthermore, gas can also furnish the cooling requirement of air-conditioning by use of the absorption system, and it can provide power generation by means of the gas turbine. This co-ordination of all the energy requirements of a given area, such as a new town, offers very considerable scope for the more efficient use of national fuel resources. This subject is dealt with further in Chapter 24 under the title of *Total Energy*.

* $\frac{1}{10}$ inch water gauge = 25 N/m² approx. Thus 100 N/m² = $\frac{4}{10}$ inch water gauge.

TABLE 6.2

FLOW OF TOWNS GAS IN STEEL PIPES
(MEDIUM GRADE TO BS 1387)

Litres per second in pipes having nominal bores, as stated

N/m² per m run	15 mm	20	25	32	40	50	65	80	100	125	150
	½ in	¾	1	1¼	1½	2	2½	3	4	5	6
1·0	0·15	0·39	0·75	1·62	2·47	4·72	9·6	14·9	30·6	54	88
2·5	0·3	0·67	1·29	2·77	4·20	7·99	16·2	25·1	51·2	90	147
5·0	0·45	1·01	1·92	4·12	6·24	11·8	23·9	36·9	75·2	133	215
7·5	0·57	1·28	2·43	5·19	7·84	14·8	30·0	46·2	93·9	166	268
10·0	0·67	1·51	2·86	6·10	9·22	17·4	35·1	54·1	110	194	313
15·0	0·85	1·91	3·60	7·66	11·6	21·8	43·9	61·6	137	241	389
20·0	1·00	2·25	4·24	9·00	13·6	25·6	51·4	79·1	160	282	454
25·0	1·14	2·55	4·81	10·2	15·4	28·9	58·0	89·2	181	318	511

Note:

1. *For Natural Gas* at constant volume flow, multiply tabulated pressure loss by 1·10.
   At constant pressure loss, divide tabulated volume flow by 1·05.
   (For approximate purposes, the Table may be used as it stands for natural gas).

2. *Single Resistances*
   For approximate purposes equivalent length may be taken as 1 m per 1 inch nominal bore for each bend, tee etc. For accurate values see *Guide*.

3. *Heat equivalent*
   Natural Gas 41 MJ/m³ = 41 kJ/litre
   thus   1 litre/s = 41 kW
   Towns Gas 1 litre/s   = 20 kW

CHAPTER 7

Low-Pressure Hot-Water Heat Emitters

IN THIS COUNTRY, WITH its unpredictable climate, hot water is usually the best medium for heating on account of its simple temperature control to meet variations in weather, and the absence of the parching effect commonly associated with steam and heated air. Warming by hot water therefore deserves detailed consideration.

Low pressures are generally employed—which means temperatures below boiling point, usually at 82° C maximum. The use of higher pressures is considered in Chapter 11.

The question of the choice and disposition of heating surfaces for any particular installation has previously been discussed, and it now remains to show how the amount of such surface necessary to meet a given heat loss may be established.

EMISSION FROM HEATED SURFACES OF VARIOUS FORMS

The total emission from a heated surface may be divided into radiant and convective components, of which the radiation is proportional to the difference of the fourth powers of the absolute temperatures of the radiating and absorbing surfaces. The convection varies considerably with the form and height of the surface, but if the heat emitted by convection at any one temperature is known, that at some other temperature will be proportional to the temperature difference to the power of 1·25.

Due to the convection currents set up by any heated element in air, no surface gives 100 per cent radiation, though a heated ceiling closely approaches it (about 90 per cent). Panels on walls may have 60 per cent radiation and 40 per cent convection, and in floors 50 per cent radiation and 50 per cent convection. Radiators of the ordinary type vary according to their convolutions, but transmit commonly 20 per cent by radiation and 80 per cent by convection. Convectors give almost 100 per cent of their output by convection and none by radiation.

It has in the past been usual to evaluate heating surface in terms of area and to give a coefficient of heat emission per unit area based on makers' tests. The coefficient multiplied by the temperature difference mean water to air gives the transmission per m², and hence the heating surface for a given heat loss in a room may be derived. Whilst the heat emission from a plane surface or a simple shape such as a cylinder can be calculated on theoretical grounds, conventional radiators and the like are not susceptible to such treatment.

It is not possible, in any event, to apply this method to a convector which contains some form of finned surface in a metal box, and the same applies to many other forms of heating appliance. Thus a British Standard has been produced, No. 3528, which stipulates a standard method of test according to which actual emissions are to be stated per section of radiator or per unit length.

Normal Design Temperatures—For low-pressure hot-water systems, recommended operating temperatures are given in Table 7.1 (see page 144)* when the external temperature is assumed to be − 1° C. As explained in Chapter 3 this does not preclude higher flow temperatures being used under colder weather conditions.

EMISSION FROM PIPES

Exposed piping as a form of heating surface is rarely used, but in any system of distribution, main and branch piping running through the spaces to be heated may contribute something thereto. For instance, if there is a system of overhead radiant panels or unit heaters in a factory, the piping serving this equipment may be run below north-light glazing, so serving to prevent downdraught.

Any piping so disposed in the heated space gives useful heat and the amount emitted may be deducted from the total required.

The same applies where piping serving radiators under a range of windows is left exposed. Greater economy is achieved than if buried in a trench, though with some sacrifice in appearance.

The theoretical heat emission from piping may be calculated from formulae given in the *I.H.V.E. Guide*. Table 7.2 (page 145), based on these formulae, is extracted from the *Guide*. Note that pipe sizes are given in inches as well as in metric, in accordance with common usage.

The reduction in emission, where pipes are used in coil form vertically over one another, may be taken approximately as follows:

2 pipes	5 %	reduction
4 ,,	15 %	,,
6 ,,	25 %	,,

In order to reduce the heat loss from mains and other piping, insulation is applied in one of the forms discussed later (see page 185). The loss is dependent upon the conductivity of the insulation and its thickness. Table 7.3 (page 145) gives the loss expressed as a percentage of the bare-pipe loss for 85 per cent magnesia insulation, having a conductivity at 80° C of about 0·6 W/m° C. The conductivity of glass fibre at the same temperature is about 0·45 and the percentage loss would be reduced pro rata.

By way of example in the use of these Tables, assume a two-pipe pumped system and main piping 30 m in length.

* Table 7.1 and all ensuing Tables dealing with heat emissions are grouped together at the end of this Chapter, on pages 144 to 152.

Plate II. A Pipe Tunnel serving a large hospital. The pipes convey steam, condense, heating and hot and cold water. Electric cables are routed through it (see p. 53)

Plate III. An Automatic Stoker-fired Boiler Plant serving blocks of flats (see p. 90)

Plate IV. Example of use of long steel radiators in a hospital ward (see p. 131)

From Table 7.1, flow at 80° C   return at 65° C

$$\begin{array}{cc} \text{air } 15 & 15 \\ \hline \text{difference } 65 & 50 \end{array}$$

bare loss from Table 7.2,

$$\begin{array}{ccccc} & 180 \text{ W/m} & & 130 \text{ W/m} & \\ \times 30 \text{ m} & 5400 \text{ W} & + & 3900 \text{ W} & = \quad 9300 \end{array}$$

If insulated 85% magnesia 25 mm thick

Mean temp. diff. hot surface to air   =   57·5°
take percentage for 55° from Table 7.3   =   21
The loss for the two pipes insulated

$$= 9300 \times \frac{21}{100} \qquad\qquad = \qquad 1953 \text{ W}$$

RADIATORS

Radiators are generally best placed under windows, both for architectural and technical reasons. Architectural, because here they do not mar an otherwise unblemished wall surface, and technical, since this is the point of maximum heat loss and dirty marks on the walls are minimized.

The placing of radiators in relation to cooling surfaces also has a bearing on temperature gradient. Thus a radiator on an inner wall will deliver a strong current of warm air to the ceiling, whereas under a window the cool downward currents from the glass will lower the rising air temperature and hence the force of the upward current. For the same reason, long windows may be better dealt with by long radiators rather than short concentrated ones.

In a high building, such as a church, radiators or pipe coils under high level glass will prevent strong downward draughts, and, under a factory north-light roof or any roof-light, the same applies.

Shelves over Radiators—It is frequently desirable to carry the sill of the window over the top of the radiator, the latter being placed in a recess in the wall so as to save floor space.

Similarly, when placed against a wall, not under a window, the provision of a shelf over a radiator reduces the smoky marks above it, which would certainly otherwise occur. The jointing to the wall should be sound, as the slightest crack will allow a black mark to form, and end shields are necessary to prevent markings at the sides. A useful shield is one made

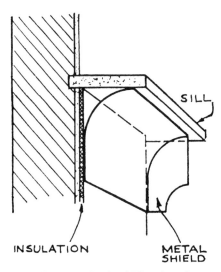

INSULATION   METAL
                              SHIELD

FIG. 7.1.—Sectional Elevation of
Metal Radiator Shield.

of one piece of metal with the angle curved, and to the top of this, timber, marble, or other material may be screwed as in Fig. 7.1.

Provided such shelves are not less than 75 mm above the radiator the reduction in transmission caused by them is so slight that it may be ignored.

Radiator Grilles—Where the appearance of radiators cannot be tolerated on architectural grounds, grilles of painted iron, bronze, hardwood or other material were at one time commonly used to enclose them. These not only reduce the heat transmission, thereby calling for increased surface, but are in themselves costly and dirty. The greater use of convectors has generally superseded this method of treatment.

Fresh-Air Inlets to Radiators—These were, in the past, often provided behind radiators. Their effect is to increase the heat transmission by some 20 to 30 per cent. When such inlets are used, a baffle plate is necessary in front of the radiator to prevent the direct in-flow of cold air, and means for closing the grating by adjustable louvres or a hit-and-miss register is provided.

These inlets, admitting air straight from outside, tend to make the radiators and their enclosures dirty; and, moreover, it is frequently found that they are stopped up on account of the draughts which otherwise occur with strong winds on that particular face. Other methods, such as the fan-assisted convector with filter, are a better solution.

Painting of Radiators—It has already been mentioned that the best radiating and absorbing surface is dead black, and that the painting of a surface white (such as a roof) will reduce the absorption and radiation considerably. Whilst this is true of high-temperature radiation, such as is received from the sun, and is important when considering the material for a roof, it does not apply at low temperatures. It has been established that radiation from various painted surfaces, glossy or matt, including black, green, brown, red, ivorine, and white, is indistinguishable at temperatures below 100° C.

Polished metallic surfaces are, however, bad *radiators* of heat even at these low temperatures (in some cases only 10 per cent as efficient as the unpolished surface), as are the metallic paints such as aluminium, bronze or gold, though the *convection* is not of course affected.

The use of such metallic paints reduces the radiation component by about 50 per cent. Assuming that an ordinary radiator emits 20 per cent by radiation, a 10 per cent total reduction may be expected. It is stated that a coat of varnish over a metallic paint brings the emissivity back almost to normal.

Cast-Iron Radiators have been in existence for nearly 100 years, though they have greatly improved in appearance over this period. The varieties comprise two- to six-column types, wall type, window type and hospital type (which has smooth external surfaces easily cleanable). Heights vary from 300 mm to 870 mm. Emissions are given in Table 7.4 (page 146) at the end of this Chapter. Fig. 7.2 shows two of the types mentioned.

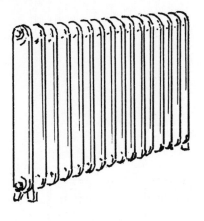
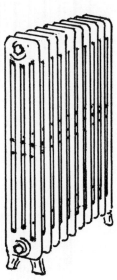

Fig. 7.2.—Cast-Iron Radiators. *Above:* Hospital Type. *Right:* 4-Column Type.

Steel Radiators are made of light steel pressings welded together, thereby being generally considered more attractive and modern in appearance, particularly for domestic applications. Furthermore, they are lighter for erection and may be supported on brackets from light partitions, which is not possible with the heavier cast iron ones. Where necessary, steel radiators may be obtained in longer lengths as in Plate IV (facing page 129). A disadvantage of steel is that it is prone to attack from any corrosive propensities in the water and hence a corrosion inhibitor is advised.

Various types of steel radiator are shown in Fig. 7.3. Emissions for some examples are given in Tables 7.5 and 7.6 (on page 147).

Temperature Difference—The basis of emissions in the Tables above referred to is 60° C between air and mean water. Where some other temperature difference is adopted, the factors given in Table 7.7 (page 148) apply. These are based on $\left(\dfrac{\text{Temp. diff. °C}}{60}\right)^{1.3}$.

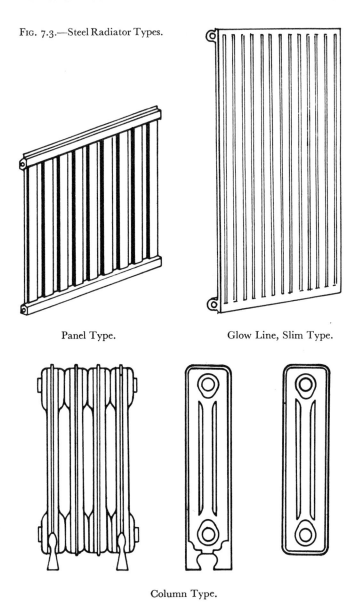

FIG. 7.3.—Steel Radiator Types.

Panel Type.   Glow Line, Slim Type.

Column Type.

CONVECTORS

A 'natural' convector comprises a finned tubular-element mounted near the bottom of a casing, such that the 'flue effect' of the rising column of heated air within the casing causes a flow of warm air to issue from the top, whilst cooler air is drawn in near the floor. The greater the height of the casing, the greater is the emission of heat for a given size.

Convectors may be free-standing as in Fig. 7.4, or recessed (see Fig. 3.12, page 57). A damper for hand control is incorporated in some types. Convectors should only be used with pumped circulation.

Where control is by adjustment of water temperature, it is to be noted that convector emission falls off more rapidly as water temperature is lowered than does that from a radiator, due to the absence of a radiation component as well as to the drop in flue effect with the decline in temperature. Hence, control of temperature with convectors should be over a more limited range than with radiators. Furthermore, it is inadvisable to mix the two forms of equipment on the same circuit. Separate circuits under different control characteristics are necessary.

The velocity of water through the tubes also has a bearing on transmission rates; thus, where the element has two or more tubes, a series arrangement offers a greater output than a parallel arrangement.

Typical heat emissions for a few sizes and heights of convector are given in Table 7.8 (page 148) based on 60° C temperature difference air to mean water. A correction for other temperature differences may be based on $\left(\dfrac{\text{Temp. diff. °C}}{60}\right)^{1.5}$; but see makers' lists also.

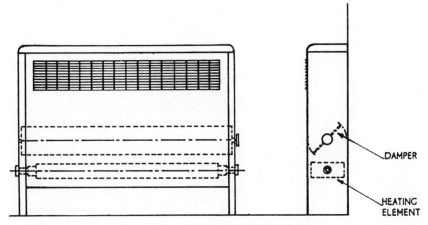

FIG. 7.4.—Natural Convector, Cabinet Type.

Continuous Convectors—This application of the convector is specially suited to the modern form of continuous glazing in a building designed on a modular basis. The finned heating element is continuous from end to end of the building (subject to certain provisions for expansion) and the steel casing is likewise continuous, see Fig. 7.5. The louvre outlets for warm air at the top are in sections, one per module, and each section is provided with a damper for control of heat. The heating element naturally runs continuously uncontrolled. Where partitions occur, a sound-proof barrier is necessary inside the casing.

An example of the use of continuous convectors treated architecturally, as a barrier rail, may be seen in Plate V (facing page 144).

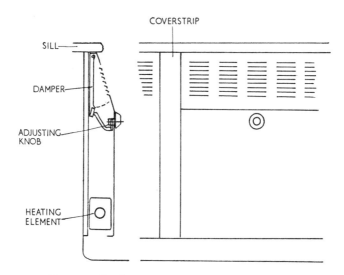

FIG. 7.5.—Continuous Convector (Copperad Sill-line).

Indirect Radiators—These are convectors composed of gilled pipes for placing below floors or in hidden recesses with grilles. They are largely obsolete.

FIG. 7.6.—Skirting Heating (Copperad Wallstrip).

Skirting Heating—This provides a neat and unobtrusive system especially for domestic use. The skirting is heated by pumped hot-water circulation and may run round one or all sides of the room.

One form is in effect a miniature convector and comprises a finned element enclosed in a steel casing, as in Fig. 7.6. Masking and dummy sections are available to fit corners, etc.

Emission for this type is given in Table 7.9 (see page 149). Another form is in cast-iron, by Crane.

Fan Convectors—One make is illustrated in Fig. 3.13 (page 57) and another is shown in Fig. 7.7. This is the basic unit having an outlet grille on the front. A base section may also be added, if required, allowing fresh air to enter from the back, as well as recirculated air from the front. The unit shown also has a filter.

In some applications a high-level outlet is required, for which purpose an extension is added to the casing.

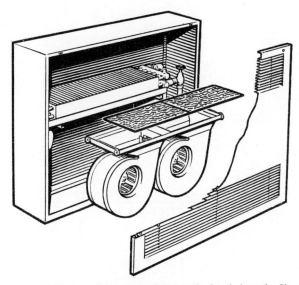

Fig. 7.7.—Fan Convector (Copperad). Note the fans below, the filter, the heating coil above, and the inlet and outlet grilles on the face of the casing.

Two-speed or three-speed control of the fan is usually included, the top or boost speed being useful for quick heat-up when the noise level is unimportant. At the lower speeds these units are generally acceptably quiet. Thermostatic control is by on–off switching or by change speed.

Typical emissions are given in Table 7.10 (page 150), from which it can be seen that relatively large spaces can be heated with few units. Thus, for instance, one unit per classroom suffices in a school and two or three units suffice in a hall or gymnasium. The system has been applied success-fully in churches. The emissions in Table 7.10 are based on 60° C tem-perature difference air to mean water. Outputs for other temperatures follow a linear relationship.

Fan convectors can be used for the heating of multi-storey flats. Hot water is circulated from a boiler serving a unit in each flat, as shown diagrammatically in Fig. 7.8. This then functions as any other gas or electric warm-air system, air being drawn into the unit and delivered through short ducts to the various rooms. Charge for heat may be on the basis of fan-running hours, for which a simple meter suffices.

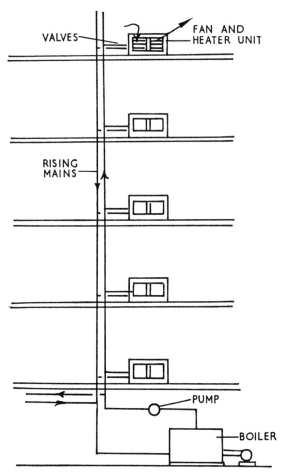

FIG. 7.8.—Fan-Convection System for block of flats.

Miniature-size fan convectors have been developed, such as in Fig. 7.9, suitable for domestic use and taking the place of radiators.

The small water content of these convectors means that they are very rapid in response and give the greatest economy with intermittent heating.

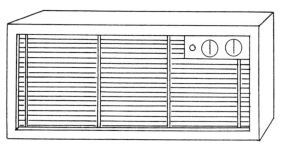

FIG. 7.9.—Minivector (Copperad).

RADIANT PANELS (METAL)

Radiant panels of cast iron are shown in Fig. 7.10, type 15. In type 35 the water-way is cast with a front plate of steel.

If they are used in the ceiling, they must be jointed to the plaster at the edges, and to prevent a crack appearing a cover strip is usual. This is not always architecturally desirable, and for this reason the embedded panel or 'acoustical ceiling' is to be preferred. Metal plates are much more adaptable to wall positions, though it will generally be found that their length considerably overruns the spaces under the windows, unless convection types such as No 44 are used.

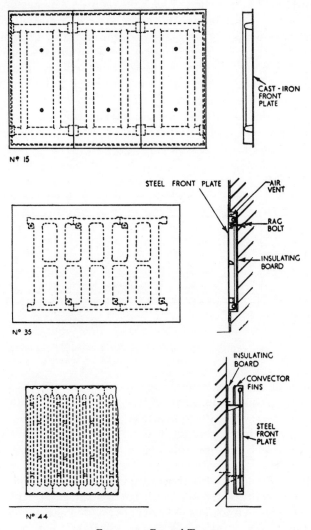

Fig. 7.10.—Rayrad Types.

Steel plate panels (of which there are various makes comprising, in principle, a flat steel plate with tubes attached or welded to the back) are adaptable to almost any shape, size and position. They are best concentrated under or over the windows, where the major heat loss occurs. This type has also been developed for warming skirtings, floor-borders and cornices, or they may be suspended overhead in work-rooms or factories—but these are matters dealt with in Chapter 11.

It is important that where panels of any type back onto exposed walls, the rear surface should be insulated to reduce direct transmission losses.

EMBEDDED PANELS

Ceiling Panels—The practice of heating rooms by means of embedded ceiling panels was common in the older type of building with heavy construction, where a long time lag was unimportant. Present-day light forms of construction and extensive glass areas require systems with rapid response. Furthermore, the greater use of false ceilings, often with some form of acoustical treatment, renders the embedded system impracticable. It is proposed, therefore, to make only brief reference to it here.

The sinuous coils, usually of ½ in bore steel or copper at 150 mm centres, are laid on the shuttering of a concrete slab prior to concreting. For a good key, slip tiles are placed between the pipes, see Fig. 7.11.

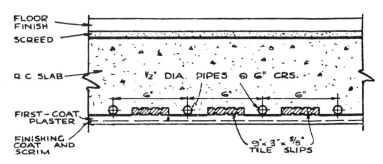

FIG. 7.11.—Detail of Embedded Ceiling Panel in Reinforced Concrete Slab.

The soffit is plastered to a special specification as issued by the Invisible Panel Warming Association. Ground floors having no heat below may require some floor heating. Roof slabs must be well insulated.

A technique for incorporating heating panels in suspended ceilings is shown in Fig. 7.12. Emissions from ceiling panels are given in Table 7.11 (page 151).

Floor Panels—Floor panels consist of ½ in or ¾ in bore pipes, usually at 225 to 300 mm centres in coil form, laid in the floor screed which is usually of 50 to 75 mm thickness. A variety of floor finishes is now commonly used over floor panels, and, subject to appropriate measures being taken

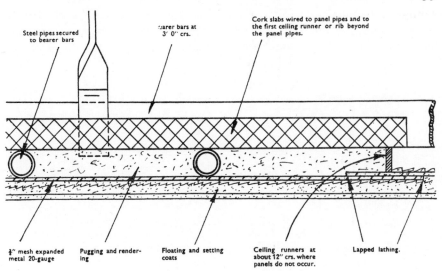

Steel pipes secured
to bearer bars

bearer bars at
3' 0" crs.

Cork slabs wired to panel pipes and to
the first ceiling runner or rib beyond
the panel pipes.

½" mesh expanded
metal 20-gauge

Pugging and render-
ing

Floating and setting
coats

Ceiling runners at
about 12" crs. where
panels do not occur.

Lapped lathing.

FIG. 7.12.—Heating Panels in a Suspended Ceiling.

to render material suitable for the temperatures involved, almost any normal floor finish may be used. Thus, wood blocks adequately kilned and bedded in suitable mastic, not softening at the temperatures involved, are found to be satisfactory. Cork tiles and a variety of thermoplastic and rubber tiles have also been used.

Most of the harder types of material for floor finishes are equally satisfactory for floor warming. Thus, marble, stone, terrazzo, Granwood and various similar proprietary materials are suitable, subject to due precautions being taken, bearing in mind the possibility of expansion which might lead to cracking. Advice on precautions and details associated with floor heating are available from the Invisible Panel Warming Association. For instance, in the laying of stone or marble paving over floor-warming panels, it is necessary to bed the material all over and not by dabs as is sometimes done, otherwise an air gap is introduced and the transmission seriously retarded.

Another important matter for consideration, where floor panels occur on the ground floor, is the insulation of the floor from the earth below. It is found that where insulation is provided a quicker heating-up rate takes place, and the downward loss, which may be as much as 25 per cent is largely eliminated. Fig. 7.13 shows a method of insulation in which it will be noted that the insulating material is carried up at the perimeter of the ground floor slab, where it abuts on the wall, so as to form an edge insulation.

The form of the material must be suitable for the conditions, and involves the choice of such materials as are capable of withstanding loads likely to occur as well as being immune from rot or other deterioration. A damp-proof course is usually carried under the insulating material. It will also be noted that the pipes forming the panel coils do not lay directly on the insulating material, which might be liable to cause external corrosion through moisture contained therein.

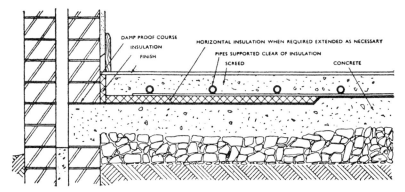

FIG. 7.13.—Panel Heating. The floor is on made-up ground with the membrane either sufficiently protected by insulation or of a type not readily affected by local heat.

It is found to be unpleasant and tiring to sit for long with the feet in contact with a warm floor. The maximum floor-temperature to give comfort with continuous use is about 24° C, and, where occupants are not stationed for long periods, 27° C. In entrance halls and the like, and within 1·6 to 2 m of the walls of a room, a temperature of 30° C is permissible.

Where floor heating is adopted, it is generally found necessary to cover the whole area of the floor with embedded pipes, varying the spacing according to the amount of heat required. Such systems have been successfully used in schools and churches (see Plate VI, facing page 145).

Floor panel coils are sometimes made of copper, to avoid rusting due to water from washing of floors, etc., entering through cracks.

Emissions from floor panels are given in Table 7.12 (page 151).

Wall Panels—Pipes may be embedded in walls and covered with plaster, as in the case of ceilings, or with marble or some other similar finish. They are, however, liable to be obscured by furniture and to cause convection-current markings. Where pipes are in external walls, insulation is desirable.

Ceiling Heating and Room Height—It is necessary to consider, in the case of ceiling heating, whether the radiant heat on the head will be unpleasant. In the case of a low ceiling height there is obviously a greater danger than in higher rooms. At the same time, by limiting the width of a panel the effect of the direct radiation can be reduced. The *Guide* gives the relationship between maximum desirable temperatures of ceiling panels at different heights and of different sizes in relation to the comfort of the occupants. From this it is clear that, for any room height of 3 m or over, the panel dimensions common in practice will not produce conditions of discomfort. For rooms of less height it may be desirable to reduce the width to 0·6 m or less, which would involve the spreading of the panel surface in a narrow band round the perimeter of the room.

The effectiveness of ceiling panels is reduced with increasing height of rooms. An arbitrary allowance for this is to add 1 per cent to the heating surface for every 0·3 m of height above 5 m. In addition it is desirable to bring the panels for unusually high rooms nearer to the centre of the ceiling so as to reduce the amount of heat falling on the exposed walls and windows. High rooms commonly have tall windows, and it is desirable to check the downdraught from these by convection heaters under the sills.

'PANELITE'

A proprietary version of panel heating under the above trade name consists of pipe coils enclosed in a split tubing of asbestos cement as shown in Fig. 7.14. This arrangement allows freedom for expansion and contraction; consequently, higher water temperatures may be used than in the solid embedded type. The coils can be installed in concrete or hollow tile slabs, or laid in floor screeds, or recessed into walls. Emissions are similar to the embedded panel type.

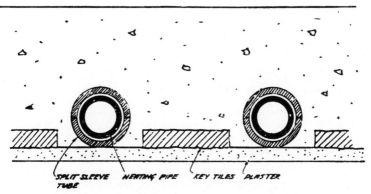

Fig. 7.14.—'Panelite' Heating as applied to Concrete Slab.

HEATED METAL ACOUSTIC CEILINGS

This type of heating combines acoustical treatment with radiant warming. In one of the proprietary makes, the ceiling is composed of thin aluminium plates (0·6 metres square) having perforations, and above this is placed a layer of thermal and acoustical insulating material, such as a 25 mm blanket of glass silk or slag wool. Fig. 7.15 shows a typical layout of a ceiling. The plates are clipped to the piping. It is usual to cover the whole ceiling of a room with this ceiling treatment, and only such parts of the coils as are necessary for heating are made alive, the remainder being blanked off or plugged during the construction.

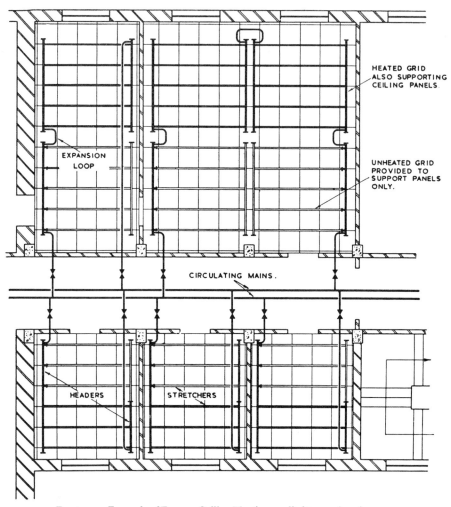

FIG. 7.15.—Example of Frenger Ceiling Heating applied to a series of rooms.
Heated sections are shown in solid black.

It is usual to equip systems of this kind with thermostatic control, which may be of the individual room-control type comprising a thermostat in each room which controls, by means of a motorized valve, the circulation of water through the coils serving that particular room. It should be noted that as the metal ceiling replaces the normal sub-ceiling, no plastering is involved. (See Fig. 7.16 and also Plate VII, facing p. 145.)

In another proprietary form a continuous hot-water pipe coil is supported above and independently of the perforated steel plates forming the acoustical ceiling, as shown in Fig. 7.17. Above the pipes is placed a thermal and acoustic insulating pad supported on wire mesh. The con-

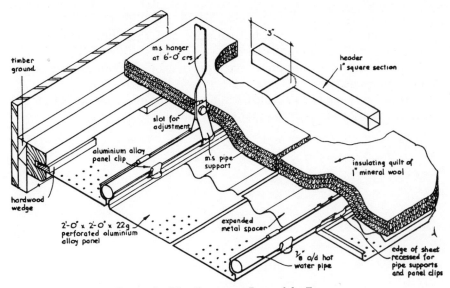

FIG. 7.16.—The Component Parts of the Frenger
Suspended Heating and Acoustic Ceiling.

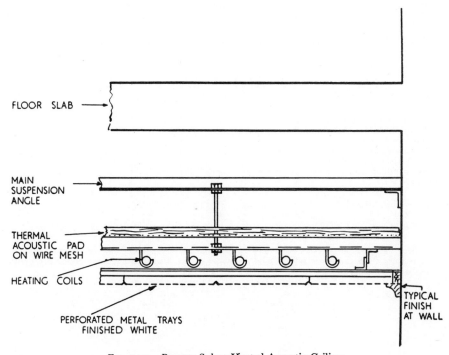

FIG. 7.17.—Burgess-Sulzer Heated Acoustic Ceiling.

struction may be applied directly below a floor slab, or with a space as in the case of a normal sub-ceiling and as shown in the Figure.

The system is of low thermal capacity and therefore lends itself to individual room thermostatic control.

Typical emissions from the above types of acoustic ceilings are given in Table 7.13 (page 152).

Further systems of suspended ceiling heating make use of plaster tile heated from above in a similar manner to that just described.

TABLES: CHAPTER 7

TABLE 7.1*
DESIGN WATER TEMPERATURES FOR LOW-PRESSURE HOT-WATER SYSTEMS

System	Temps. °C at Boiler or Calorifier	
	Flow	Return
Radiators (including pipe coils)		
Gravity circulation - - - - - -	80	60
Pump circulation 2-pipe - - - - -	80	65
Pump circulation 1-pipe - - - - -	80	70
Metal Plate Panels		
Pump circulation 2-pipe - - - - -	80	65
Convectors and Unit Heaters		
Pump circulation - - - - - - -	80	70
Embedded Panel Coils		
Pump circulation 2-pipe		
Ceiling and wall - - - - - -	53	43
Floor - - - - - - - -	43	35
Sleeved panel coils, all positions - - - - -	70	60

Where radiators are served off a single pipe, successive radiators receive water at a lower temperature than those preceding. The following temperatures are recommended for design purposes:

Flow temperature at radiator	Minimum temperature at outlet from radiator
80	70
75	65
70	60
65	55

* See page 128.

*Plate V. A Continuous Convector forming a barrier rail. An example of the integra-
tion of the heating system with architecture (see p. 134)*

Plate VI. Floor Heating: embedded hot-water coils at St Bride's, Fleet Street (*see p. 140*)

Plate VII. A Heated Acoustic Ceiling during construction showing grid heating pipes,
perforated aluminium panels and insulation above (*see p. 142*)

TABLE 7.2*

THEORETICAL EMISSIONS FROM HORIZONTAL STEEL PIPES†

Nominal Bore		Heat emission Watts/metre run						
		°C Temperature difference surface to air. Ambient air temperature between 10° C and 20° C						
inches	mm	40	45	50	55	60	65	70
¼	8	28	32	35	39	43	46	50
⅜	10	33	37	41	46	50	55	60
½	15	40	46	53	59	66	73	80
¾	20	48	56	62	70	78	87	95
1	25	58	68	78	88	98	110	120
1¼	32	71	82	93	110	120	130	150
1½	40	78	92	105	120	130	150	160
2	50	96	112	130	150	170	180	200
2½	65	120	140	160	180	200	220	240
3	80	140	160	180	210	230	250	280
4	100	170	200	230	260	290	310	350
5	125	200	240	270	310	350	380	420
6	150	230	280	320	350	400	440	480

* See page 128.
† Steel pipes to BS 1387 and ISO/R 65.

TABLE 7.3*

EMISSIONS FROM INSULATED PIPES EXPRESSED AS A PERCENTAGE OF BARE PIPE LOSS
For 85% magnesia for temp. differences stated

Temp. diff. hot surface— ambient air °C	Thickness of insulation in mm	Nominal bore of steel pipe to BS 1387						
		in. ¾	1	1½	2	3	4	6
		mm 20	25	40	50	80	100	150
40	19	32	31	29	28	27	26	26
	25	29	27	25	23	22	22	21
	38	24	22	20	19	17	17	16
	50	21	19	17	16	14	14	13
55	19	29	28	26	25	24	24	24
	25	26	24	22	21	20	20	19
	38	21	20	18	17	16	15	15
	50	19	17	15	14	13	13	12
85	19	26	25	23	22	21	21	21
	25	23	22	20	19	18	17	17
	38	19	18	16	15	14	13	13
	50	16	15	14	13	12	11	11

* See page 128.
† *Note:* The effect of any hard-setting composition is usually ignored.

TABLE 7.4*

EMISSIONS FROM CAST-IRON RADIATORS FOR TEMPERATURE DIFFERENCE
AIR TO MEAN WATER OF 60° C (20° to 80°)

Type	Height without feet mm	Depth mm	Length of Section mm	Heating Surface per Section m²	Emission		
					Watts per m²	Watts per section	kW per m length of radiator
Beeston 2 Column	409	67	50	0·070	555	39	0·76
	509	67	50	0·095	555	52	1·04
	709	67	50	0·125	555	69	1·38
4 Column	409	142	50	0·14	510	71	1·42
	559	142	50	0·19	510	97	1·94
	722	142	55	0·24	510	122	2·22
	871	142	55	0·30	510	153	2·78
6 Column	409	219	50	0·20	480	96	1·92
	559	219	50	0·28	480	134	2·68
	722	219	55	0·38	480	182	3·30
	871	219	55	0·46	480	221	4·00
Hospital 76 mm	390	76	51	0·70	555	39	0·76
	543	76	51	0·93	555	52	1·02
	695	76	51	0·121	555	67	1·31
146 mm	390	146	67	0·14	474	66	0·99
	543	146	67	0·18	474	86	1·29
	695	146	67	0·23	474	109	1·74
	848	146	67	0·28	474	132	2·10
185 mm	390	185	67	0·18	450	81	1·58
	542	185	67	0·23	450	104	2·04
	695	185	67	0·29	450	121	2·37
	848	185	67	0·35	450	157	3·07
Window	330	330	57	0·27	465	125	2·20
New Royal Wall	457	54	203	0·28	525	148	0·72
	609	54	203	0·37	525	195	0·94
	762	54	203	0·47	525	247	1·20

* See page 131.

TABLE 7.5*

EMISSIONS FROM STEEL-PANEL RADIATORS FOR TEMPERATURE
DIFFERENCE AIR TO MEAN WATER OF 60° C (20° to 80°)

Type	Height mm	Depth mm	Length of 12 corrugations mm	Heating Surface per 12 corrugations m²	Emission		
					Watts per m²	Watts per 12 corrugations	kW per m length of radiator
Steirad Super Panel Single	300	28·6	480	0·336	690	230	0·48
	440	,,	,,	0·480	660	310	0·64
	590	,,	,,	0·636	630	400	0·83
	740	,,	,,	0·792	620	490	1·02
Double	300	66·7	,,	0·672	540	360	0·75
	440	,,	,,	0·960	520	500	1·04
	590	,,	,,	1·272	510	550	1·23
	740	,,	,,	1·584	500	790	1·64
Treble	300	—	—	—	500	—	—
	440	—	—	—	480	—	—
	590	—	—	—	470	—	—
	740	—	—	—	455	—	—

Note: 1. The above emissions adjusted for 60° C temperature difference from published data for 55·6° C.
2. Standard lengths are in multiples of 4 corrugations i.e.: 12, 16, 20 etc. .... 80.
3. For fuller details see makers' lists.

* See page 131.

TABLE 7.6*

EMISSIONS FROM STEEL-COLUMN RADIATORS FOR TEMPERATURE
DIFFERENCE AIR TO MEAN WATER OF 60° C (20° to 80°)†

Type	Height without feet mm	Depth mm	Length of Section mm	Heating Surface per Section m²	Emission		
					Watts per m²	Watts per Section	kW per m length of radiator
Hull 2 column	432	76	37	0·07	500	35	0·95
	584	—	—	0·10	—	50	1·35
	736	—	—	0·13	—	65	1·75
	990	—	—	0·17	—	85	2·30
3 column	304	127	45	0·07	490	39	0·8
	432	—	—	0·12	—	59	1·3
	584	—	—	0·15	—	74	1·6
	736	—	—	0·20	—	98	2·2
	990	—	—	0·28	—	137	3·1
5 column	304	190	45	0·12	480	57	1·3
	432	—	—	0·18	—	85	1·9
	584	—	—	0·24	—	115	2·5
	736	—	—	0·31	—	148	3·3
7 column	304	280	45	0·19	470	90	2·0

* See page 131.
† Based on Hull Steel Radiator Ltd. list—converted to SI.

TABLE 7.7*

$$\text{Values of } \left(\frac{\text{Working Temp. Diff. }°C}{60}\right)^{1\cdot3}$$

Working Temp. Diff. °C	0	1	2	3	4	5	6	7	8	9
20	0·24	0·25	0·27	0·28	0·30	0·32	0·34	0·36	0·37	0·38
30	0·41	0·43	0·44	0·46	0·48	0·49	0·51	0·53	0·55	0·57
40	0·59	0·61	0·63	0·65	0·67	0·69	0·71	0·73	0·75	0·77
50	0·79	0·81	0·83	0·85	0·87	0·90	0·92	0·94	0·96	0·98
60	1·00	1·02	1·04	1·07	1·09	1·11	1·13	1·16	1·18	1·20
70	1·23	1·25	1·27	1·30	1·32	1·34	1·37	1·39	1·41	1·43
80	1·45	1·48	1·50	1·52	1·55	1·57	1·59	1·62	1·65	1·67
90	1·69	1·72	1·74	1·77	1·80	1·82	1·84	1·87	1·89	1·92
100	1·94	1·96	1·99	2·02	2·05	2·07	2·10	2·12	2·15	2·18

* See page 131.

TABLE 7.8*

EMISSIONS FROM NATURAL CONVECTORS FOR TEMPERATURE
DIFFERENCE AIR TO MEAN WATER OF 60° C (20° to 80°).
Water-temperature drop 10° C

Type	Height	Depth	Approximate Emission kW per m length of Convector
2 tube	460	115	0·75
	610	,,	0·84
	760	,,	0·90
	910	,,	0·98
3 tube	460	165	0·96
	610	,,	1·12
	760	,,	1·24
	910	,,	1·3
4 tube	460	210	1·10
	610	,,	1·37
	760	,,	1·60
	910	,,	1·78

Note: Lengths are from 0·5 m to 1·5 m in
steps of 50 mm.
For fuller details and factors for
other temperatures see Makers' lists.
Based on Flexaire.

* See page 133.

TABLE 7.9*

EMISSIONS FROM CONTINUOUS CONVECTORS AND SKIRTING HEATING
FOR TEMPERATURE DIFFERENCE AIR TO MEAN WATER OF 60° C
(20° to 80°). FRONT OUTLET

Type	Height	Depth mm		Emission kW per m finned length	
	mm	1″ tube	2″ tube	1″ tube	2″ tube
Copperad Sill Line Single tube	310	73	133	0·57	0·80
	410	,,	,,	0·66	0·93
	510	,,	,,	0·69	1·02
	610	,,	,,	0·72	1·08
	710	,,	,,	0·75	1·12
Double tube	410	73	133	0·74	1·06
	510	,,	,,	0·86	1·21
	610	,,	,,	0·94	1·36
	710	,,	,,	1·02	1·49
Copperad Skirting Convector (Wallstrip)	231	76		0·47	

Note: 1. Above emissions for Sill Line water-flow rates—
1″ tube 0·5 kg/s   2″ tube 2·0 kg/s
For Wallstrip   –   0·06 kg/s.
2. For fuller details, other temperature differences and water-flow rates consult Makers' lists.

* See page 134.

TABLE 7.10*

EMISSIONS FROM FAN CONVECTORS (COPPERAD)
FOR TEMPERATURE DIFFERENCE AIR TO MEAN WATER OF 60° C (20° to 80°).

Temperature drop 10° C
Height of basic unit, top outlet, 685 mm
Depth   229 mm

Type	Length m	Air volume litre/s Normal speed	kW emission	
			Normal speed	Boost speed
3 row	0·610	70	2·8	3·4
	0·915	140	5·9	7·2
	1·220	210	9·2	11·0
	1·525	280	11·9	14·3

Minivector Metal case	Height	298 mm
	Depth	160 mm
	Length	762 mm

Type	kW emission	
	Normal speed	Boost speed
210/240	3·1	4·5
M 14	2·9	4·1

Note: 1. For full details and emissions at other temperatures consult Makers' list.
2. In addition to the two speeds stated, a third low speed is available.

* See page 135.

TABLE 7.11*

EMISSIONS FROM CEILING PANEL HEATING WITH WATER AT 50° C MEAN

Construction	Downward emission W/m²		Upward emission W/m²	
	Room Temperature °C		Room Temperature °C	
	15	18	15	18
1. Hollow Tile 150 mm				
(a) cement screed and 25 mm hard finish above	175	160	80	70
(b) Screed and wood block finish above	190	175	70	55
2. Solid Concrete 150 mm				
(a) Screed and 25 mm hard finish above	175	160	100	80
(b) Screed and wood block finish above	175	160	80	70
3. Suspended ceiling 50 mm pugging, 25 mm insulation over	190	175	55	40
4. Roof. Hollow Tile or concrete with 50 mm insulation over	190	175	30	

Note: 1. Where upward emission is 80 W/m² and over with room temperature of 18° C, the floor area over the panel may be too warm for comfort.
2. Roof loss is to be taken in estimating boiler power and pipe sizing but not as contributing to room heating.
3. Data based on Invisible Panel Warming Association data (converted to SI).

* See page 138.

TABLE 7.12*

EMISSIONS FROM FLOOR-PANEL HEATING

Floor Finish	Centres of nominal ½ in bore pipes mm	Upward emission W/m² floor surface for temperature differences air to mean water °C				
		20	25	30	35	40
Screed 50 to 75 mm thick and 25 mm tile or other hard material finish	150	90	112	136	155	—
	225	83	104	124	142	—
	300	77	97	112	130	—
Screed as above but with 25 mm wood block finish	225	—	—	75	85	95
	300	—	—	70	80	90

Note: Based on Invisible Panel Warming Association data (converted to SI).

* See page 140.

TABLE 7.13*

EMISSIONS FROM HEATED ACOUSTICAL CEILINGS

Metal plates clipped to pipes (as Frenger):

Air temperature	20° C
Mean water temperature	80° C
Emission	175 W/m²

Metal plates with pipes above, but not attached
(Burgess Sulzer):

Temperatures	(as above)
Emission	170 W/m²

Upward emission (heated room over):

Both types	15 W/m²

* See page 144.

CHAPTER 8

Pipe Sizing for Hot-Water Heating

HAVING SELECTED THE TYPE, position, and extent of the heating surface, and marked this on the plans of the building in question, the next stage is to determine how this shall be connected to the boiler with pipes in the most economical and effective manner, and how the water shall be circulated.

Water Circulation—Water circulation in a system of pipes may either be by thermo-syphon (gravity), or by mechanical means (pump). A thermo-syphon or gravity system is one in which the circulation is produced by difference of temperature. Water at a higher temperature, such as issues from a boiler (flow), is less dense than the cooler water which has passed round a circulation (return), and advantage is taken of the difference of mass of rising warm-water column and dropping cool-water column to cause the circulation through the system.

In the case of circulation by mechanical means this usually implies the employment of a centrifugal pump, which is a convenient method of impelling water since it may be easily connected to an electric motor. A centrifugal pump is preferable to a plunger type of pump owing to the absence of pulsations whereby noise is produced. The centrifugal pump can be rendered to all intents and purposes noiseless. Furthermore it is more foolproof; for instance, even the closure of valves in the circulation will cause no harm.

Pump circulation has supplanted gravity circulation for all medium and large installations, and is now common even for the smallest installations, such as in houses. The advantages of pump circulation are:

(*a*) The circulation is independent of temperature.

(*b*) It gives a quicker response.

(*c*) Pump circulation allows the use of smaller bore pipes, thus saving in first cost and enabling pipes to be fitted inconspicuously more easily. The thermal capacity of a system using smaller pipes is naturally less than one with large pipes, and this again affects the rate of heating up and response to controls.

(*d*) Many forms of heat emitter are now suitable only for pump circulation, such as forced air convectors and metal plate radiant panels. Other types such as skirting heating, and normal convectors, are also much more advantageously served by mechanically circulated water. Temperature drops

153

may be less, and hence the size of equipment may be reduced due to the higher mean-water temperature in the apparatus.

New kinds of pumps have become available (as described later) designed to reduce the maintenance and cost of fixing, and some are particularly suitable for small installations.

Although the cost of the pump or pumps and the cost of power for running have to be allowed for, the advantages of forced circulation greatly outweigh the consideration of cost, which is usually a minor item compared with the overall capital and running cost of the installation.

In view of the importance of forced or pump circulations, it is proposed to consider the design of these in greater detail first, and to refer to gravity circulation second. Before doing so, however, it is necessary to describe some of the systems of piping commonly used.

SYSTEMS OF PIPING

Fig. 8.1: Single-Pipe Circuit—In this case the circulating piping leaves the boiler and returns to it in one continuous loop. Heat emitters are teed off the single-pipe flow and return and are therefore in shunt with a short section of the piping. The circulation from the pipe into the heat emitter and out again is thus largely independent of the circulation produced by the pump, and it is necessary in this case to size the connections of the branches as if they were gravity operated. The first radiator receives water at full boiler temperature, and the cooler return water from the outlet mixes with the water in the main which has bye-passed the radiator. So radiator No 2 receives a mixture at some lower temperature than that received by radiator No 1. The temperature drop from radiator to radiator is progressive, the last one receiving water at little more than boiler-return temperature. To preserve heat output, radiator sizes are therefore graded according to their calculated mean temperature. A small

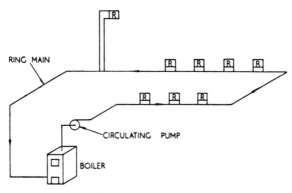

Fig. 8.1.—Single-Pipe Circuit.

overall-system temperature drop—boiler flow to boiler return—reduces the effect. The pump is shown in this case in the flow, but may equally well be in the return to the boiler.

Fig. 8.2: Parallel Single-Pipe Circuits—In this type of system there are a number of single-pipe circuits arranged in parallel, either in the form of a ladder or in the form of a series of single-pipe drops. Such an arrangement is frequently the most economical for serving a multi-storey building, or a school, as there are any number of ways in which similar parallel circuits may be arranged. Like the previous case, radiators are only slightly assisted as to their circulation by the action of the pump, being in shunt from a short section of the pipe in each case, and they are similarly subject to progressive mean temperature decline.

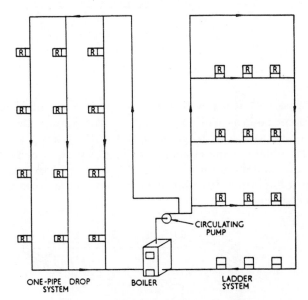

FIG. 8.2.—Single-Pipe Circuits in Parallel.

Fig. 8.3: Two-Pipe System—This system is sometimes called a 'tree' system, as it starts from a main near the boiler and branches out in all directions. Each branch conveys water at boiler temperature to the heat emitter, less any temperature drop due to heat given out by the mains, and returns the water direct to the boiler. Thus, the progressive lowering of temperature to successive heat emitters of the single-pipe system is largely avoided and, due to this, the two-pipe system is of greater application, being widely used on extensive systems. It may be applied to multi-storey buildings with risers, or to single-storey factories etc. with long horizontal flow and return mains, and drop-pipe connections to radiators or heat emitters in any one of a vast variety of ways. All radiators or other heat emission surfaces connected to the system, receive the full benefit of the pump in promoting circulation through the equipment in question. Many types of heater offering a high resistance to flow are only suitable for two-pipe application.

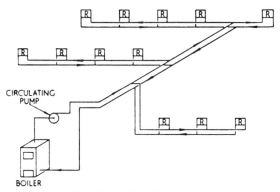

FIG. 8.3.—Two-Pipe System.

Fig. 8.4: Reversed Return System—This is a modified form of the two-pipe system and, as will be noted from the Figure, the pipework is arranged such that the total travel from the boiler to any one heat emitter and back to the boiler is the same in each case. This is clearly a method of simplifying the balancing-up of resistances, and is an admirable system where conditions permit. It is, however, usually more expensive than the previous tree system.

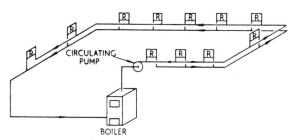

FIG. 8.4.—Reversed Return System.

Fig. 8.5: Panel Systems—These are always designed on the two-pipe principle. Panel coils offer a considerable resistance to flow, and assist in establishing a differential throughout the system, which aids in balancing up. The resistance of the water flowing through the panel coil must be allowed for in establishing the pump head of the system. It will be noted that the circulation is upwards through the panels, so as to remove accumulation of air to the highest points where it may be easily vented. An alternative method of piping panel systems on the drop-system has been used, but has fallen into disfavour due to difficulties of air venting.

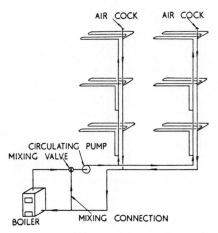

Fig. 8.5.—Panel System.

FLOW OF WATER IN PIPES

The mass flow of water to convey the required quantity of heat from the boiler or water heater to the heat emitters, in unit time, depends on the temperature drop from flow (t_f) to return (t_r) and the specific heat capacity of water $(\mathrm{J/kg°C})$ at that temperature. Thus

$$\text{mass flow kg/s} = \frac{\mathrm{J/s}}{(t_f - t_r) \times \mathrm{J/kg\ °C}}$$

Over the range of low-pressure hot-water systems, the specific heat capacity may be taken as $4.2\ \mathrm{J/kg°C}$.

This is a tedious sum to work out a great many times for various sections of a pipework complex and shows up the disadvantage of the SI system compared with a calorimetric unit, such as the British Thermal Unit, in which the specific heat of water is taken as unity.

Tables of mass flow in piping are to be found in the *Guide* based on kg/s. Heating loads, on the other hand, are worked out in kW; but $1\ \mathrm{kW} = 1\ \mathrm{J/s}$, so it would be more convenient if the data were presented direct in kW per °C, thus:

$$\mathrm{kW/°C} = \text{mass flow kg/s} \times \mathrm{J/kg\ °C}.$$

It then only remains to multiply by the temperature difference flow to return to obtain the heat flow in question. The properties of water do not enter into the matter.

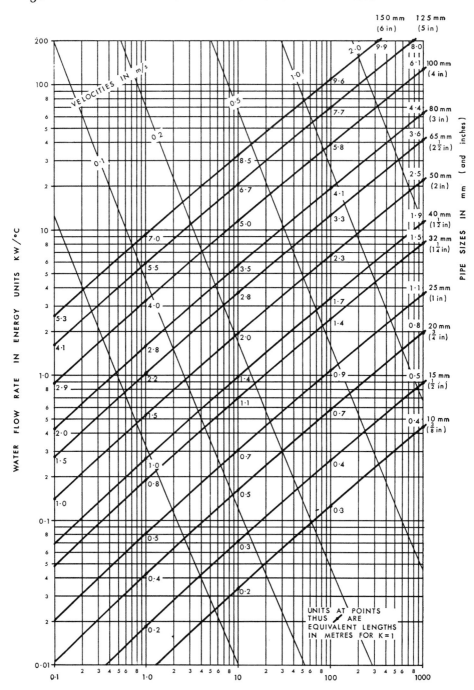

FIG. 8.6.—Pipe-Sizing Chart for Water at 75° C. (steel).

This book presents for the first time Fig. 8.6, which has been prepared from *I.H.V.E.* data by computer* in the form of a chart. The pipe size lines give kilowatts of flow per degree C temperature drop. Pressure loss is in N/m^2 per metre: i.e, N/m^3. The chart covers the range $\frac{3}{8}$ in to 6 in. Larger sizes are less commonly used and *I.H.V.E.* Tables are available.

By the use of this chart it is possible to retain the kilowatt throughout all heating calculations: heat loss, proportioning of heat emitters, summation of boiler power and pipe sizing.

It is as well to remember that the kilowatt is no more than a measure of energy flow. Thus we see in the above steps, first, how the rate of energy dissipation through the fabrics is calculated in heat losses, and second, the output of energy to counteract this loss room by room or space by space. Third, all come together at the boiler, or other heat source, where energy is to be produced in sufficient quantity to supply all demands. Finally, the system of piping has to be sized so that water circulating through the boiler and carrying the energy which it generates is split up into perhaps a great many sub-circuits and branches, and hundreds of terminal connections, to heat-emitting appliances. The object of pipe sizing is to achieve a balance throughout such that each appliance receives its proper share of the total energy distributed.

It is as well to keep this picture in mind and to remember that pipe sizing is only a means to an end and not an academic exercise. Thus, extreme meticulosity is entirely unnecessary and a waste of time. The limitations of commercial pipe sizes and the tolerances to which they are made have to be accepted; furthermore, for one reason or another the practical fitting of the system into the building may not go exactly to the drawings: more bends may be introduced, tees may branch in a different way and rough joints may occur, particularly if welded. A separate study might be made of the relative importance of things—some surprises might be in store!

Single Resistances—The method of making allowance for the resistance of bends, tees and other fittings is to add to the measured length of piping an *equivalent length* (*EL*) for each fitting. The values of *EL* are marked on the graphs in Fig. 8.6 at intervals and, as will be seen, they vary according to velocity and pipe size. They are, in fact, velocity-pressure equivalents and require to be multiplied by a factor, K. As given in the *Guide*, the factors vary from 0·3 for a long radius bend over 4 inch in size to 0·9 for a return bend of one inch or under. Tees and normal bends and elbows are in the range $K=0·5$ to 0·7.

Bearing in mind the above dictum on the relative importance of things, and that resistances are of minor significance, it is excusable in the interests of simplicity to allow all such fittings as having $K=1$.

---

* The Authors are indebted to Oscar Faber and Partners for the use of their computer for this purpose.

This is certainly on the right side and contributes something to unknown factors such as have been mentioned.

The allowance for single resistances thus becomes simply a matter of counting up bends, tees, elbows, reducers and the like, and multiplying by the *EL* for the particular size taken off the chart. As a first shot, when neither size nor velocity are known, no more than a guess can be made, but this can be checked later.

PIPE SIZING

Single-Pipe Circuit—Referring again to Fig. 8.1 (page 154), the single pipe must carry sufficient energy, via the water flowing in it, to supply the total emission from all the radiators plus the emission from the main. This total will equal the output of the boiler.

$$
\begin{array}{ll}
\text{Assume Radiator emission} & = 60 \text{ kW} \\
\text{Pipe emission (taken as 2 in)} & = 10 \text{ kW} \\
\hline
& 70 \text{ kW} \\
\hline
\end{array}
$$

$$
\begin{array}{ll}
\text{Temp. drop flow to return at boiler} & 12° \text{ C} \\
\text{kW/}°\text{C} = \dfrac{70}{12} & = \quad 6
\end{array}
$$

$$
\begin{array}{ll}
\text{Length of main} & 90 \text{ m} \\
\text{Allow for single resistance} & 10 \text{ m} \\
\hline
& 100 \text{ m}
\end{array}
$$

There is now a choice, using Fig. 8.6:

$$
\begin{array}{lll}
2 \text{ in} & 85 \text{ N/m}^3 \times 100 \text{ m} = & 8500 \text{ N/m}^2 \\
1\tfrac{1}{2} \text{ in} & 300 \text{ N/m}^3 \times 100 \text{ m} = & 30\ 000 \text{ N/m}^2
\end{array}
$$

If $1\tfrac{1}{2}$ in is used, radiators would be increased, as emission from 2 in is assumed as contributing to room heating. On the other hand, 2 in gives a very low pump head, but there would be little significant difference in the cost of a pump if the higher pressure loss of $1\tfrac{1}{2}$ in is selected.

Probably, overall, $1\tfrac{1}{2}$ in is preferable: the energy-flow rate is quite out of the range of $1\tfrac{1}{4}$ in.

Having established the size, the allowance for single resistances can now be revised. The pressure to be set up by the pump will be the sum of the circuit-pressure loss and the boiler resistance. The latter may be found from makers' data (where high velocities may be used to promote efficiency).

The radiator connections, being in shunt from a short length of main, derive a pressure difference corresponding to that of the length of pipe carrying the bye-pass water. This can be evaluated and added to the gravity head referred to later.

Parallel Single-Pipe Circuits—It will be apparent from Fig. 8.2 (page 155) that each single-pipe element can be treated as set out above, as if a boiler served it individually. The mains to which they connect will, however, be sized as the two-pipe system considered next. Note that the pump pressure will be that of the longest or index circuit plus boiler resistance. To adjust for out-of-balance of parallel circuits, offering less resistance than the index circuit, regulating valves are desirable.

Two-Pipe system—By way of a simple example, the system shown in Fig. 8.3 (page 156) is reduced to one element of the circulation, as in Fig. 8.7.

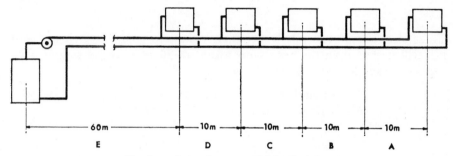

| ⊢——60m——⊢ | ⊢—10m—⊢ | —10m— | —10m— | —10m—⊢ |
| E | D | C | B | A |

FIG. 8.7.—Apportionment of Mains Emissions.

First approximation—Assuming the emitting units are 20 kW fan convectors:

Five fan convectors of 20 kW each $\quad=\quad$ 100 kW
The mains loss may be say 10 per cent $\quad=\quad$ 10 kW
$\quad\quad\quad\quad\quad\quad\quad\quad\quad\quad\quad\quad\quad\overline{}$
$\quad\quad\quad\quad\quad\quad\quad\quad\quad\quad\quad\quad\quad$ 110 kW

The overall temperature drop may be taken at $12°$ C. Thus:

$$\text{kW/°C} = \frac{110}{12} = 9 \cdot 1$$

From Fig. 8.6, assume a unit pressure drop of say 100 N/m³. The flow for various pipe sizes can then be set down thus:

$$\text{kW/°C} \times 12 = \text{kW}$$

$\frac{1}{2}$ in	0·25	3
$\frac{3}{4}$ in	0·6	7
1 in	1·2	14
$1\frac{1}{4}$ in	2·4	29
$1\frac{1}{2}$ in	3·4	41
2 in	6·7	80
$2\frac{1}{2}$ in	13·0	156

Using the same overall mains heat loss we may then give progressive totals thus:

	Section A	11 kW
	B	22 kW
	C	33 kW
	D	44 kW
	E	55 kW

Sizes will then be:

	A	1 in
	B	$1\frac{1}{4}$ in
	C	$1\frac{1}{2}$ in
	D	$1\frac{1}{2}$ in
	E	2 in

These sizes being generally larger than strictly needed, the unit pressure loss will be less than the 100 N/m³ assumed. If taken at 90 average, we can arrive at the mains pressure loss thus:

Total length flow and return	200 m
Allow single resistances	20 m
	220 m

$$90 \times 220 = 19\ 800\ \text{N/m}^2*$$

A fresh run-through with a higher pressure loss could be tried:

> e.g. if N/m³ is taken at 200 instead of 100 N/m³, the largest pipe size will be $1\frac{1}{2}$ in not 2 in, but the overall mains pressure loss will be about 40 000 N/m²;* which, with loss through boiler and heater added, may be considered a rather high pump head for an installation of this modest size.

'Accurate' Sizing—The previous example may serve to show a quick method of pipe sizing, sufficient perhaps for an estimate of cost; but, when the project is to proceed further, a more accurate approach is necessary. The following steps, related to Fig. 8.8 (which is a diagram of the system shown in Fig. 8.3 with dimensions added), will illustrate one method by which 'accurate' pipe sizing may be accomplished.

1. The process of allocating guessed preliminary pipe sizes to a system in order to establish the heat emitted by each section, is tedious and use may be made of Fig. 8.9 as a short-cut. From this figure, a near approximation to heat emission for a pipe may be read from the right-hand scale, for a number of system-temperature drops, against the energy-flow

---

* Those more familiar with the rating of pumps in feet head of water will recall that 1 ft head = approx. 3000 N/m². Thus 19 800 = approx. 6·6 ft head; 40 000 = approx. 13·3 ft head.

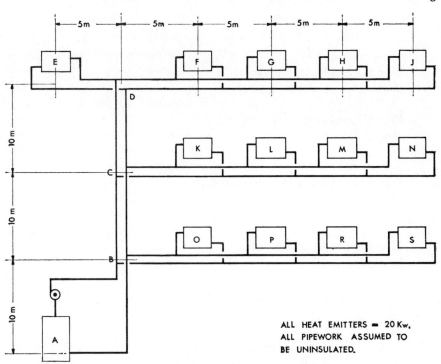

FIG. 8.8.—Example of Two-Pipe Sizing.

rate shown on the bottom scale. Thus, for the system in Fig. 8.8, a list may be made as follows for a 12° temperature drop:

Pipe section key	Nett Energy flow (kW)	Pipe heat emission W per (m) Fig. 8.9	length (m)	Total (kW)
AB	260	188	30	5·64
BC	180	178	20	3·56
CD	100	148	20	2·96
DE	20	92	10	0·92
DF, CK, BO	80	135	10	1·35
FG, KL, OP	60	125	10	1·25
GH, LM, PR	40	112	10	1·12
HJ, MN, RS	20	92	10	0·92

2. The apportionment of these heat emissions, section by section, must now be carried out systematically on a 'compound interest' basis. A first step would be to consider the three identical main circuits on the right-hand side of the Figure, taking the upper identifications as follows:

	Heaters			
	F	G	H	J
Rated emission, kW	20·00	20·00	20·00	20·00
Pipe HJ is allocated to heater J alone				0·92
	20·00	20·00	20·00	20·92
Pipe GH is allocated to heaters H and J			0·55	0·57
	20·00	20·00	20·55	21·49
Pipe FG is allocated to heaters G, H and J		0·40	0·41	0·44
	20·00	20·40	20·96	21·93
Pipe DF is allocated to all four heaters	0·32	0·33	0·34	0·35
	20·32	20·73	21·30	22·28

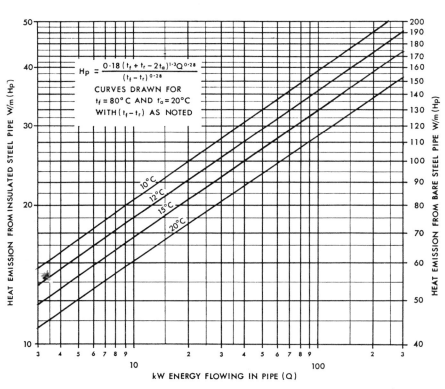

$$H_p = \frac{0.18 \, (t_f + t_r - 2t_a)^{1.3} \, Q^{0.28}}{(t_f - t_r)^{0.28}}$$

CURVES DRAWN FOR
$t_f = 80°\text{C}$ AND $t_a = 20°\text{C}$
WITH $(t_f - t_r)$ AS NOTED

HEAT EMISSION FROM INSULATED STEEL PIPE W/m (Hp)

HEAT EMISSION FROM BARE STEEL PIPE W/m (Hp)

kW ENERGY FLOWING IN PIPE (Q)

FIG. 8.9.—Heat Emission related to Energy Flow.

3. Proceeding similarly, pipe CD is allocated to heaters E, F, G, H and J, pipe BC to heaters E, F, G, H, J, K, L, M and N, and lastly, pipe AB is allocated to all the heaters shown. This process produces the total loadings shown below and it will be seen that, although the *average* mains loss is just over 10 per cent, that for heater J is 18·9 per cent whilst that for heater O is only 3·5 per cent.

Heater	Gross load (kW)	Mains loss (%)
O	20·71	3·5
P	21·15	5·7
R	21·74	8·7
S	22·73	13·6
K	21·09	5·4
L	21·53	7·1
M	22·12	10·6
N	23·14	15·7
E	22·35	11·7
F	21·69	8·4
G	22·14	10·7
H	22·76	13·8
J	23·79	18·9

The effect of the mains losses upon the heat emitters is that the full *system* temperature drop of 12° C will not be available at the individual flow and return connections. Taking the two extreme examples:

*Heater O, temperature drops**

$$\text{In flow main} \quad = 12 \times \frac{0·35}{20·71} = 0·25°$$

$$\text{Across heater} \quad = 12 \times \frac{20·00}{20·71} = 11·50°$$

$$\text{In return main} = 12 \times \frac{0·35}{20·71} = 0·25°$$

*Heater J, temperature drops**

$$\text{In flow main} \quad = 12 \times \frac{1·89}{23·79} = 0·95°$$

$$\text{Across heater} \quad = 12 \times \frac{20·00}{23·79} = 10·10°$$

$$\text{In return main} = 12 \times \frac{1·89}{23·79} = 0·95°$$

* These are not strictly correct since the flow pipe is hotter than the return, and will thus have a slightly higher heat loss.

The object of the exercise is to maintain the *mean* temperature at the heating constant throughout the system. Naturally, this small simple example does not show up the relative importance of mains loss as would be the case in an extensive plant, but no doubt it will serve to show the principle.

It will be seen that, for a large installation involving a great many branches of different lengths and sub-circuits of varying size, the apportionment of mains losses, if pursued to its ultimate refinement, can be a very laborious process. Various methods have been devised to simplify this task, such as have appeared in previous editions of this book. By way of rough compromise, if the total mains loss of one circuit from the boiler is calculated and divided by the number of branches on that circuit, even if not of uniform load, some attempt at apportionment can generally be made by sight to achieve a percentage basis which is probably not far from reality.

However arrived at, the mains loss for each section is added to the emitter load of the branch, and these are added progressively back to the boiler, or to the headers if there are several main circuits.

4. We now have loads which each section of main must carry, and the size can be judged from a starting basis of say 100 N/m³ unit pressure drop. The length of each section, flow plus return, plus single resistances can be set down and a table prepared, thus, for heater J in the example:

Section of Pipe (1)	Load Carried kW/°C (2)	Total Length (L + EL) m (3)	Pipe Size mm (4)	Unit Pressure N/m³ loss (5)	Section Pressure Loss (3) × (5) (6)
AB	23·85	41·1	80 (3″)	140	5760
BC	16·70	23·6	80 (3″)	70	1652
CD	9·37	21·3	65 (2½″)	58	1235
DF	7·52	11·1	50 (2″)	120	1332
FG	5·72	11·2	50 (2″)	70	784
GH	3·88	11·1	40 (1½″)	120	1332
HJ	1·98	14·0	32 (1¼″)	60	840
Heater	1·98	—	—	Catalogue	5000

Total for circuit = 17 935 N/m²

5. Next come the other branches and sub-circuits. At each off-take from the index circuit there will be some surplus pressure available. This must be dissipated in the branch connections, otherwise short circuiting will occur.

Each branch taken in turn then becomes a fresh exercise to be re-tabulated as above, sizes being adjusted to absorb surplus pressure.

There often comes a stage where this is impossible within the limits of commercial pipe sizes, and hence all such branches are provided with regulating valves which can be adjusted to take up over-pressure. Alas, in the hurry of completion of site work this final regulation is often scamped and the circuits are never properly balanced. The saving grace, however, is a generous pump capacity above that strictly necessary so that the ill-effects of the short-circuiting hazard are minimized!

6. Having established the total pressure loss of the system based on a number of initial assumptions, there is always the question as to whether, by selecting a different temperature drop or a greater or lesser pump pressure, a more economical solution would have emerged. Furthermore, as main sizes are changed, so are their heat losses, and the question of optimum insulation thickness also arises where appropriate.

Pipe Sizing for the Reversed-Return System—It will be obvious that with this system (see Fig. 8.4) the loads to be carried by the piping are added progressively forwards for the return and back from the index run to the boiler on the flow. Branches will be dealt with, however, in the normal way as for a two-pipe system. A word of caution is necessary, calling for a careful checking of the available pressure (flow and return) at each branch, as a condition may be found where a greater than average pressure drop has occurred in, say, the flow and a less than average one has occurred in the return. In such a case there might be no differential pressure available for the branch, which would then fail to work.

Pipe Sizing for the Embedded-Panel System—The normal two-pipe method of sizing (see Fig. 8.5) is used except that, due to low water temperatures, mains losses are low and may be taken on an overall basis of say 10 per cent. Furthermore, as the pressure loss in each coil is high relative to the pressure loss in mains and the temperature drop in the system is low, involving large water quantities, the system tends to be self-balancing.

The task of pipe sizing may be simplified by noting that, as panel coils are all usually of much the same length, each may be assumed to take the same quantity of water. The progressive totalling of loads for the mains is then only a matter of adding up the numbers of coils branch-by-branch and length-by-length.

It is necessary to recall that the heat emitted by an embedded panel has an upward and a downward component, whether in ceiling, roof or floor. Thus the total in each case must be allowed for, and not simply the emission to the room which the panel is designed to heat.

Design by Computer—To obtain the most refined solution to the problem of pipe sizing by long-hand arithmetical methods, having regard to the large number of variables each of which is mutually interdependent, is virtually impossible. Methods in the past have followed well-defined principles and limitations which, in the main, have worked in practice.

It is obvious, however, that with the terrific advantage of a new tool, namely the computer, a technique can be developed to assimilate all these variables and to evolve therefrom solutions to problems which are impossible to solve by any other means. It is clear, therefore, that the future of pipe sizing will no longer be in the sweating-through innumerable pages of abstruse calculations, and the pushing of a slide rule until it is red hot, but in the setting-up of programmes and data for computer processing.

A number of suitable *programs* (to use the transatlantic spelling unfortunately adopted by the computer industry) are already in existence. Experience is showing that a project of suitable size can be processed in one-quarter of the time and at one-third of the cost which would apply to the equivalent manual exercise. Pipe sizing *per se* is, however, only a part application of a fully developed computer approach: a full suite of programs will accept building dimensions and system criteria as input and will provide, as output, not only pipe sizes but also radiator dimensions and bills of quantities. The development of a sub-program to compose letters of appreciation from satisfied building owners does not appear yet in any catalog (sic); but this refinement will no doubt appear in due course of time!

PIPE SIZING WITH GRAVITY CIRCULATION

Circulating Pressure—Fig. 8.10 shows a simple single-pipe gravity circulation. The force creating circulation is that due to the difference in weight of positive column P_1 at temperature t_2 and negative column N_1 at temperature t_1.

The force available will be that of unit mass, subject to acceleration due to gravity, which may be taken as 9·81 m/s. If ρ_2 and ρ_1 are densities of water at temperature t_2 and t_1 respectively in kg/m³, H is the height of P_1 and N_1 in metres, and CP is the circulating pressure in N/m³, then

$$CP = H \times 9\cdot81\ (\rho_2 - \rho_1)$$

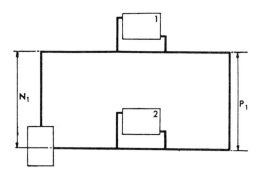

FIG. 8.10.—Simple Single-Pipe Gravity Circulation.

For convenience, the *I.H.V.E. Guide* gives a Table of circulating pressures for various flow temperatures and temperature drops, from which Table 8.1 is extracted.

TABLE 8.1

CIRCULATING PRESSURES FOR GRAVITY HOT-WATER SYSTEMS

Flow Temperature °C	CP in N/m² per metre height for following temperature differences °C flow minus return						
	8	10	12	14	16	18	20
40	27·63	33·73	39·50	44·90	49·94	54·59	58·83
50	33·35	41·01	48·37	55·44	62·20	68·64	74·74
60	38·38	47·36	56·10	64·57	72·78	80·72	88·37
70	42·96	53·14	63·09	72·80	82·28	91·52	100·50
80	47·57	58·54	69·61	80·46	91·09	101·50	111·68
90	51·25	63·57	75·64	87·61	99·32	101·82	122·11

In order to determine the temperature drop from N_1 to P_1, it is necessary to take the heat emission of the top radiator 1 and the top length of the circulating pipe and that of the bottom radiator 2 and the bottom length of pipe. The overall temperature drop from boiler flow to return multiplied by the ratio of emissions top-total gives the temperature drop to P_1. If Q_1 is emission at the upper level and Q_2 that at the lower level, then

$$\text{Temp } P_1 = \text{Temp } N_1 - \left\{ \frac{Q_1}{(Q_1 + Q_2)} \right\} \times \text{overall temp. drop.}$$

Heat loss from riser and drop are ignored.

Resistance to Flow—The circulating pressure assessed in the manner described is the means of creating and maintaining a circulation through the system, and if this is a closed circuit, as in Fig. 8.10, it will cause just such a velocity that its force is balanced by the resistance or friction encountered in the pipes, boiler and radiator.

For a given quantity of heat to be transmitted from the boiler to radiators 1 and 2, water must flow through the system, the mass depending on the temperature drop. For a high temperature drop t_1 to t_2 each kg of water will carry more heat and, therefore, less water will be required than at a lower temperature. At the same time, the greater the difference between t_1 and t_2 the greater will be the circulating pressure, and obviously the greater the flow of water.

With a constant heat output from the boiler, these effects strike a balance in such a way that the temperatures t_1 and t_2 adjust themselves to produce just that circulating pressure which will be absorbed in impelling through the circuit a quantity of water to an amount equal to the energy delivered, divided by the temperature difference $t_1 - t_2$.

Thus the circulating pressure balances the sum of the resistances, or in other words:

$$CP = \Sigma R$$

where ΣR is the sum of all the resistances to flow of water throughout the circuit.

Of these, the most important is the resistance of the piping, bends, tees, and valves. The lesser resistance is that in the boiler.

Available CP per Metre Run—It is necessary to determine the total metres travel 'T' of the circulation, including an allowance for the single resistances, which must therefore be assumed beforehand, both for size and velocity and corrected later. In the case of a branched system we take the travel for which $\dfrac{CP}{T}$ is the least, and this is called the *index circulation*.

The flow and return temperatures will be as selected from Table 7.1 (page 144), such as 80° C flow and 60° C return in a typical case.

Centre of Boiler—In measuring the height of the column producing circulation it is convenient to assume the boiler to be concentrated at one central point. This is the mid-point between flow and return connections.

Radiators on Single Pipe—The matter of radiator connections from a single pipe has been mentioned. Fig. 8.11 illustrates the method often

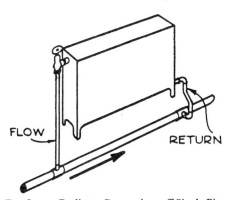

FIG. 8.11.—Radiator Connections off Single Pipe.

employed. The gravity circulation through the radiator is calculated as if there was a boiler at the centre of the main pipe. The *CP* available is then that from the centre of the radiator to the centre of pipe. The *T* will be say one metre in the connecting pipes plus the equivalent length for the valve and radiator resistance and other fittings. The *K* factor for an angle valve is 5·0 and for a radiator also 5·0. The bends in the return and the tees may total three: a grand total of thirteen. The EL for $K=1$ may be taken from Fig. 8.6 and the total length assessed. $\dfrac{CP}{T}$ is then obtained and hence the kW carried per 1° C. The actual temperature drop across the radiator must be decided, trying this perhaps on 8° C at first. The lower the drop the higher the mean-radiator temperature, but, on the other hand, the larger the connections.

Where the system is pumped, as mentioned earlier, some additional pressure is available corresponding to the differential loss through the length of main between the flow-and-return connections carrying the bye-pass water. This extra pressure is added to the gravity *CP*.

Two-Pipe (Gravity) System—In a two-pipe system, as Fig. 8.12, the *CP* (centre of radiator to centre of boiler) will obviously be greatest for radiator 1 and least for radiator 3, which is the index and the worst placed. This latter circuit must then be sized first and the resistance to the branches to radiators 2 and 3 deducted from the *CP* in each case. The risers to radiators 2 and 3 are then sized to absorb the available *CP*.

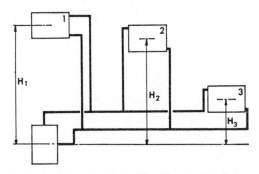

Fig. 8.12.—Example of Two-Pipe Gravity Circulation.

Having established $\dfrac{CP}{T}$ for the index and other circuits, sizing then proceeds as with a pumped system by reading, from Fig. 8.6 (page 158), the kW of energy flow for each pipe size for the available pressure drop per metre run. The loads will have been totalled progressively, including mains losses as before in the case of the two-pipe system duly apportioned.

Sizes of pipes necessary for the flow in question will then be selected and, if different from the first assumption, some revision of emissions may be necessary followed by a final run-through again.

Gravity Systems in Perspective—The above brief reference to sizing of gravity systems is intended to give some idea of the principles involved. Being no longer in common use for anything but the smallest installations, there is no point now in pursuing the subject in great detail. Those who wish to do so for historical interest may consult earlier editions of this book and other works. There were, for instance, many more types of system— the one-pipe drop, the two-pipe drop, the irregular and so on. With the tiny forces available, large pipes were the rule and, indeed, considerable

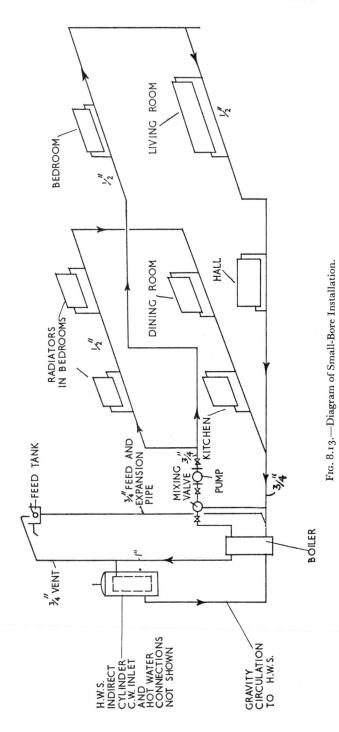

Fig. 8.13.—Diagram of Small-Bore Installation.

experience was called for to ensure that a system would circulate in all its parts. For one reason or another some circuits might be sluggish, only responding when the boiler was forced unusually hard. Then there was always the menace of an air lock. All this is a thing of the past, due to the great flexibility now offered by the pump.

THE SMALL-BORE SYSTEM

This is a special case of a single-pipe system developed for residential property using a silent submerged rotor pump. Fig. 8.13 is a diagram of a typical layout; in this case two single-pipe loops are in parallel, which is necessary where the number of radiators becomes greater than can be carried on one loop, or there may be three or four loops. Also, it will be noted that an indirect hot-water supply cylinder is served by gravity circulation from the same boiler, Fig. 8.14.

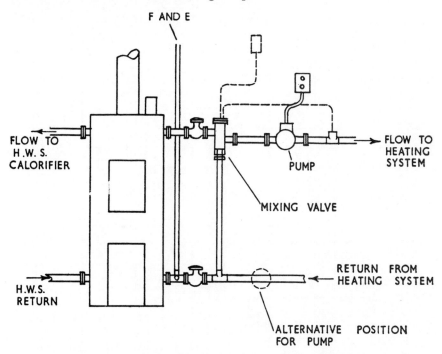

FIG. 8.14.—Small-Bore System: pump and mixing arrangement.

Use is made of small-bore copper piping, usually $\frac{1}{2}$ in nominal bore for the loops and $\frac{3}{4}$ in or over where serving several loops, thus enabling it to be unobtrusively fitted into new or existing houses. Radiators are served by tees from the single pipe as shown.

Pipe sizing is a matter of assessing the maximum heat output which can be carried by a $\frac{1}{2}$ in loop, devising further loops as necessary. Piping common to the multiple circuits is then sized to carry the total load.

As copper piping is not covered by Fig. 8.6 (page 158), reference may be made to Fig. 8.15 which, as before, is based on *Guide* data but in terms of kW per °C drop.* A temperature drop of 10° C is commonly adopted. Water velocity is limited by what is permissible without creating noise, usually considered to be about 1 m/s. Single resistances are dealt with as already stated, the values for equivalent length being given in Fig. 8.15.

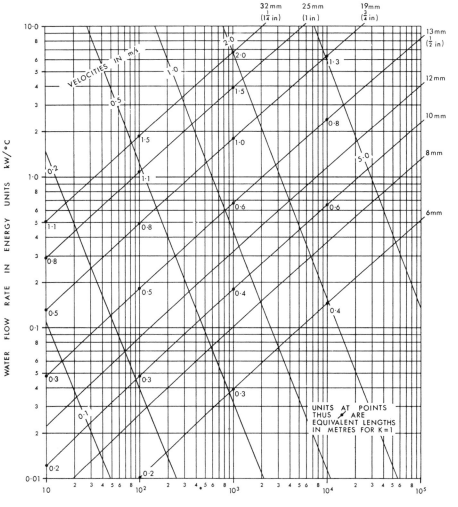

FIG. 8.15.—Pipe-Sizing Chart for Small Copper Pipes. *Note:* 6 to 12 mm pipes are outside diameters for use with minibore. ½ in pipes, and above, are nominal inside diameters to B.S. 659. For larger sizes see Fig. 14.4.

* This chart was prepared in advance of the issue of a British Standard for metric copper pipes (still pending).

THE MICRO-BORE OR MINI-BORE SYSTEM

In this system, the possibilities of small-bore piping are further exploited using the two-pipe principle and water velocities up to 1·8 m/s. Each radiator, convector or other form of heat emitter receives a separate flow and return from headers or manifolds strategically placed, as in Fig. 8.16. (See also Plate VIII, facing page 208).

It is found under these conditions that copper piping no more than 6 mm outside diameter may often suffice or, for larger duties, 8, 10 or 12 mm outside diameter. Alternatively, piping of stainless steel and nylon has been used. The common header or manifold may be ¾ in, or as required, and a number of types of prefabricated unit are currently available.

Radiators or other emitting elements may be connected with two small-bore valves as normally, or use may be made of a special single valve carrying an extension pipe which runs inside the waterway to the far end, the valve having two connections—one for flow and one for return.

Pipe sizes being so small, heat losses therefrom can be ignored. A 10° C drop may be assumed.

Where the piping is to be of copper, Fig. 8.15 gives the necessary data. Pressure loss in the header or manifold may be 2·5 to 5·0 kN/m²: if of proprietary make, the makers' information should be available.

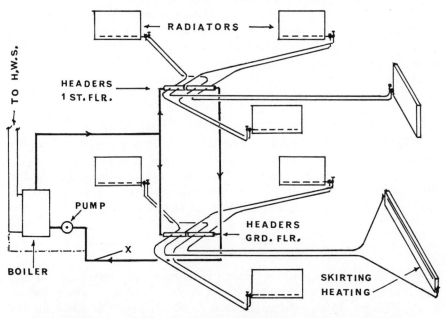

Fig. 8.16.—Diagram of Micro-bore (or Mini-bore) System. Note that the feed and expansion pipe at X connects to either (a) a small tank or (b) a sealed pressure cylinder (see Chapter 9). The hot-water supply indirect cylinder is not shown.

CHAPTER 9

Hot-Water Heating Auxiliaries

IN THE PRESENT CHAPTER, the various accessories associated with a hot-water heating system will be described.

The general use of pumped circulation has already been mentioned, and this is the first item to be considered. *The pump* has now become a vital and essential part of the installation and with small-bore and micro-bore heating systems for houses, this applies even to the smallest installations.

It is then necessary to consider how the system is filled with water and what happens to the water when it expands due to heating.

Next, it will be appreciated that the system itself becomes larger as its temperature is raised, and provision for expansion of piping has to be carefully arranged and designed for.

The control of the heating system is of the greatest importance both from the point of view of comfort of the users and economy of fuel consumption.

Another matter to be considered is the safety of the system under a variety of fault conditions, and the various fittings which are mounted on boilers for checking their operation.

For small installations, the pump, boiler, burner and various accessories are now to be obtained in one combined packaged unit. The tendency in larger installations is also in the same direction, but whether or not this is the case, the designer will always want to know the basis on which the capacities and sizes have been determined.

This chapter attempts to deal with these problems.

PUMP CIRCULATION—CAPACITY OF PUMP

The capacity of the pump is determined from the total kW emission of the system. Pump duties are measured by volume per unit time which, in SI units, is litres per second. The mass flow of water per second is thus kW/specific-heat capacity × system-temperature drop. Taking specific-heat capacity of water as 4·2 kJ/kg °C and 1 kg of water = 1 litre,

$$\text{litre/s} = \frac{\text{kW}}{4 \cdot 2 \times \text{°C drop}}$$

To limit the pump to this duty would mean that every circuit and radiator is expected to take exactly its correct volume of water and no more. This is obviously impracticable, and a margin to allow for the hazard of short circuiting is desirable of the order of 20 to 30 per cent.

In the case of single-pipe or ladder systems where the water is not subdivided, or is split into some three or four circuits only, 10 per cent margin would be sufficient.

Taking as an example:

Total emission of radiators and mains - 1000 kW
Temperature drop - - - - 15° C

$$\text{Litre/s} = \frac{1000}{4 \cdot 2 \times 15} = 16$$

A pump of, say, 20 litre/s would be suitable. The pressure drop would be that previously determined for the pipe sizing plus pressure loss in boiler and in heat emitter on index run.

Taking a total pressure loss of 30 kN/m², the power delivered by the pump to the water will be

$$20 \times 30 = 600 \text{ W}$$

Pump efficiencies vary between 50 per cent and 75 per cent. If the former, the power absorbed at the pump shaft would be

$$600 \times \frac{100}{50} = 1200 \text{ W} = 1 \cdot 2 \text{ kW}$$

The electric motor for this duty would require to have a margin over this to allow for the possibility of error in the estimation of the head. Probably a motor of 1·5 kW output would be suitable.

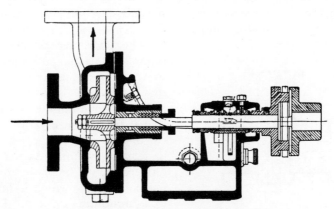

FIG. 9.1.—End-Suction Type of Centrifugal Pump.

CIRCULATING PUMPS

Centrifugal Pumps—The centrifugal type of pump is most suitable for the purpose. This consists of an impeller rotating in a fixed casing of volute shape. The impeller may be of the end suction type, as in Fig. 9.1, although a split casing type is sometimes preferred, in which the top half of the casing is removable for inspection.

N

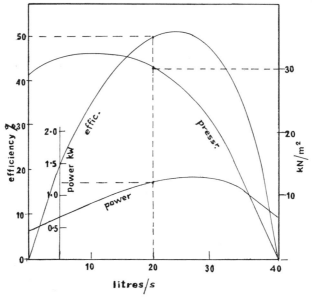

FIG. 9.2.—Typical Characteristic Curves of Centrifugal Pump.

Centrifugal pumps of either type are more suitable for heating circulations than other types of pump since they are very conveniently and simply driven by electric motors, are low in cost, require little maintenance and may be made silent in running.

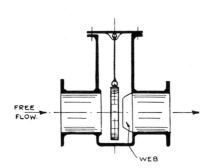

FIG. 9.3.—Automatic By-pass Valve.

The characteristic curves of a pump such as that in the example is shown in Fig. 9.2. It will be noted that the selected duty is near the point of maximum efficiency, and that if the pressure varies from that calculated so does the performance in terms of volume. As the pressure reduces, power increases to a limiting point, after which it falls away to a minimum on free discharge. Motors are usually selected to cover the power limit, thus being 'non-overloading'.

Pump Arrangement—Direct coupling to an electric motor is the simplest form of drive, and a flexible coupling is usual, to allow for possible distortion of the base plate and misalignment of the bearings. An alternative arrangement is one in which the motor is separate, with a V-rubber belt drive to the pump. This permits the motor speed and pump speed to be independent of one another, which is sometimes an advantage.

Automatic By-pass—This fitting—of the type shown in Fig. 9.3—is placed in a connecting pipe between suction and delivery of the pump to allow partial gravity circulation should the pump stop. With solid-fuel boilers and systems of low pressure loss, this was a common feature, but with the current use of higher velocities and automatic firing the tendency is to omit such provision.

Self-Contained Pumps—This is a class of pump incorporating in one unit, pump, motor, and by-pass, and in some cases includes isolating valves to the pump.

Fig. 9.4 shows a typical example with V-belt drive to the motor. A vertical spindle type (Fig. 9.5) is so designed as to require no by-pass, the water flowing freely through the pump when stopped. The merit of these self-contained pumps is compactness and convenience in operation.

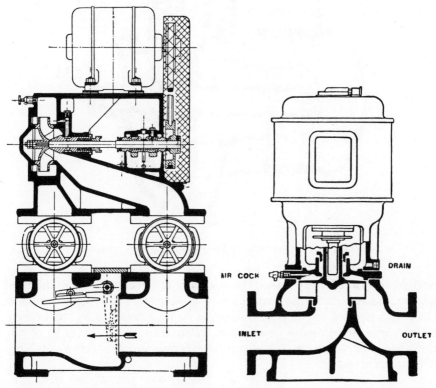

FIG. 9.4.—Self-contained Pump Unit. FIG. 9.5.—Vertical Spindle Pump.

Where a pump is relied on as the sole means of circulation, the gravity effect being too small to be useful, in larger installations it is usual to arrange for duplicate sets, with valves, so that either pump may be in commission with the other as standby.

Submerged Rotor Pumps—An interesting development has come about in pumps specially designed for heating systems in which the rotor of the driving motor is contained within the circulating water. The rotor is surrounded by a non-magnetic water barrier and the field coils are exterior to this. Rotating parts require no lubrication, and there is no gland, so that this type of pump is put forward as requiring no maintenance of any kind, and is generally silent. One make of this type is illustrated in Fig. 9.6. It will be noted that it is mounted in the pipeline, and requires no base or other fixing. It is suitable only for alternating current.

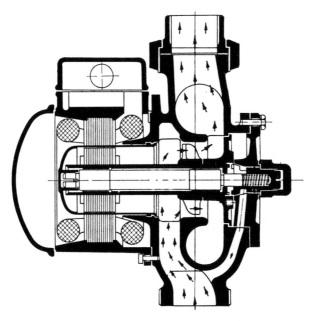

Fig. 9.6.—Submerged Rotor Pump. (Sigmund 'Thermopak')

Self-Accelerated Systems—Mention should perhaps here be made of accelerated systems of the past in which a pump was not employed. Most of these depended for their operation on steam generated by the heating boiler, or independently, and have fallen into disuse because of the simplicity and cheapness of the centrifugal pump, and the wide availability of electric current.

DUAL TEMPERATURE SYSTEMS

It is sometimes necessary to supply water at two different temperatures from the same boiler plant, such as when a constant temperature circuit is desired for fan convectors, unit heaters and possibly hot-water supply, and, in addition, a radiator heating system is also to be served but with a variable water temperature to suit the weather.

This can conveniently be achieved by a mixing arrangement whereby part of the return water from the low-temperature circuit is re-circulated direct into the flow without passing through the boiler. The amount of water re-circulated may be controlled by hand, or by thermostatically-operated mixing valve. (See Fig. 9.7.) A separate thermometer is desirable in the low-temperature flow.

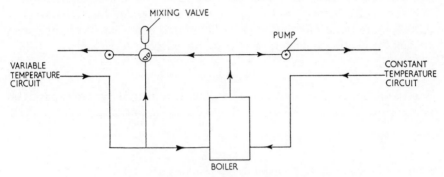

FIG. 9.7.—Diagram of Dual Temperature Circuits.

The same method of variable flow may be extended to a number of circuits by using a ring circuit for boiler circulations, maintained by a separate pump, or pumps, and each circuit with its own pump and mixing arrangements connected by means of 'accept-reject' connections (see Fig. 9.8). The capacity of the ring-circuit pump must of course exceed the total of the zone-circuit pumps.

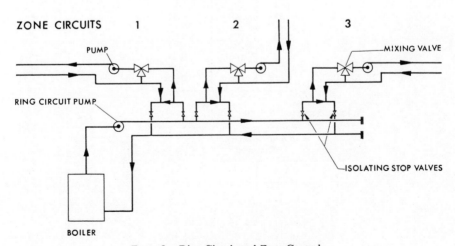

FIG. 9.8.—Ring Circuit and Zone Controls.

BOILER-CIRCULATING PUMPS

In order to avoid corrosive effects of oil firing, referred to on page 77, due to low water temperatures, boiler-circulating pumps are being increasingly used and are standard fittings on some makes of boiler in the larger range of sizes.

A boiler-circulating pump acts independently of other pumps and circulations. It draws hot water from the top of the boiler and delivers it into the cooler bottom, so raising its temperature. Another advantage is that it ensures some circulation in the boiler in the case where mixing valves serving the heating system are liable to shut-off against the boiler.

GENERAL

Feed and Expansion Tanks—The water in a heating system expands on being heated, and the purpose of the feed and expansion tank is to receive

TABLE 9.1

CONTENTS OF HEATING APPARATUS

(a) Boilers, cast-iron sectional type -	- -	- 1·5 to 2 litre/kW
(b) Boiler, steel types—vary greatly -	- -	- 0·5 to 3 litre/kW
(c) Radiators, steel panel types	- - -	5 litre/kW
(d) Radiators, hospital and column types	-	- 10 to 15 litre/kW
(e) Piping, steel to B.S. 1387, medium grade:		

Nominal Internal Diameter		Litre/m run
in	mm	
$\frac{1}{2}$	15	0·21
$\frac{3}{4}$	20	0·37
1	25	0·58
$1\frac{1}{4}$	32	1·01
$1\frac{1}{2}$	40	1·37
2	50	2·20
$2\frac{1}{2}$	65	3·71
3	80	5·11
4	100	8·70
5	125	13·26
6	150	18·90

(f) Piping, Copper to B.S. 659:

Nominal Internal Diameter		Litre/m run
in	mm	
$\frac{1}{2}$	13	0·14
$\frac{3}{4}$	19	0·28
1	25	0·53
$1\frac{1}{4}$	32	0·81
$1\frac{1}{2}$	38	1·18
2	51	2·10

this water when the system is hot and return it when it cools down. In so doing the water in the system is not changed, and encrustation or corrosion which might otherwise occur with a constantly changing supply is avoided.

For the same reason it is inadvisable to empty the system in the summer or to change the water at all, except when repairs or alterations call for it.

The expansion of water from 7° C to 100° C is one twenty-third of its volume at the initial temperature. In order to determine the size of expansion tank for a certain system it is therefore necessary to estimate its total water contents. For boilers and radiators, makers' catalogues may be consulted. As an approximation is all that is necessary, however, Table 9.1 will be found to give a fair average. This table also gives the contents of piping.

In addition, about 100 mm of water is necessary permanently in the tank to float the ball valve, and a fair margin of space above, before the overflow is reached. This will call for a tank capacity about double that estimated.

The method of connection of the tank by means of a 'feed and expansion pipe' to the system requires some consideration. In a gravity system the feed and expansion pipe simply connects into the boiler return; a vent off the boiler or off the system is turned over the tank and there is no problem.

With pump circulation, which is now usual, there are three possible alternatives (see Fig. 9.9):

A. pump in return, feed and expansion pipe into suction;
B. pump in return, feed and expansion pipe into delivery;
C. pump in flow, feed and expansion pipe into boiler return.

In case A, the vent pipe must be carried above the tank water line to a height exceeding the pump head to prevent water discharging. With high pump heads this is often impossible.

In case B, the vent pipe is not subject to the pump head and no extra height is necessary, but there is a tendency to draw air in at the air cocks on top-floor radiators so that, in order to release air, the pump must be stopped.

In case C, with pump in flow, the feed and vent are in balance; the heating system being under pressure from the pump, air release presents no difficulty.

The feed and expansion tank is usually made of galvanized iron and is provided with a ball-cock connected to the cold-water supply. The lever of the ball-cock is bent so as to keep the ball near to the bottom, and the valve itself is above overflow level. The overflow should be of large.dimensions such as 1¼ in for small systems and 2 in or more for large ones. If in

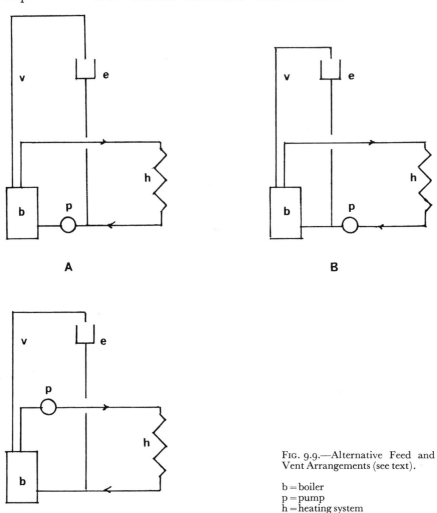

FIG. 9.9.—Alternative Feed and Vent Arrangements (see text).

b = boiler
p = pump
h = heating system
e = feed and expansion tank
v = open vent

an exposed position, or in a cold roof-space, the tank and its connections should be protected against frost.

Expansion Taken in Pressure Cylinder—The open expansion tank suffers, particularly in small installations, from being out of sight and out of mind, and not infrequently is found to have run dry due to the ball valve having stuck. Also it is frequently placed in a position liable to frost, and so constitutes a danger should it become frozen with the boiler alight. The use of a pressure cylinder removes this distant object from the scene altogether; the pressure cylinder may be placed alongside the boiler, and thus be under constant observation.

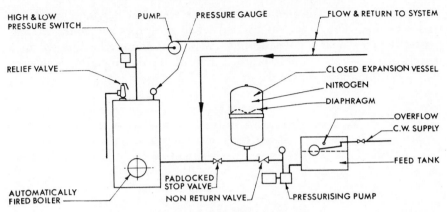

FIG. 9.10.—Pressurized System.

Fig. 9.10, shows the arrangement of such a system and it comprises a closed cylinder containing nitrogen above a neoprene diaphragm. As the water expands the gas is compressed and pressure rises. A safety valve prevents undue pressure being reached. Proprietary forms of this equipment of continental origin exist which include a special form of safety valve, pressure gauge and non-return valve for filling. The size of vessel is dictated by the amount of expansion water to be accommodated within the limits of rise and fall of the diaphragm.

In small domestic systems the filling may be checked annually using a removable hose from the water main attached to the non-return valve.

For larger systems a filling unit is provided comprising a small tank with ball-cock and an electrically driven pump as in Fig. 9.10. A pressure switch connected to the system brings on the pump to maintain a minimum water pressure sufficient to ensure that the highest point is filled. A second pressure switch may be arranged to cut off the firing unit on over-pressure before the safety valve operates.

This kind of system is suitable only for automatically-fired boilers, not for hand-fired solid-fuel boilers where undue temperature swings may occur.

Insulation of Pipes—Piping is insulated to reduce heat loss to a minimum, where such heat would be wasted or a nuisance. This applies, for example, particularly to main piping run in trenches or ducts, and through basements or other unheated spaces. Table 7.3 (p. 145) gives heat losses from pipes insulated with 85 per cent magnesia, expressed as a percentage of bare-pipe loss. It will be seen that the larger the pipe, the less the percentage loss for a given thickness of insulation—due to the increase in outer radiating surface being proportionately greater for small pipes than large.

Conductivities of insulating materials may be compared in Table 9.2, the values being applicable to the range of temperature used in low-pressure systems.

As to the materials themselves, dry or sectional laggings avoid wet trades on site. They may be applied when the pipes are cold, whereas plastic requires heat for drying out. All materials may be finished in a variety of ways with enamels, fabric, bituminous paint, metallic sheathing, etc., according to position and cost.

TABLE 9.2

THERMAL CONDUCTIVITIES OF PIPE INSULATING MATERIALS
At Pipe-Face Temperature 60° to 90° C

Material	W/m° C
Amosite Asbestos	0·049
Blue Asbestos	0·062
Plastic Asbestos	0·094
Glass Fibre	
(rigid sections)	0·046
85% Magnesia	0·062
Mineral Wool	0·041

Expansion of Pipes—From Table 1.3 (page 9), it will be noted that the coefficient of linear expansion of steel is $11·3 \times 10^{-6}$ per 1° C. Thus for a steel tube 100 metres long raised from 0° C to 100° C, the expansion will be roughly 0·11 m or 110 mm.

In the practical installation, flexibility in pipework is achieved by changes of direction and intentional offsets, but where long straight runs are involved flexible bellows of the type shown in Fig. 9.11 are introduced. For larger mains, *horseshoe loops* are sometimes used—for which, see makers' data.

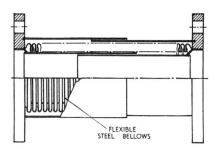

FLEXIBLE
STEEL BELLOWS

FIG. 9.11.—Bellows Expansion Joint. ('The Power Flexible Tubing Co., Ltd').

HOT-WATER BOILER MOUNTINGS

Boiler mountings for a hot-water heating boiler comprise:

(a) Safety valve.

(b) Thermometer.

(c) Altitude gauge.

(d) Draw-off cock.

(*a*) **Safety Valve**—The safety valve or relief valve is a device to relieve pressure should this rise above normal for any reason. Thus, if boiler valves have been shut and feed and vent perhaps frozen, the safety valve is the last resort to prevent an explosion should the boilers have been fired through an act of stupidity. Safety valves are either of deadweight or spring type. Fig. 9.12 shows one form of the latter approved by insurance companies: it will be noted that the spring is enclosed and the setting cannot be altered. Sizes are given in Table 9.3 according to boiler rating (based on BS 779.1961).

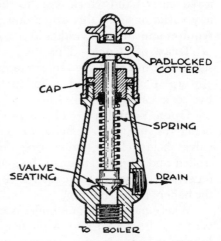

FIG. 9.12.—Typical Spring-Loaded Relief Valve.

TABLE 9.3
RELIEF VALVE SIZES

	in	mm
Up to 270 kW	$\frac{3}{4}$	20
270 to 360 ,,	1	25
360 to 450 ,,	$1\frac{1}{4}$	32
450 to 540 ,,	$1\frac{1}{2}$	40
540 to 750 ,,	2	50

(*b*) **Thermometer**—The thermometer is a most important accessory to the boiler, and may be a plain mercury type in a brass case or one of dial pattern. The latter is desirable where the boiler is either a large one or a high one.

In order that the thermometer (of whatever type) may be removed for repair, an immersion well is always provided (see Fig. 1.2, page 4). This is screwed into the boiler and forms a water-retaining joint. The well may be of steel or of brass; a steel well may be filled with mercury, but a brass well should be filled with oil for the reason that mercury might form an amalgam with the parent metal.

Thermometers are also useful on other parts of the system, such as after a mixing valve, and on individual circuit returns: they should assist in regulation—the main object of which is to ensure that all returns are at the same temperature.

(*c*) **Altitude Gauge**—An altitude gauge is simply a pressure gauge, and is calibrated so as to read in bar units, or alternatively in metres head of water. One bar equals nearly 10 m head of water. The gauge is fitted with a red index hand which is set to the normal pressure in the system, and any variation from this up or down will at once indicate trouble which requires immediate investigation.

(*d*) **Draw-off Cock**—This item needs little comment. It is usually a standard fitment of the boiler manufacturers. Sizes are $\frac{3}{4}$ in up to 50 kW, 1 in up to 150 kW and $1\frac{1}{2}$ in up to 300 kW. Larger sizes are 2 in, or two cocks can be used of $1\frac{1}{4}$ in.

(*e*) **Open Vent Pipe**—This is simply a pipe taken from the top of the boiler (as in Fig. 9.9), being turned over the feed tank. In the event of valves being closed, preventing flow, or the boiler running on uncontrolled and causing steam to be generated, the worst that can happen is for steam and water to be discharged over the tank, the water returning to the boiler via the feed pipe.

Where more than one boiler occurs and to save the cost of a separate vent pipe from each, use is made of three-way 'safety' cocks as shown in Fig. 9.13. In the position shown, each boiler is connected to the common vent. If, however, the boiler valves are shut for draining down the boiler, the cock is turned to the opposite position connecting to the open end in the boilerhouse.

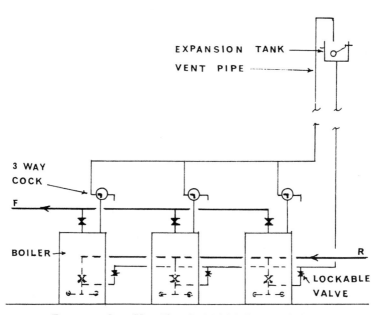

FIG. 9.13.—Open Vent Pipes for Multi-Boiler Installations.

Sizes of open vent pipes are given in Table 9.4 (from BS 779.1961 converted to SI). Open vent pipes are not of course required with closed pressure cylinder systems.

TABLE 9.4

SIZES OF OPEN VENT PIPES

kW	in	mm
45 to 150	$1\frac{1}{4}$	25
150 to 300	$1\frac{1}{2}$	32
300 to 600	2	40
over 600	$2\frac{1}{2}$	65

WATER TREATMENT

In the newer forms of heating system, often using thin steel radiators, small-bore copper pipes and other metals, there is a danger of corrosion internally which can be overcome, however, by dosage of the initial filling of water with a suitable corrosion inhibitor.

With larger systems it is piously hoped that the water will not be changed; but for one reason or another, such as leakage from pump glands and alterations necessitating draining down, water is subject to change. The problem is probably one more of preventing scale formation than of

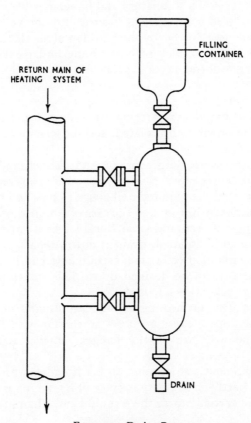

FIG. 9.14.—Dosing Pot.

preventing corrosion, or it may be both. It is necessary, then, for the treatment to be repeated periodically and use may be made of a dosing pot for this purpose, as shown in Fig. 9.14. The usual addition is sodium metaphosphate, in order to raise the alkalinity, sometimes combined with tannin. The effect of such treatment is to destroy the scale-forming properties of calcium carbonate and other products of a hard water district by raising the pH value to about 10 or 11.

The use of the dosing pot is as follows:

The chemical dose is placed in the top vessel. The valves connecting the pot to the system are closed, and the small cylinder is drained off. The chemical dose is then admitted to the small cylinder by the opening of the top valve. This is then closed, and the two valves connecting the cylinder to the system, usually in the return main, are opened, so that the liquid can pass into the general circulation. Such a method of treatment requires careful and periodical control in order to check the pH value, this being usually done by a simple indicator test which any ordinary boilerman can carry out. It is desirable to obtain advice in regard to such treatment and certain proprietary firms undertake this function.

Where the water is of a soft nature and liable to cause corrosion, there may equally be a need for treatment in order to remove the acidity and raise the pH value, to 10 or 11, by the addition of an alkali in some form, so as to avoid corrosive attack of steel piping and particularly of steel radiators, should they form part of the system.

VALVES

Valves in a heating system are provided for two purposes: one is to enable parts of the system to be isolated, and the second, to enable regulation to be carried out.

These two purposes are quite separate and distinct, and it is indeed a fact that a good isolating valve usually makes a very poor regulating valve.

Isolating valves will be required on boilers, pumps and throughout the system on all main branches, and, in the case of a multi-storey building, at the base of each pair of risers and so on. Such valves are to enable a section of the installation to be isolated, drained down, repaired and altered as may be required without affecting the rest of the system. In order to enable the section or sections to be drained down, it is therefore necessary to provide emptying cocks at each pair of valves.

Valves for isolation on heating systems are usually of full-way gate type, which, when open, cause virtually no resistance to flow. Valves of 3 in (80 mm) and over are usually flanged; smaller sizes are usually screwed.

Emitting equipment also requires to be provided with valves, partly for isolation and partly for regulation, since, if regulation is necessary, it is much more easily carried out at the terminal ends than in a general way on mains.

Radiators, convectors, ceiling panels, unit heaters and the like on any extensive installation are usually provided with two valves, one in the flow and one in the return, so that each piece of apparatus may be disconnected and removed, if required, without affecting other parts of the system. At the same time, for purposes of regulation, usually the valve in the return is fitted with a lockshield, whereby the valve can be adjusted by a key, and this valve is then called upon to serve the double purpose of isolation and regulation. The ordinary screw-down valve is virtually useless for regulation, as it is far too critical. A curve of flow against amount of opening for this type of valve is of the form indicated in Fig. 9.15. Various attempts have been made to devise a valve form which will give nearer to a straight line, and the acorn disc nearly achieves this object. The amount of heat emitted, however, is dependent on the overall temperature drop of the water and will be greater the smaller the flow. The ideal valve curve for straight line heat-output is shown dotted.

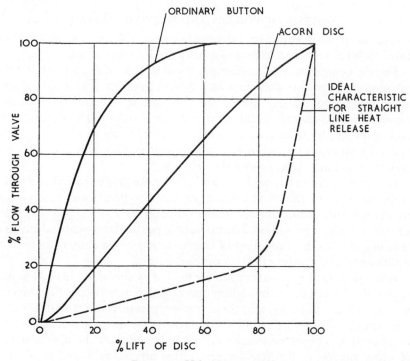

FIG. 9.15.—Valve Characteristics.

Where such a valve is used for regulation, there is always the danger, when it is used for isolation purposes, that it will not be reset back to its original setting after it has been shut off for some reason. To overcome this disadvantage, various designs of double regulating valve have been devised which can be used for isolation, but which, on being opened up, can do so no further than by a given amount.

What has been said above refers mainly to two-pipe installations. In the case of a single-pipe system such as a small bore installation for a house, a single valve on a radiator may be all that is necessary, since, with all radiators virtually in series, no question of regulation as between one and another arises. The only purpose of a valve on a radiator in such case is so that the heat can be shut off or turned on as required, and there are now a number of neat forms of such valves incorporated in the radiator unit as a standard. If in such an installation there is need to remove a radiator for painting or repair, it becomes a matter of emptying down the circuit complete, which is of no great disadvantage seeing that the water content is so small.

In the micro-bore system there is much to be said for the single connection valve to radiators, etc., for the two-pipe entry as referred to on p. 175.

CONTROL OF BOILERS AND HEATING SYSTEMS

A separate book could be written about control systems, as they have tended to become more and more sophisticated as time goes by.

The elements of control are simple enough; first, there is control of the heat producing equipment, such as a boiler, in order that it may deliver just so much heat as is required by the heating system and no more; second, there is control of the heat emitters in the various rooms of a building or in the different parts of a factory, so as to ensure that the required temperature is achieved without overheating, which might cause discomfort and in any event would waste fuel.

Whereas in the past, with a hand-fired coke boiler, a crude form of damper-regulator sufficed to control the air admitted below the grate for combustion, this meant that the boiler often ambled along at much below its rated output. This is bad for the boiler, particularly with oil-firing or gas-firing, due to the likelihood of condensation being caused under such conditions with corrosive effect on the boiler itself.

It is now considered better practice to control the boiler in such a manner that it runs at a constant high temperature, so as to reduce the tendency for condensation to occur, and then to mix the return water with the flow, by means of a mixing valve, for the purpose of serving the heating system. Where heat at a constant temperature is required from the boiler, such as hot-water supply, or for heat for ventilation air, or for forced convectors, unit heaters and the like, this would be served direct from the boiler with a separate pump, as indicated in Fig. 9.7 (p. 181).

The control of the boiler may also be under the influence of a time switch, by means of which the firing is started up at some predetermined time in the morning and shut off at night. Alternatively, it may be considered preferable simply to cut the temperature down at night; which involves a second thermostat, and the clock switch, in effect, changes over from one stat to the other. Time switches may have special features to cut out heating at weekends, or for a certain period in the middle of each day.

Control by Weather—A general control of temperature of water supplied to the heating system may be by means of a compensator system, one version of which is shown diagrammatically in Fig. 9.16. This comprises an external temperature-sensing element, and a similar temperature-sensing element in the flow main to the heating system. By means of a control box, in cold weather a balance of the circuit is achieved only by the element in the water flow being at a high temperature; similarly, in warm weather a balance is achieved with the water at a lower temperature. At any point between the two extremes the water temperature is adjusted proportionately to the outdoor temperature. Refinements have been introduced into this system by way of a heating element in the external unit, so that the effect of wind is taken into account. This type of compensator control may be used to control directly an oil burner or a gas-fired boiler; but if, as stated above, and as is preferable, the boiler is run at a constant temperature, the compensator system will then control the mixing valve. If there are a number of main circuits to various parts of the building, it is possible for each one to have its own compensator and circulating pump with mixing valve.

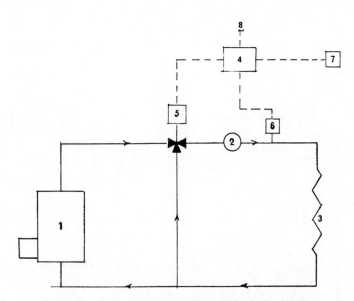

FIG. 9.16.—Diagram of Compensator Control (Satchwell).

1. Boiler-burner unit.	5. Modulating motor to mixing valve.
2. Pump.	6. Water-temperature detector.
3. Heating system.	7. Outside temperature detector.
4. Compensator control box.	8. Electric supply.

Zone Control—Control of the heating system for a building or series of buildings by zone control, is based, in effect, on the assumption that certain areas are thought to have common characteristics and hence similar heat requirements. For instance, in a rectangular building with north and south aspects on the long face, it is probable that rooms on the south aspect will often require less heat than on the north, and therefore might well be served off a different zone with a different mixing valve, as referred to above.

In a tall building, likewise, due to the greater exposure of the upper floors, it might well be desirable to zone the upper half of the building separately from the lower half. Again, if there are two or more main elevations, this might involve four or more zones.

Individual Room Control—The greatest refinement is to be achieved by control of the heat emitter in each room. In buildings where some rooms are sparsely occupied and others crowded or with their own internal heat gains, it is probably only by individual room control that uniform temperatures can be maintained. The alternative is, of course, to assume that the occupants themselves shut off the heat when it is not wanted and turn it on again later. In practice, however, this seldom seems to happen, but,

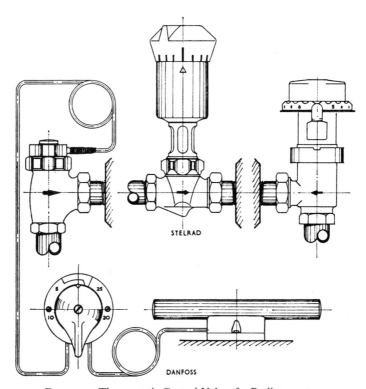

Fig. 9.17.—Thermostatic Control Valves for Radiators, etc.

rather, the window is opened if the room becomes too hot, with obvious wastage of heat.

Individual room control appears to be particularly necessary in ceiling-heated systems, where the ceiling is one of the forms of suspended metal trays.

A combination of compensator control and individual room control has considerable merit, as the thermostatic equipment in the various rooms then has less work to do and there would be much less tendency for fluctuations of temperature to occur.

Control Systems—The simplest form of controller is direct-acting, comprising a sensing element in the room or in the water flow, and, by liquid expansion or vapour pressure through a capillary, power is transmitted to a bellows or diaphragm operating a valve spindle. The valve spindle may be a mixing valve, mixing water from the boiler and return water on the suction or delivery of a pump, as referred to previously, or the valve may be in the supply pipe to a radiator, convector or other heat emitter (as in Fig. 9.17), so serving the purpose of an individual room controller.

Direct-acting thermostats have little power, and their control band is sometimes somewhat wide, but they have been considerably improved of recent years and are applied extensively, particularly to the smaller systems. Direct-acting thermostatic equipment may be said to be modulating or gradual-acting. An application to a domestic small-bore system is shown in Fig. 8.14 on page 173.

For larger installations, electrical systems of control are by far the most common, and can be devised to perform any of the functions desired in a great variety of ways. The simplest electrical system comprises an on-off thermostat connected to an electrically-operated valve or damper, such that, when the thermostat calls for heat, the valve or damper is opened and, likewise having become satisfied, the valve or damper is closed. An on-off thermostat may also be used to control the stopping and starting of a motor driving a fan, as in the case of forced convectors and unit heaters. This type of control also appears to be satisfactory for controlling ceiling heating. Indeed, it is probably true to say that most heat emitters in rooms may be satisfactorily controlled on an on-off method, particularly if the heat supplied through the water has been adjusted according to weather.

Electrical modulating controls enable, for instance, a mixing valve to be so adjusted automatically that it finds some mid-position and so supplies the desired flow-temperature to the heating system. For example, for a quick heat-up in the morning, the control system may be so arranged as to call for full heat for the first hour or so according to external weather, after which the temperature will be dropped proportionately to the external temperature. The system shown in Fig. 9.16 incorporates provision for this kind of variation. Furthermore, buildings vary in their response to heating according to their construction and exposure: at medium external temperatures, some may require higher water temperatures than others. This kind of adjustment or compensation can be made in the control box.

Electrical control systems now make use of transistors and thermistors which, with electronic circuits, eliminate mechanical moving parts to a large extent, and hence tend to make for greater reliability.

Controls using compressed air as a means of actuating valves and dampers are also available, but are more used in air-conditioning and industrial applications, so need not be considered here.

Control Panels—Where a comprehensive system of controls is to be provided, it is advantageous to locate the whole of the equipment at one point on a control panel, which can be pre-wired, and greatly facilitates installation. On the same panel would be mounted controls to boilers and starters for pumps, etc., together with any time-switching equipment, temperature-indicating dials, pressure indicators, if required, and combustion control equipment such as CO_2 instruments, flue-gas thermometers, and so on. Plate XXIV, facing page 465, illustrates a control panel of this type.

CHAPTER 10

Heating by Steam

STEAM AS A MEDIUM FOR HEATING in radiators and the like is a thing of the past. Hot water, with its flexibility to meet variable weather conditions and its simplicity, has supplanted it in all new residential, commercial and public buildings.

Steam is, however, often used for the heating of industrial buildings where steam-raising plant occurs for process or other purposes. It is also used as a primary conveyor of heat to calorifiers such as in hospitals, where again steam-boiler plant may be required for sundry duties such as in kitchens, laundry and for sterilising. Heating is then by hot water served from the calorifiers, further sets providing hot-water supply.

The present Chapter will mainly be confined to the principles involved in the generation of steam and to its utilisation for the purposes mentioned.

Generation of Steam—It will be as well to recapitulate here what was stated briefly in Chapter 1: namely that steam is produced when heat is applied to water in a partially filled closed vessel; and that, when boiling point is reached (100° C at atmospheric pressure), the addition of further heat causes a change of state to occur from water to steam. The quantity of heat involved in the process—the *latent heat of evaporation* (see page 14)—is considerable: 2258 kJ/kg at atmospheric pressure, compared with 420 to raise it to boiling point from 0° C.

In a closed vessel, the steam has no means of escape and the addition of further heat causes the pressure to rise. Means for preventing the rise from being above the strength of the vessel are provided by safety valves. As the pressure rises so does the temperature. Thus, this further heat entering the water is in the form of sensible heat in the liquid and steam, the latent heat falling as the pressure rises.

Steam in contact with water is termed *saturated*. If it carries some water droplets in suspension, it is referred to as *wet*.

If steam is removed from the vessel in which it is generated and further heat is added out of contact with water, the steam is said to be *superheated*. In this condition it tends to behave as a dry gas and is of little use for heating until it has cooled and become saturated.

The utilisation of steam for heating involves the process of condensation, in which the latent heat is removed by the heat-emitting surfaces of the heating system and reverts to water at the same temperature as the steam. This hot water or *condense* must be removed as soon as it is formed, or the heating apparatus will become water-logged and useless. The condense is, however, under pressure and, as it is released to atmospheric pressure, it suffers a reduction in temperature. In effect, part of the heat in the condense above atmospheric boiling point goes to re-evaporate a proportion of the liquid into steam at the lower pressure, this being termed *flash-steam*. Use can be made of this in a variety of ways, but if it is not usefully employed it constitutes a loss, as also does the remaining heat in the condense if this is run to waste. Thus, in practice, condense is collected and returned to the boiler for re-use, which at the same time affords a supply of distilled water and saves on water consumption.

Unfortunately, condense also may carry with it uncondensible gases such as chlorine, CO_2 and O_2 which are liable to cause severe corrosion of condense lines.

STEAM TABLES

Tables of properties of saturated steam in SI units are to be found in the *I.H.V.E. Guide*. Table 10.1 at the end of this chapter gives extracts from this data. Fig. 10.1 gives a graphical representation thereof. From this data, the following may be noted:

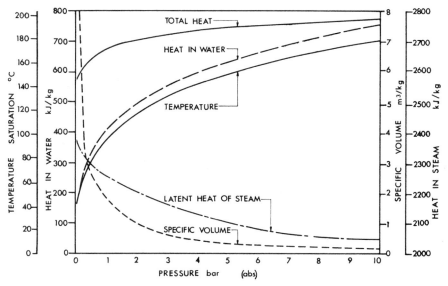

FIG. 10.1.—Graphical Representation of Properties of Saturated Steam.

Pressure. This is stated in *bar absolute* (1 bar $= 10^5 N/m^2$).* Atmospheric pressure very nearly equals one bar (1·013); and thus, as pressure-gauge readings are relative to atmosphere, gauge pressures are equivalent to absolute pressures minus 1·0. Sub-atmospheric pressures are given in the same units, though it is well to recall that, in the past, vacuum has been

---

* Those more familiar with lb f/sq. in. will find that 10 lb f/sq. in. is equivalent approximately to 0·7 bar (0·6895).

referred to in *inches of mercury column*—a descendant of the mercury baro-meter (1 in. mercury = 3·386 × 10³N/m² = 0·034 bar). Pressures over 10 bar need not concern us for heating purposes.

Temperature. Atmospheric boiling point is taken as 100° C. The 'stan-dard' atmosphere is 1·013 bar, and thus the boiling point at 1·0 bar is less (99·63° C). Steam temperatures at sub-atmospheric pressures fall to levels comparable with hot water, which feature was made use of in the vacuum and vapour systems now obsolescent in this country. Steam at temperatures above atmospheric will be seen from Table 10·1 to advance to 180° C at 10 bar. The higher the temperature the greater the output from a given heating surface. On the other hand, the higher the pressure, the greater the potential heat losses from mains and from flash steam and condense. Furthermore, the higher the pressure the greater the cost of pipework and apparatus.

Enthalpy (or heat content). The next three columns give the heat content or *specific enthalpy* in terms of kJ/kg for the water, for the latent heat, and for the sum of the two (total heat of saturated vapour). The enthaply of water is calculated from nought at 0° C. The decline in the latent heat with the increase of pressure will be noted; thus, per unit mass of steam, the lower the pressure the greater the ratio of latent heat to total heat, and consequently the less the loss from flash steam.

Volume. The last column quoted in Table 10.1 gives volumes in m³/kg, these being of use in pipe sizing where velocity of flow is used as a criterion.

Other values. The steam Tables in the *I.H.V.E. Guide* include in ad-dition the Prandtl number, which is the ratio of viscosity to thermal diffusivity and is a factor involved in heat-transfer problems under forced convection. It need not concern us here.

Another column in the *Guide* Tables gives values of specific-heat capacity for steam at each pressure. The value varies from 2·01 kJ/kg° C at 1 bar (roughly atmospheric pressure) to 2·62 at 10 bar absolute.

Use of Steam Tables—By way of example, assume that a heat load of 1000 kW (= 1000 kJ/s) is to be served from a steam boiler at a distance.

The mass flow of steam involved is obtained by dividing the heat energy content by the latent heat, which in turn is dependent on pressure. Thus,

at 1 bar abs. $\frac{1000}{2258} = 0·44$ kg/s

" 5 " " $\div 2108 = 0·47$ kg/s

" 10 " " $\div 2015 = 0·49$ kg/s.

In terms of volume, these mass flows represent

at 1 bar 0·44 × 1·69 = 0·75 m³/s (or 750 litre/s)

" 5 " 0·47 × 0·38 = 0·18 m³/s (or 180 litre/s)

" 10 " 0·49 × 0·19 = 0·09 m³/s (or 90 litre/s).

Thus it is clear that the higher the pressure the smaller the pipe size to carry the load.

When the steam is condensed at the distant point to which it has been conveyed, assuming no losses on the way, we have:

heat in liquid

$$\text{at} \quad 1 \text{ bar} = 417 \text{ kJ/kg}$$
$$\text{,, } \ 5 \ \text{ ,, } = 640 \text{ kJ/kg}$$
$$\text{,, } 10 \ \text{ ,, } = 762 \text{ kJ/kg}$$

At 1 bar absolute, being almost atmospheric, the condense issues as boiling water and may be conveyed back to the boiler as such.

At 5 bar, when released to atmosphere,

$640 - 417 = 233$ kJ/kg of heat re-evaporates part of the parent water (as explained previously);

i.e. $\frac{223}{2108} = 0.105$ kg/kg as flash steam.

At 10 bar: $762 - 417 = 345$ kJ/kg of heat;

i.e. $\frac{345}{2015} = 0.17$ kg/kg as flash steam.

Thus it is apparent that the higher the pressure the greater the potential loss due to flash steam: at atmospheric pressure, nil; at 5 bar, 10·5 per cent; at 10 bar, 17 per cent. In practice, the re-condensing of flash steam may be achieved in equipment for the purpose, or, if it is passed into return-condense lines within the building the loss can be turned into useful heating. Otherwise, the overall efficiency of the system in terms of heat input to heat usefully employed is reduced to the extent of the percentages above.

Entropy—This is a conception which is much used when considering steam used in a heat engine. For ordinary heating problems it is not important, and need only be considered very briefly.

Entropy does not correspond to any physical property of heat of which we can have direct knowledge. It is simply a mathematical ratio for expressing the availability of heat, and the entropy of any *closed* system involving heat-transfer increases as the transfer takes place, e.g. from steam at 150° C to secondary water at 10° in a calorifier, where finally we are left with condensate at, say, 95° and water at 80°. By no known method could the heat be taken out of the water and put back into the condensate to make steam (without adding external energy as in a heat pump), and this irreversible process takes place in the direction of increase of entropy. Similarly, the expansion of steam through a throttling valve is an irreversible process which causes an increase in entropy of the steam.

Entropy may be depicted graphically by drawing a curve of the heat content of a closed system, the curve having ordinates of absolute temperature. If the area under the curve represents heat, then the abscissae represent entropy.

For the development of this subject the reader is referred to any of the standard works on Heat Engines.

Steam Boiler Evaporation Ratings—are commonly listed *from and at* *100° C.*

This denotes that if the water in the boiler is at atmospheric pressure the steam generated at that pressure will be the quantity given, i.e. it is assumed that only latent heat is added.

Such a rating, however, is of little practical use as it stands. The feed water is seldom exactly 100° C and the pressure is usually above atmospheric.

For any other set of conditions the *actual evaporation* can be derived as follows:

Actual evaporation

$$= \frac{\text{Evaporation from and at } 100° \text{ C} \times 2258}{\left\{\begin{array}{c}\text{Total heat at pressure}\\ \text{required}\end{array}\right\} - \left\{\begin{array}{c}\text{Heat in water at}\\ \text{feed-water temp.}\end{array}\right\}}.$$

It will generally be found that the actual evaporation is less than the 'from and at' rating.

STEAM GENERATION IN PRACTICE

A whole bibliography exists on this subject but, within the limits of the future use of steam in heating projects, the scope of this chapter may be limited to considering firstly the question of pressure and secondly the types of equipment available for this purpose.

It will be clear from the foregoing that high pressures are not necessarily desirable. Low ones such as 1·5 to 2 bar abs. may suffice for a canteen with a steam-cooking load and for some heating in the canteen itself. In such case, a cast-iron sectional steam boiler, as in Fig. 10.2, would be

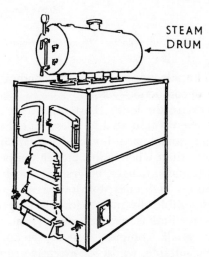

STEAM DRUM

FIG. 10.2.—Cast-Iron Sectional Steam Boiler for low pressures (Ideal Britannia).

possible. It will be noted that the steam separating space is contrived by means of a drum external to the boiler. The limitation of this type of boiler as to pressure is 2 bar absolute.

Higher pressures are dealt with by means of steel boilers, now generally of packaged welded construction, automatically fired by gas or oil. A boiler of this type is shown in Fig. 10.3 covering a range from 1 MW to 10 MW. Beyond this output, water-tube boilers are employed, but these are not relevant in the present context.

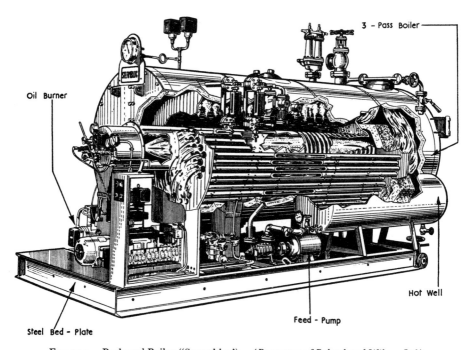

FIG. 10.3.—Packaged Boiler ('Steambloc'). (*By courtesy of Babcock and Wilcox, Ltd.*)

Boiler Pressure—The selection of the most economical boiler pressure depends on a number of conflicting factors.

The lower the pressure the lower the temperature and, hence, losses from piping surfaces, even if insulated, are less. But the volume of steam is greater and larger mains are involved. Heat-emitting apparatus requires larger surfaces than when using higher temperatures. Pressure-retaining equipment (boilers, flanges, valves, pipework etc.) is less costly for low pressures than for high ones; but low-pressure steam, when conveyed over any considerable distance, becomes progressively wetter in losing temperature until a limit is reached where it is no more applicable.

Thus—except for small, compact systems—steam pressures of the order of 4 bar may be suitable, whilst for extensive runs of main 7 to 10 bar abs. is often used.

Pressure Reduction—It is often necessary to generate steam at one pressure but desirable to serve equipment at a lower pressure. For instance, sterilizing and laundry plant may require 7 bar absolute, but, for reasons stated above, the rest of the system can be run more economically at some decreased pressure, which at the same time reduces losses.

This is done by means of a pressure-reducing valve as shown in Fig. 10.4. In fact, if the steam entering such a valve is dry-saturated, it will be

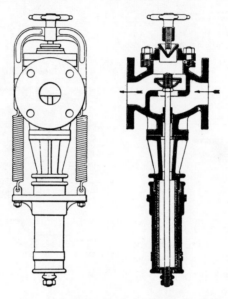

Fig. 10.4.—Steam Reducing Valve (Royles).

apparent that it becomes superheated when pressure is reduced. Some initial wetness in the steam and some losses following pressure reduction may, however, be enough to restore its condition to saturation.

Flash Steam Recovery—This has been mentioned already and it only remains to refer to means for taking advantage of it. In a calorifier system, a separate coil may be used to receive the flash steam. In a unit heater system, one or more of the units may be designed to receive flash steam.

Condense Return—Having been condensed in the heating equipment it is grossly wasteful to discharge condense to drain. It may represent as much as 20 per cent. on the fuel bill. Thus it is normally returned for re-use in the boiler, as explained already.

At the drain point of the heat emitting apparatus or calorifier, some means is required to allow water to pass but not steam. This device is termed a *steam trap*. Various types exist, some depending on a float

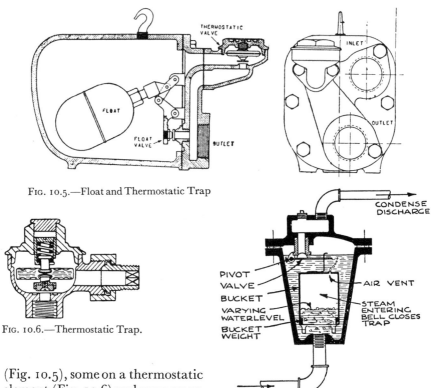

FIG. 10.5.—Float and Thermostatic Trap

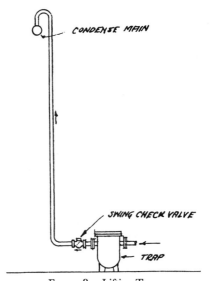

FIG. 10.6.—Thermostatic Trap.

FIG. 10.7.—Bucket Trap (Drayton Armstrong).

(Fig. 10.5), some on a thermostatic element (Fig. 10.6) and some on an inverted bucket (Fig. 10.7). Where the condense outlet is below the level of the return main, then use is made of a 'lifting trap' (Fig. 10.8).

Where condense mains are too low to allow a gravity fall back to the boilerhouse, and the steam pressure is too low to operate a lifting trap, the condense is taken to a receiver at a low point from which it is pumped back to the boilerhouse either by an electric pump, as in Fig. 10.21 (page 218), or by a pumping trap (Fig. 10·22, page 219) using steam taken from a high-pressure main.

Air—One of the bugbears of steam is air. Its presence in heating equipment acts as a blanket preventing the steam from condensing

FIG. 10.8.—Lifting Trap.

on the heating surfaces: this is particularly so at low pressures. Means for air removal are therefore essential. Types of steam trap with a thermostatic element allow air to pass through, due to the air causing a lowering of temperature. Certain types of float traps contain a separate thermostatic element for this purpose. Air is more dense than steam and collects at the lower portions of equipment. Thus, another fitment is an air eliminator which is connected near the bottom of the equipment and is also thermostatically controlled.

At higher pressures, air is usually purged by hand-opening of the blow-through cocks on mains and equipment, after which the steam traps may be expected to cope with it mixed with the condense.

By 'air', in this connection, is meant the mixture of uncondensible gases already alluded to.

Boiler Feeding—The simplest method now common is to provide, near the boiler, a collecting tank or 'hot-well' into which condense return is delivered. In addition, the tank receives cold-water make-up through a ball valve from the mains, which water may in turn have passed through a water-treatment plant.

From the tank, water is pumped to the boiler through a non-return valve by means of an electrically-driven or steam-driven feed pump under the control of a water-level regulator attached to the boiler. This comprises a float in a vessel connected to the boiler at the water line, such that if the level drops the pump is started. When the level is restored the pump is stopped.

The condense return generally serves to keep the feed-water hot; but where this is not the case, a steam coil in the tank is used to pre-heat the feed so as to avoid the injection of cold water into the boiler, which is liable to cause thermal shock.

In the packaged boiler (see Fig. 10.3, page 202) the feeding equipment forms part and parcel of the unit: i.e; feed tank, pump and level control.

Steam Boiler Mountings—The essential fittings of any steam boiler are:

>*water gauges*: to give a visual indication of water level in the boiler;
>*pressure gauge*: to give indication of steam pressure;
>*safety valves*: to limit the pressure to that for which the boiler is designed;
>*feed valve*: for admittance of feed water, and incorporating a non-return valve;
>*steam stop-valve*: to shut off the outlet;
>*blow-down valve*: for draining-off and clearing sludge.

In addition, it is common practice to add:

>*high and low-water alarms*: these give an audible alarm on the levels rising or falling below safe limits; at the same time, with automatic firing, the burner may be cut-off. They are usually electrical;

pressure control: this acts in the same way as a thermostat on a hot-water boiler by simply opening contacts when the set pressure is reached, so shutting off the firing equipment. On fall of pressure below a minimum setting, the reverse occurs. For duties over about 300 kW, high-low or modulating control is used (as explained in Chapter 5).

EMISSION FROM STEAM PIPING

Piping Systems—It is unlikely that exposed steam piping will be used as a heating surface. It was a common and cheap method in the older type of factory. It may, however, occur for drying or other special purposes and emissions are given in Table 10.2 (see page 222).

Mains—Piping used for mains will be insulated and the losses must be taken into account in determining the steam flow to be carried. Table 10.3 (see page 223) gives the emission where insulated by means of a commonly used material—glass fibre. This Table has been compiled in a form convenient for reading emissions direct according to the hot-face temperature, the equivalent steam pressure being also given.

Emission from condense piping may be taken as at 100° C temperature difference and according to whether bare or insulated. Whilst this is on the high side it is usually of little account and ignored.

UTILISATION OF STEAM

Probably the most general method of utilising steam for the heating of industrial and similar buildings is through the medium of warm air.

Plenum System—The Plenum System accepts steam at a central point in a set of heating coils or 'battery', and a fan is arranged to blow air over the coils, delivering it warmed into a system of ducts for distribution throughout the space to be heated. This system, however, lacks flexibility —the ducts are often cumbersome, and distribution of heat is liable to be upset by wind pressure and the opening of doors.

Unit Heaters—The unit-heater system makes use of warm air as the final distribution medium but avoids the use of ducts. Thus, steam is conveyed to a series of units throughout the space to be heated, each unit being equipped with a fan discharging the heated air into a limited zone such that, with a suitable lay-out of units, fairly uniform distribution of temperature is achieved. Local cold spots such as at doors can be dealt with by units directed to such areas, and local control of temperature is achieved by on-off switching thermostatically of individual units or groups of units.

Various types of unit heater have been devised: the horizontal type (see Fig. 3.14, page 58), the vertical-downward type (Fig. 10.9) and the projector type (Fig. 10.10). Ratings vary as follows:

horizontal type:	10 kW to	30 kW
vertical type:	30 kW to	200 kW
projector type:	100 kW to	300 kW

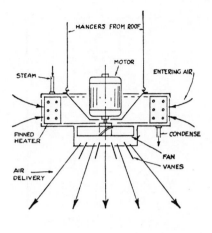

FIG. 10.9.—Downward Discharge
Unit Heater.

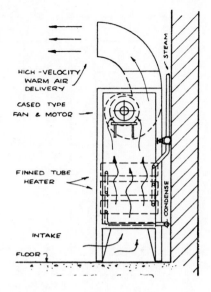

FIG. 10.10.—Projector Type Unit Heater.

For desirable air volumes, discharge temperatures and mounting heights, makers' lists should be consulted. Lower air-delivery temperatures such as 40° C to 50° C, with correspondingly larger air volumes, are preferable to small air quantities at higher temperatures.

Typical lay-outs of horizontal-type unit heaters are shown in Fig. 10.11. The projector type relies on an unimpeded path for the discharge into the heated space, the air-flow pattern being as in Fig. 10.12.

Where quiet running is essential, the selection of type and speed should take this into account. Some noise in industrial applications is often unimportant.

Fresh Air Inlets—It is often necessary to provide artificial inlet ventilation to a building where unit heaters are used. This may conveniently be done by connecting the suction side of the unit by means of a duct to external air, terminating either with a roof cowl or with a louvred opening in the wall as in Fig. 10.13. The basis of design must take into account the volume of fresh air introduced, and that the unit inlet will be at 0° C or below in cold weather. This, of course, affects the rating of the unit. When some units in a building are provided with fresh air inlets and some are recirculating, it is desirable that their final air temperatures should be as nearly the same as possible.

It is sometimes arranged that the fresh air inlet is controlled by a damper, so that in cold weather the unit may be made to recirculate.

It should be noted that the battery of a unit connected by vertical ducting to a roof cowl will induce a strong reverse air current when the fan is stopped, with risk of overheating of the motor. It is generally arranged

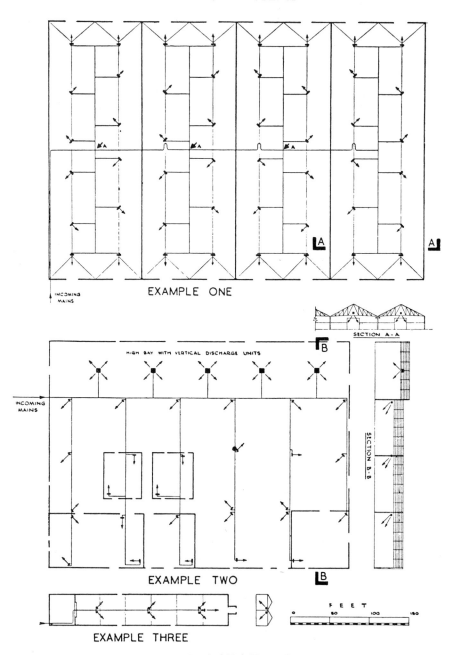

EXAMPLE ONE

EXAMPLE TWO

EXAMPLE THREE

Fig. 10.11.—Typical Unit Heater Layouts.

Example One—Storeshed.
  „   Two—Workshops.
  „   Three—Canteen.

Plate VIII. Above: A Micro-bore System during installation. Note the flow and return headers in the floor trap. Below: a radiator in a room, similar to the one above, served by a micro-bore system. Note the temperature limiting valve controlling the district-heating supply (see p. 175)

Plate IX. A Steam Calorifier chamber (see p. 214)

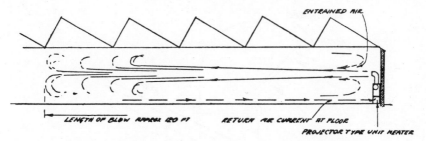

FIG. 10.12.—Air Circulation with 'Projection' Type Unit Heaters.

that these units should be not under thermostatic control by switching off the fan. If they are switched off, the steam should be shut off also.

Motors—The sizes of motors for unit heaters vary from about 60 W to 1000 W for horizontal and vertical types, and up to 4 kW for the projector type. Single-phase motors are usually of the 'Capacitor start-run' type which have no brush gear to give trouble, but sometimes the condensers break down, due to overheating from the radiant heat. Where it is possible, it is preferable to use three-phase current, though protection against single-phasing seems to be desirable. Motors are usually best if totally enclosed, and the bearings should be suitable for long running without attention.

Margin—The required output of unit heaters is sometimes not achieved in practice, due to steam pressure drop, dirt in the fins of the heater, direct escape of warm air through doors before mixing, unsatisfactory distribution and other causes. It is customary to make the installed load somewhat larger than the calculated heat losses. A margin will furthermore allow the thermostatic control to take charge even during coldest weather when the units would otherwise be running continuously.

Such a margin must be purely empirical and is generally taken at from 15 to 20 per cent of the heat losses. This margin will also be available for sub-design temperature and quick heating-up, and the piping and boiler power should be installed to include for it.

Piping Connections—The large concentrated steam loads of unit heaters call for care in the piping connections. The steam inlet requires a valve, preferably of the fullway gate type. The condense outlet requires a trap capable of passing large volumes of condense rapidly as, on heating up a cold building, the condensation rate will be much higher than when warm.

For small units, up to about 30 kW on low pressure steam, a thermostatic trap will suffice; for larger units a float and thermostatic trap is necessary. For medium and high-pressure units an inverted bucket and certain types of float trap are suitable. Unions are necessary for disconnection of steam and condense.

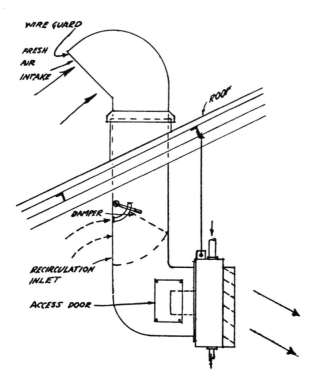

Unit Heater with Fresh Air Intake through Roof.

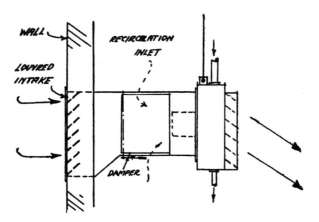

Unit Heater with Fresh Air Intake through Wall.

FIG. 10.13.

Fig. 10.14 shows a typical arrangement of connections.

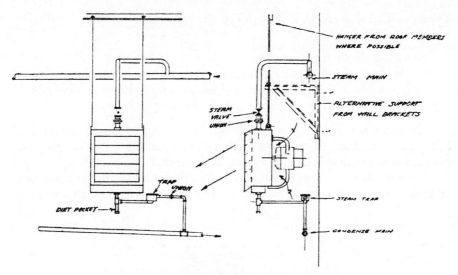

FIG. 10.14.—Typical Unit Heater Connections.

CONVECTORS

Convectors of the same general type as those already described (see page 57) may be used with steam as the heating medium, and while it is physically possible to operate them with the same range of pressures as, for instance, unit heaters, it is preferable to avoid the use of pressures over about 2 bar absolute.

Steam convectors are an economical way of heating factory offices particularly where these occur singly or in small groups over a wide area. For larger office layouts it is preferable to install a steam-water calorifier serving hot-water radiators, either by gravity or pump circulation.

Fan Convectors—Fan convectors illustrated in Fig. 3.13 (page 57) may satisfactorily be used with steam, particularly where large outputs are required. Fan convectors may be used with fresh air inlets behind and a change-over recirculating damper to meet cases such as a canteen where inlet ventilation is required to replace kitchen exhaust.

Radiant Metallic Panels using Steam—These are an alternative method of heating factories and similar large spaces, and are dealt with under *High-Pressure Hot Water* in Chapter 11. When used with steam each will be fitted with a valve and steam trap.

HEATING CALORIFIERS

A calorifier is, in effect, a heat exchanger; that is to say, a device whereby heat from a medium at a high temperature is transmitted to a second medium at a lower temperature. The high-temperature medium we are concerned with here is steam, and the low-temperature medium hot water to be used for heating.

Types of Calorifier—A horizontal steam calorifier for heating is shown in Fig. 10.15 and consists of:

(a) An outer shell.
(b) An internal battery of piping.
(c) A chest in which the ends of the tubes terminate so as to admit the heating medium.

The steam is generally passed through the tubes, the water to be heated being outside. This gives a lower temperature on the outer casing and consequently less heat loss than if steam is in the shell.

The outer shell is either of cast-iron or steel. Inlet and outlet connections or flanges are necessary on the casing. The thickness of shell depends on

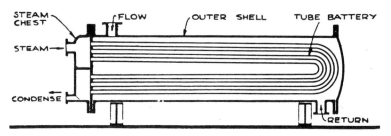

FIG. 10.15.—Steam Calorifier with U tubes.

the head pressure and on the diameter of the vessel: a minimum test pressure should be not less than one and a half times the working pressure.

The tube battery may be of steel, brass or copper. Usually the latter is to be preferred, as it may be thinner than steel, allowing a more compact arrangement of the tubes, apart from which it has a slight advantage in conductivity.

The tube battery may be made removable, as shown in Fig. 10.15, with the tubes of U (or hairpin) form, or a 'floating header' may be adopted, the tubes then being straight. An alternative construction is as shown in Fig. 10.16 in which the tube battery is not removable without complete

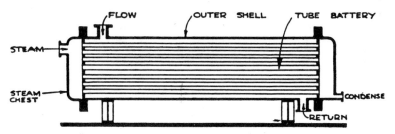

FIG. 10.16.—Steam Calorifier with Fixed Tube-Plates.

dismantling. A further type is vertical (Fig. 10.17), being economical in floor space but requiring height for withdrawal of the shell.

The chamber for the admission of the heating medium is called *the steam chest*, and it is generally of cast-iron with inlet and outlet connections for the steam and condense.

Rating of Calorifiers—The subject of heat transfer is a separate study. Within the limited scope of a steam calorifier, the parameters include:

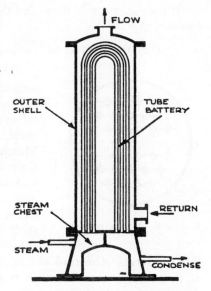

FIG. 10.17.—Vertical Steam Calorifier.

(*a*) difference of temperature between the steam and water, inlet and outlet;

(*b*) the velocity of water over the tubes;

(*c*) the material of the tubes and thickness;

(*d*) the transfer coefficient which is itself dependent on temperature;

(*e*) water quality: i.e., whether clean or liable to foul the tubes.

The *Guide* gives data under the above headings from which calorifiers may be designed. Table 10.4 (see page 223) has been prepared from these data and gives in a simplified form the essential information for a conventional set of conditions.

The user will, however, as with a boiler, refer to makers' lists for actual outputs and sizes, particularly so as by the use of controlled flow paths and high velocities within the shell of a calorifier a much enhanced performance is possible; but this can only be related to a particular design.

Example of Calorifier Sizing—

Output required	1000 kW
Steam pressure	3 bar abs.
Steam temperature	133° C
Water temperature mean	60° C
Difference	73° C
From Table 10.4 for 80° diff. transfer	8·5 kW/m
For 80° mean water, difference = 53	
$8 \cdot 5 \times \frac{73}{53}$	= 10·0 kW/m

Length 1 in (25 mm) tube required $= \frac{1000}{10} = 100$ m run

If the tubes are U form and the average length of one U tube $(2\frac{1}{2} \times 2)$ equals 5 m,

then the number of U tubes = 20.

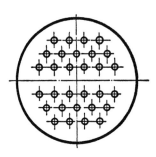

FIG. 10.18.—'Staggered' Tubes.

The tubes would be arranged in staggered formation, as in Fig. 10.18. Allowing a reasonable pitch, the shell might be about 0·7 m diameter and 3·2 m long, plus 0·3 m for the steam chest.

Mountings for Calorifiers are much the same as for boilers. Safety valve, thermometer, and drain cock are essential on the water side, and a steam pressure gauge and trap on the steam side. An altitude gauge connected to the shell is desirable. Open vents direct from calorifiers are as for boilers.

Thermostatic Control to calorifiers is a necessary provision. If control at a constant temperature is required, as when the heating system is divided into a number of zones each with its own pump and mixing valve, a simple direct-acting thermostatic valve may be adequate. Where the calorifier is controlled by compensator, an electrical system will be required using a modulating steam valve on the steam inlet. A word of warning here, as, on shut-off, a vacuum may be formed preventing escape of condensate which must in any event go out by gravity. The addition of a vacuum breaker on the steam side may be necessary. A trap giving ready air release is then a corollary.

General—Other items required in connection with calorifiers are supporting cradles, runways or other means for tube withdrawal, and lagging. These do not call for comment. (An installation of heating and H.W.S. calorifiers is shown in Plate IX, facing p. 209).

PIPE SIZING FOR STEAM SYSTEMS

For the sizing of steam mains and piping it is necessary to determine:

(a) The mass flow of steam in grammes per second. This may be derived in the case of a heating system directly from the kW transmitted, by dividing by the latent heat at the appropriate pressure. The value at atmospheric pressure is 2258 kJ/kg (2·258 kJ/g).

In the case of kitchen or laundry apparatus the steam condensed in g/s may be arrived at by calculation if the duty is known, or from manufacturers' data for each item.

Where the mains are lagged and comparatively short their heat loss may be ignored in estimating the steam flow. If such is not the case due allowance should be made from Tables 10·2 or 10·3 (pages 222 and 223).

(b) The initial pressure of the steam.

(c) The pressure drop permissible between the two ends of the system. This may be 0·07 to 0·14 bar for low pressure, and 0·35 to 1·4 bar for high pressure, depending on size of system.

(d) The resistances in the pipe line due to bends, tees, valves, etc.

With SI units, in order that flow rates may be represented by whole numbers, it has been necessary to adopt the unit of *grammes of steam per second*. This may seem strange to engineers used to pounds per hour. But if it is remembered that 1000 grammes equals 2·2 pounds and one hour equals 3600 seconds, we have

$$1 \text{ lb/h} = \frac{1000}{2·2 \times 3600} \text{ g/s}$$
$$= 0·126 \text{ g/s}$$

i.e: roughly 8 lb/h = 1 g/s.

Pipe sizing for steam may be based on a pressure-drop method or on a velocity method.

Sizing by Pressure Drop—This method, first developed by Reischel and followed in earlier editions of this book and by the *I.H.V.E.* in the *Guide*, makes use of a pressure factor Z, where $Z = P^{1·929}$ and where P is pressure in bar absolute. Table 10.5 (p. 224) gives Z values up to 10 bar.

Table 10.6 (see page 224) gives the flow of saturated steam in steel pipes in terms of difference between initial and final Z values per metre length of pipe. By way of example, assume

Heat to be transmitted	= 1000 kW
Steam pressure initial	= 5·0 bar
Steam pressure final	= 4·5 bar
Latent heat	= 2164 kJ/kg
Steam flow $\dfrac{1000 \text{ kJ/s}}{2164}$	= 0·46 kg/s
	= 460 g/s
Length of main (including single resistances)	= 50 m
Z_1 initial	22·3
Z_2 final	18·2
$Z_1 - Z_2$	= 4·1
$\therefore \dfrac{Z_1 - Z_2}{L}$	= 0·08

From Table 10.6, $2\frac{1}{2}$ in carries 468 g/s.

The required main will thus be $2\frac{1}{2}$ in (65 mm).

The above pressure drop of 0·5 bar in 50 m equals 0·01 bar/m.

Sizing by Velocity—Piping in boilerhouses such as main headers, offtakes, short runs of mains, connections to calorifiers etc. are more conveniently sized on the basis of velocity. Table 10.7 (page 225) gives some conventional steam velocities and also pipe areas in square metres. For a given pressure, the volume of steam in m³/kg can be read from Table 10.1. This multiplied by the mass flow per second will give the volume to be handled, which divided by the velocity selected from Table 10.7 then gives the pipe area necessary, and hence the pipe size.

Alternatively, the velocity lines on Table 10.6 may be used for the smaller sizes, though probably with less facility.

Sizing of Condense Returns—The presence of air and flash steam, as pressure is relieved on passing the steam trap, has been mentioned. Any precise calculation of pipe sizes is consequently impossible and various rule-of-thumb methods are commonly employed. One is to make the condense pipe one size smaller than the steam pipe.

The *Guide* postulates two conditions: one where air is released by other means and the trap discharge is at low pressure, when the condense lines may be sized using the hot-water-flow data. The pressure in N/m² can be derived from the head causing flow (1 metre head = 9804 N/m²). The second condition is that referred to above, where gases and re-evaporated steam add greatly to the resistance to flow and it is recommended that a pressure drop ten times that of ordinary hot water be used.

Where condense is collected in a receiver and pumped back to the boilerhouse, the pumping main can be assumed to run full and hence may be sized as for normal hot water.

STEAM ACCESSORIES

Provision for Expansion—Reference has been made in Chapter 9 to the need to make provision for expansion, but, as the temperatures employed with steam are much greater, the problem is more severe.

As before, the amount of expansion to be accommodated in metres is simply length in metres × coefficient of expansion per °C × temperature rise.

By way of example:

Initial temperature	10° C
Steam pressure	5 bar abs
Steam temperature	152° C, thus the rise = 142° C
Coefficient expansion of steel	$11 \cdot 3 \times 10^{-6}$
Length of main	50 m
Expansion	$= 50 \times 11 \cdot 3 \times 10^{-6} \times 142$
	$= 0 \cdot 08$ m $= 80$ mm or $(3\frac{1}{8}'')$

Methods of providing for expansion are:

1. By changes of direction of the piping in the manner indicated in Fig. 10.19 (and see also Plate X, facing page 224). It will be noted that certain points are fixed or anchored. This is to prevent successive

creep which might otherwise eventually rupture the pipe. The U expansion type or the Lyre type are standard fittings for which data on movement allow able per unit is available.

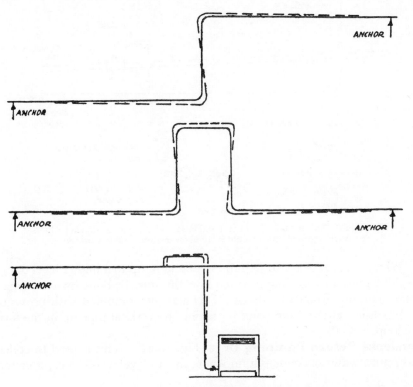

Fig. 10.19.—Use of Changes of Direction for Expansion.

2. By flexible bellows of the type shown in Fig. 9.11 (page 186). These are also used in an off-set to rack sideways in various forms, being restrained from extension due to pressure by external pivotted links.

3. By sliding expansion joints. These are, however, prone to leakage if not well maintained and have largely gone out of favour.

The provision for expansion in whatever form needs to be considered in conjunction with the form of supports: hangers, rollers, guides, spring suspensions and the like. But this is beyond the scope of the present book.

Steam Main Drainage (Fig.10.20)—Where possible, steam mains should fall in the direction of steam flow at about 1 in 200. Reducers should be eccentric to allow no pocket for condense to collect. At the end of a run the main should be drained with a steam trap discharging to condense main. Mains longer than about 100 m should be drained intermediately.

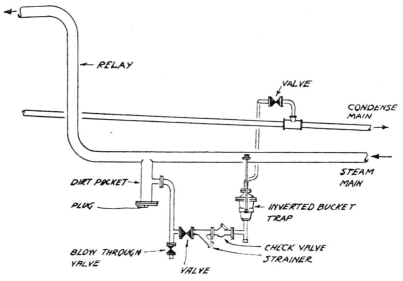

FIG. 10.20.—Steam Main Drip Point and Relay. Flanged joints shown for high-pressure; screwed joints would be used for medium and low pressures.

Where it is necessary to step the steam main up to a higher level due to the fall having brought the pipe too low, this may be done by arranging a 'relay' suitably drained as shown. This may be combined with provision for expansion at the same point by taking the vertical pipe up in the form of a loop.

Condense Return Pumping Unit (Fig. 10.21)—This is used to collect and return water of condensation when, due to level or distance, it cannot

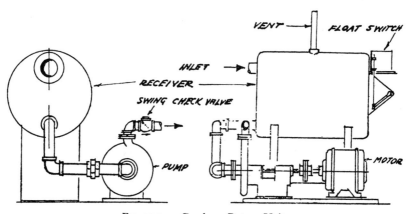

FIG. 10.21.--Condense Return Unit.

be returned by gravity. It consists of a receiver containing a float switch, and a centrifugal pump, motor-driven, stopped and started by the float switch. The rating of the pump is usually about three times the condense

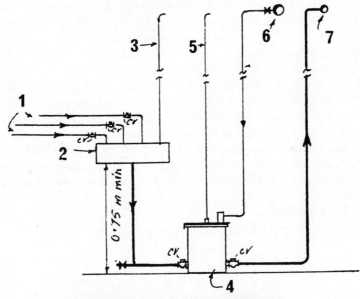

DIAGRAM OF CONNECTIONS

KEY
1. Low-pressure con-
dense returns
2. Receiver
3. Vent
4. Super lifting trap
5. Exhaust
6. High-pressure
steam
7. High-level return
line
8. Inlet check valve
9. HP steam inlet
10. Strainer
11. Exhaust valve
12. Exhaust outlet
13. Toggle
14. Float in top
position
15. Outlet check valve

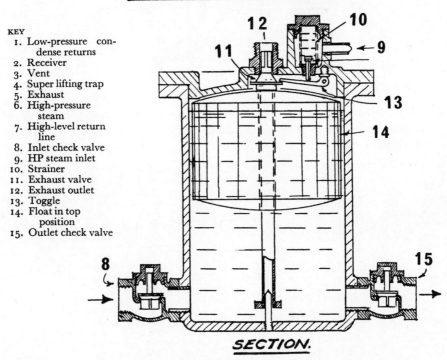

SECTION.

FIG. 10.22.—Super Lifting Trap.

return rate. The receiver requires a large vent to deal with the vapour from re-evaporation. It is necessary for the pump to be lower than the receiver to provide a 'head' on the pump suction.

Super-Lifting Traps (Fig. 10.22)—As an alternative to the pumping unit, where the condense is from steam at too low a steam pressure to lift the condense the required height, use may be made of the 'super-lifting' trap. The steam inlet is connected to a high-pressure steam main at 3 bar abs or over. A small vented receiver is necessary to accommodate the condensate during the period of discharge.

Steam Equipment Generally—In addition to the above brief list of steam apparatus particularly concerned with heating systems, there are numerous other fittings and equipment to which it is not possible to refer in detail.

These include:

> Feed water-treatment plant.
> Feed pumps and injectors.
> Hot-well arrangements and fittings.
> Boiler mountings for high pressures.
> Steam valves of various types.
> Condense valves.
> Steam separators.
> Flanged joints.
> Welding and screwing.
> Cast-iron and cast-steel fittings.
> Steam and feed-water meters.
> Instruments and Controls.

For further information on these matters, reference should be made to a textbook on Steam Engineering, and to the various makers' catalogues and handbooks.

TABLES: CHAPTER 10

TABLE 10.1*

PROPERTIES OF SATURATED STEAM

Pressure bar absolute	Temperature °C	Specific Enthalpy kJ/kg			Specific Volume m³/kg
		Heat in Water	Latent Heat	Total Heat	
Sub-atmospheric 0·2	60	252	2358	2610	7·65
0·4	76	318	2319	2637	3·99
0·6	86	360	2293	2653	2·73
0·8	94	392	2274	2666	2·09

Table 10.1—*Contd.*

Pressure bar absolute	Temperature °C	Specific Enthalpy kJ/kg			Specific Volume m/³kg
		Heat in Water	Latent Heat	Total Heat	
1·0	100**	417	2258	2675	1·69
2	105	439	2244	2683	1·43
4	109	458	2232	2690	1·24
6	113	475	2221	2696	1·09
8	117	491	2211	2702	0·98
2·0	120	505	2201	2706	0·89
2	123	518	2193	2711	0·81
4	126	530	2185	2715	0·75
6	129	541	2178	2719	0·69
8	131	552	2170	2722	0·65
3·0	133	561	2164	2725	0·61
2	135	571	2157	2728	0·57
4	138	580	2151	2731	0·54
6	140	589	2145	2734	0·51
8	142	597	2140	2737	0·49
4·0	144	605	2134	2739	0·46
2	146	612	2129	2741	0·44
4	147	620	2123	2743	0·42
6	149	627	2118	2745	0·41
8	150	634	2113	2747	0·39
5·0	152	640	2109	2749	0·38
2	153	647	2104	2751	0·37
4	155	653	2099	2752	0·35
6	156	659	2095	2754	0·34
8	157	665	2090	2755	0·33
6·0	159	670	2087	2757	0·32
2	160	676	2082	2758	0·31
4	161	682	2078	2760	0·30
6	163	687	2074	2761	0·29
8	164	692	2070	2762	0·28
7·0	165	697	2066	2763	0·27
2	166	702	2063	2765	0·27
4	167	707	2059	2766	0·26
6	168	712	2058	2767	0·25
8	169	716	2052	2768	0·25
8·0	170	721	2048	2769	0·24
2	171	725	2045	2770	0·24
4	172	730	2041	2771	0·23
6	173	734	2038	2772	0·22
8	174	738	2034	2773	0·22
9·0	175	743	2031	2774	0·21
2	176	747	2028	2775	0·21
4	177	751	2025	2776	0·21
6	178	755	2021	2776	0·20
8	179	759	2018	2777	0·20
10·0	180	763	2015	2778	0·19

Note: The above Table is abstracted from the *I.H.V.E. Guide*, to the nearest whole numbers.
* See page 198.
** Accurately, 99·63°C. 100°C occurs at standard atmospheric pressure (1·013 bar).

TABLE 10.2*

THEORETICAL HEAT EMISSION FROM BARE
SINGLE HORIZONTAL STEEL PIPES IN AMBIENT AIR AT
TEMPERATURES BETWEEN 10°C AND 20°C

Pipes to BS 1387

Nominal Bore		Heat Emission W/m run					
		Temperature Difference Surface to Surroundings °C					
in.	mm	100	120	140	160	180	200
½	15	130	170	200	240	290	340
¾	20	150	200	250	300	350	420
1	25	190	240	300	370	430	510
1¼	32	230	300	370	450	540	630
1½	40	260	330	410	510	600	710
2	50	320	410	500	620	740	860
2½	65	390	500	610	750	900	1100
3	80	450	580	710	900	1100	1300
4	100	550	710	900	1100	1300	1500
5	125	660	850	1100	1300	1600	1900
6	150	770	1000	1200	1500	1800	2100
With ambient air at 15°C the above temperatures correspond to steam at pressures bar:		1·7	3·1	5·4	9	14	21

Note: 1. The above Table is extracted from the *I.H.V.E. Guide.*

2. For 2 pipes in bank, factor = 0·95
   For 4 pipes in bank, factor = 0·85

* See page 206.

TABLE 10.3*

HEAT EMISSION FROM PIPE INSULATED WITH MATERIAL OF
THERMAL CONDUCTIVITY 0·055 W/m°C—e.g. GLASS FIBRE

Nominal Bore		Heat Emission W/m run											
		Temperature Difference Hot Face to Surroundings °C											
in.	mm	Thickness of Insulation 25 mm						Thickness of Insulation 50 mm					
		100	120	140	160	180	200	100	120	140	160	180	200
½	15	25	30	35	40	45	50	20	25	30	35	35	40
¾	20	30	35	40	45	50	60	20	25	30	35	40	45
1	25	35	40	45	55	60	65	25	30	35	40	45	50
1¼	32	40	50	55	60	70	80	30	35	40	45	50	55
1½	40	45	50	60	70	75	85	30	40	45	50	55	60
2	50	50	60	70	80	80	100	35	40	50	55	60	70
2½	65	60	70	80	90	105	115	40	50	55	60	70	80
3	80	65	80	95	105	120	135	45	50	60	70	80	85
4	100	80	95	110	130	145	160	50	60	70	80	90	100
5	125	100	115	135	155	175	195	60	70	85	95	110	120
6	150	110	130	155	175	200	220	70	80	95	110	120	135
With ambient Air at 15°C the above hot-face temperatures correspond to steam pressures bar:		1·7	3·1	5·4	9	14	21	1·7	3·1	5·4	9	14	21

Note: 1. Values are derived from *I.H.V.E. Guide* factors and are rounded to the nearest 5.
2. Other thicknesses between 25 mm and 50 mm may be interpolated with slight inaccuracy.
3. For other insulating materials and conductivities, the above values may be multiplied in ratio of the factors in the *Guide* to 0·055.

* See page 206.

TABLE 10.4*

STEAM HEATING CALORIFIERS
Heat Transfer with Free Convection at 0·4 m/s Water Velocity

Steam Pressure bar absolute	Steam Temperature °C	Heat Transfer to clean water at 80°C of 1 inch internal dia. (25 mm) plain Copper Tubes: kW/m run tube
1	100	3·2
2	120	6·4
3	133	8·5
4	144	10·3
5	152	11·5
6	159	12·6
7	165	13·6

Note: 1. Factor for water velocity 0·2 m/s = 0·6
Factor for water velocity 0·6 m/s = 1·26
2. For tubes of other internal diameters, heat transfer is in direct ratio of diameter.
3. Transfer rates are based on tube wall thickness of 1·6 mm (16 swg).
4. Tubes are assumed to be suitably pitched.
* See page 213.

TABLE 10.5*

VALUES OF Z FOR USE WITH TABLE 10·6

P = steam pressure in bar absolute

P	Z	P	Z	P	Z
1·1	1·2	2·2	4·58	4·5	18·2
1·2	1·42	2·4	5·41	5·0	22·3
1·3	1·66	2·6	6·32	5·5	26·8
1·4	1·91	2·8	7·29	6·0	31·7
1·5	2·19	3·0	8·32	6·5	37·0
1·6	2·48	3·2	9·43	7·0	42·7
1·7	2·78	3·4	10·60	7·5	48·8
1·8	3·11	3·6	11·83	8·0	55·2
1·9	3·45	3·8	13·13	9·0	69·3
2·0	3·81	4·0	14·5	10·0	84·9

* See page 215.

TABLE 10.6*

FLOW OF SATURATED STEAM IN STEEL PIPES TO BS 1387.

HEAVY GRADE IN GRAMMES PER SECOND (g/s).

Z_1 and Z_2 = initial and final pressure factors from Table 10.5. L = length of pipe in metres. P = absolute pressure in bar corresponding to 30 m/s.

Z_1-Z_2 mm:	15	20	25	32	40	50	65	90	100	125	150	
L in:	½	¾	1	1¼	1½	2	2½	3	4	5	6	
P												P
0·001	1	2	4	8	12	23	46	72	146	262	425	
0·002	1	3	5	11	17	33	66	103	210	379	613	2
0·004	2	4	8	16	25	47	96	149	303	547	885	3
0·0063	2	5	9	21	31	60	122	190	386	695	1130	
0·008	3	6	11	23	36	68	138	215	438	789	1280	
0·01	1	3	7	12	26	40	76	156	242	492	888	1440	5
0·02		4	10	18	35	60	110	224	349	711	1280	2070	
0·04	2	6	14	25	55	84	159	324	514	1030	1850	2990	
0·063		8	17	32	70	106	202	412	641	1300	2350	3310	12
0·08	3	9	20	36	80	121	230	468	728	1481	2670	4320	
0·1		10	22	41	89	136	258	526	819	1670	3000	4860	
0·2	5	14	32	59	128	196	373	760	1180	2140	4340	7020	
0·4		20	46	85	185	283	538	1100	1710	3470	6360	10 100	
0·63		26	59	108	236	360	684	1390	2170	4420	7960	12 900	
0·8	12	29	67	123	267	408	777	1580	2460	5020	9030	14 600	
1·0		33	75	139	301	460	874	1780	2770	5640	10 200	16 500	
2·0		47	109	200	435	664	1260	2570	4000	8140	14 700	23 700	
4·0		68	157	289	628	958	1820	3710	5770	11 700	21 200	34 300	
6·3		87	200	367	799	1220	2320	4720	7340	14 900	26 900	43 600	
8·0		99	226	417	906	1380	2630	5350	8330	17 000	30 600	49 500	

Note: 1. The above is abstracted from the *I.H.V.E. Guide*. Mass flow rates are rounded to the nearest whole number
2. Equivalent length of single resistances in metres for K = 1 given in the *Guide* Table may be approximated as follows:

½ in.: EL = 0·5	1½ in.: EL = 2	4 in.: EL = 6
¾ in.: EL = 0·8	2 in.: EL = 3	5 in.: EL = 8
1 in.: EL = 0·1	2½ in.: EL = 4	6 in.: EL = 10
1¼ in.: EL = 1·5	3 in.: EL = 4·5	

3. The value of 30 m/s velocity has no quantitative significance. Quantities corresponding to other velocities may be obtained by simple proportion and, to other initial pressures, by interpolation.

* See page 215.

Plate X. Showing High Pressure Steam Distribution and Condense Return Mains from boiler-house to factory buildings (*see p. 216*)

Plate XI. (a) above: H.P.H.W. Nitrogen-Pressurized System (Ashmore, Benson, Pease and Co.), with pressure cylinder on left, boiler control panel in centre, pressurizing panel on right and spill tank at top right. (b) below: the circulating pumps, header and mixing valve of a similar system (see pp. 234 and 237)

TABLE 10.7*

CONVENTIONAL STEAM VELOCITIES (SATURATED STEAM)

	m/s
Boiler outlet connections (LP) - - - - - -	5 to 10
Boiler outlet connections (HP) - - - - - -	15 to 20
Headers (LP and HP) - - - - - -	20 to 30
Steam mains (LP and HP) - - - - -	30 to 50
Vacuum steam mains - - - - - -	100
Calorifier connections - - - - - -	25 to 30

CROSS-SECTIONAL AREA OF PIPING

Steel to BS 1387 heavy weight and BS 806 (over 150 mm)

Nominal Bore		Area m²
in.	mm	
2	50	0·0020
2½	65	0·0036
3	80	0·0050
4	100	0·0085
5	125	0·014
6	150	0·018
8	200	0·032
10	260	0·05
12	300	0·08

* See page 216.

CHAPTER 11

Heating by High-Pressure Hot Water

THE FATHER OF ALL HIGH-PRESSURE hot-water systems was Perkins, whose patent was filed in 1831. In his system, shown in Fig. 11.1, the piping was extremely strong, about 22mm bore, and formed one continuous coil, part of it passing through the boiler and the remainder forming the heating surface in the rooms to be heated. Expansion of the water was allowed for in a closed vessel at the top. Two or more coils could be used

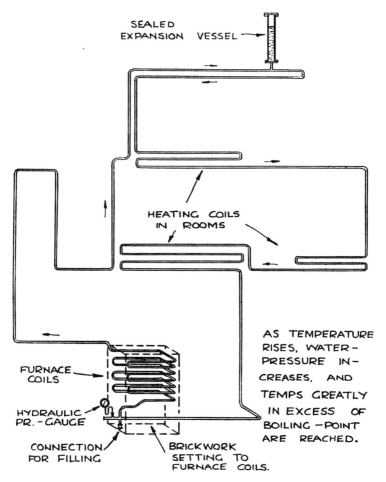

SEALED EXPANSION VESSEL

HEATING COILS IN ROOMS

AS TEMPERATURE RISES, WATER-PRESSURE IN-CREASES, AND TEMPS GREATLY IN EXCESS OF BOILING-POINT ARE REACHED.

FURNACE COILS

HYDRAULIC PR.-GAUGE

CONNECTION FOR FILLING

BRICKWORK SETTING TO FURNACE COILS.

FIG. 11.1.—Diagram of Perkins System.

where one was insufficient. As the water was heated, its expansion compressed the air in the expansion vessel and considerable pressures were
reached. The principle was that the formation of steam was prevented by
the pressure to which the water in the system was subjected. The system
is now obsolete, though examples may still be found at work in remote
spots.

Present-day Systems—The same principle is the basis of all modern
systems of this type. Hot water is the medium in a closed system and is
subjected to pressure either by steam or by a gas such that its temperature
is raised well above atmospheric boiling point. The essential differences
between this and the method used by Perkins is that circulation is by pump,
and the working pressure is controlled so that boilers and heat-emitting
equipment of various types suitable for moderate pressures may be used.
There are, in fact, two ranges of pressure in common use for space heating:

high pressure   -   -   -   5 bar to 10 bar abs.
            (approx. 150° C to 180° C)
medium pressure -   -   -   2 bar to 3 bar abs.
            (approx. 120° C to 133° C)

Temperature Ranges—The higher the initial temperature, obviously the
greater the temperature drop which can be allowed between flow and
return, and, correspondingly, the larger the heat content per unit mass
flow of water, requiring smaller pipe sizes for a given load. On the other
hand, the higher the pressure the more costly the boilers and other
equipment. Temperature drops are usually of the following order:

high-pressure system   -   -   45° C to 65° C
medium-pressure system   -   -   28° C to 30° C

Thus, a high-pressure system, with 50° C drop and a flow temperature of
160° C, has a temperature difference at the emitting apparatus, mean
water to air at 20°C, as follows:

$$\left(\frac{160+110}{2}\right) - 20 = 115° \text{ C}$$

A medium-pressure system, with 30° C drop and flow at 120° C, has a
difference of

$$\left(\frac{120+90}{2}\right) - 20 = 85° \text{ C}$$

The lower pressure system requires more heating surface for a given
output—in this case about one-third more. For smaller installations,
however, there are many compensating advantages.

For this and for other reasons, it has come to be accepted that there is
no point in designing for high pressures for systems under say 2000 to

3000 kW, below which ratings medium-pressures are most suitable. The high-pressure system comes into its own with increasing size and extensive runs of mains. Furthermore, where process equipment has to be supplied, as in many industrial applications, the temperature of operation is controlled by the needs of the process and may be up to 200° C (15 bar abs).

Comparison with steam—Hot water in a closed system under pressure may be run at any temperature up to its design maximum. Where serving space-heating apparatus, the temperature of the water can be varied according to the weather, so saving on mains heat losses and by better control generally. Variability of temperature is not possible with steam, which must be either on or off and any attempt at throttling is liable to cause water logging at the remote ends. (This assumes that the vacuum steam system with its further complications is ruled out anyway).

Hot water requires no steam traps. The potential loss of heat through flash steam and condense return has been referred to in Chapter 10 and may amount to ten per cent or more of the fuel bill. Traps also require maintenance, as do other steam accessories.

Hot-water mains may be run with complete freedom as to levels, whereas steam mains require careful grading and draining. Also corrosion of condense lines is avoided.

In terms of pipe sizes and cost it can be shown that, taken overall, there is little difference in the two systems. By way of example:

Steam, at 5 bar pressure

$$\frac{Z_1 - Z_2}{L} = 0.04$$

$$\text{main size} = 6 \text{ in. } (150 \text{ mm})$$
$$\text{g/s} = 2990$$
$$2990 \text{ g/s} \times 2109 \text{ kJ/kg (latent heat)} = 6300 \text{ kJ/s (kW)}$$
$$= 126 \text{ kW/°C.}$$
$$\text{condense size} = 3 \text{ in. } (80 \text{ mm})$$

Hot Water (50° C drop)
From Fig. 8.6 (page 158) a load of 126 kW/° C is carried by 5 in. (125 mm) flow and return at a pressure loss of 380 N/m³, or by 6 in. (150 mm) at 160 N/m³.

The saving on the smaller condense main will, however, be eaten up by traps and drainage fittings, etc.

PROPERTIES OF WATER AND PIPE SIZING AT HIGH TEMPERATURES

Table 11.1 gives the specific heat capacity and density of water from 75° C (low-pressure hot water) to 200° C. It will be noted that, whilst the specific heat capacity increases, density decreases; but, when the two are multiplied together giving heat per unit volume, there is only a small correction needed where velocity of flow is the criterion, as in pipe sizing.

It can also be shown that as temperature increases viscosity decreases and a correction in pressure loss for a given flow rate is required on this account. Table 11.2, abstracted from the *Guide*, shows the order of these corrections, which are quite small. Corrections to equivalent length factors are also required, but these are still smaller and may be ignored.

Thus, there is little error if Fig. 8.6 (page 158) is used as it stands for high-pressure and medium-pressure pipe sizing. For accurate use, the corrections may be made in accordance with Tables 11.1 and 11.2.

TABLE 11.1

SELECTED PROPERTIES OF WATER AT HIGH TEMPERATURES

Temperature °C	Pressure bar absolute	Specific Heat Capacity kJ/kg	Density kg/m³	Heat Content per litre	Correction Factor
75	—	4·19	974·9	4·094	1
100	1·0	4·22	958·3	4·043	0·99
110	1·4	4·23	951·0	4·023	0·98
120	2·0	4·25	943·1	4·007	0·98
125	2·3	4·26	939·1	4·000	0·98
130	2·7	4·27	934·8	3·987	0·97
140	3·6	4·29	926·1	3·973	0·97
150	4·8	4·32	916·9	3·961	0·97
160	6·2	4·35	907·4	3·948	0·96
170	7·9	4·38	897·3	3·930	0·96
175	9·0	4·40	892·2	3·926	0·96
180	10·0	4·42	886·9	3·919	0·96
190	12·5	4·46	876·0	3·906	0·95
200	15·5	4·51	864·7	3·900	0·95

TABLE 11.2

CORRECTION FACTORS FOR USE WITH FIG. 8.6
FOR HIGHER TEMPERATURE WATER (150° C)

Pressure Loss N/m³ from Fig. 8.6	Pipe Size mm				
	15	25	50	100	150
2	0·81	0·84	0·87	0·90	0·91
5	0·83	0·86	0·89	0·92	0·93
10	0·85	0·88	0·91	0·94	0·95
20	0·87	0·90	0·93	0·96	0·97
50	0·90	0·92	0·95	0·98	0·99
100	0·92	0·94	0·97	0·99	1·00
200	0·95	0·96	0·98	1·00	1·01
500	0·97	0·98	0·99	1·02	1·02
1000	0·98	1·00	1·01	1·02	1·02

PRESSURIZATION BY HEAD TANK AND BY STEAM

Head Tank—An elevated tank at sufficient height above the highest point of the system is rarely possible, though the simplest way of obtaining pressure. 10 metres head of water is approximately equal to 1 bar (0·984). For a medium-pressure system at 2 bar abs, some 20 metres height would be required plus an excess to prevent ebullition at the highest point of the system.

Where the medium-pressure system is confined to low level, possibly using calorifiers for low-pressure heating to the buildings, an open expansion tank may be practicable, as referred to in Chapter 9.

Pressurization by steam—Steam is generated above the water in a steam boiler, as shown diagrammatically in Fig. 11.2. The water for circulation to the heating system is taken from below the water-line of the

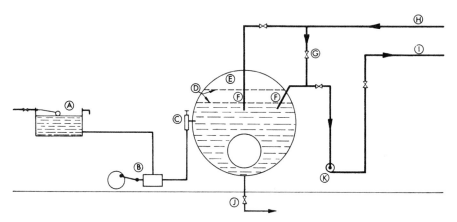

FIG. 11.2.—Diagram of Steam-Pressurizing System.

A. Feed tank
B. Feed pump
C. Check valve
D. Water line, upper and lower

E. Steam space
F. Dip pipes
G. Cooling water by-pass

H. Return
I. Flow
J. Blow down
K. Circulating pump.

boiler and returned thereto. The danger of draining the boiler, should a serious leak occur in the system, is overcome by using dip-pipes as shown.

It will be clear that the steam and water are at the temperature of saturation and that any reduction of pressure on the water side will thus cause water to flash into steam. To overcome this, a cooling supply from the return is injected into the flow outlet as it leaves the boiler, so bringing the temperature below ebullition point. Also, the flow outlet is taken down to low level, so increasing static pressure.

Where more than one boiler occurs, their respective water levels are kept uniform by steam and water balance pipes, the former from the steam space and the latter by dip pipe. Fig. 11.3 shows the various connections to a boiler of shell type.

Alternatively, the high-velocity tubular type of boiler has been much used with these systems, as shown in Fig. 11.4. The boiler tubes connect to a drum in which the steam space is formed. Where a number of such boilers occur, a common drum or drums are used, as in Fig. 11.5.

In order to feed the steam-pressurized system, a normal feed pump is used, drawing its water from an open feed tank. It is usually hand-controlled.

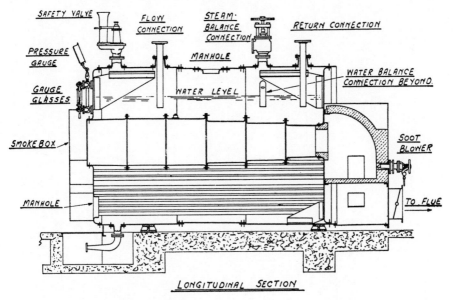

LONGITUDINAL SECTION

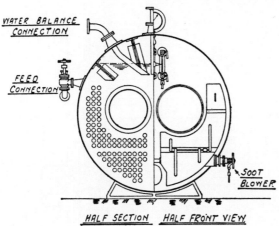

HALF SECTION   HALF FRONT VIEW

Fig. 11.3.—Super-Economic Boiler arranged for high-pressure hot-water system.

Water of expansion is to some extent contained in the rise in water level in the boiler or drum—sufficient to contain diurnal variations; but the greater volume of expansion from cold is discharged to drain.

The disadvantages of steam pressurization are that the system tends to be unstable and requires skilled operation and maintenance. Varying rates of output from boilers causing pressure changes and close interdependence of pressure and temperature tend to fickleness in behaviour not conducive to unskilled attention.

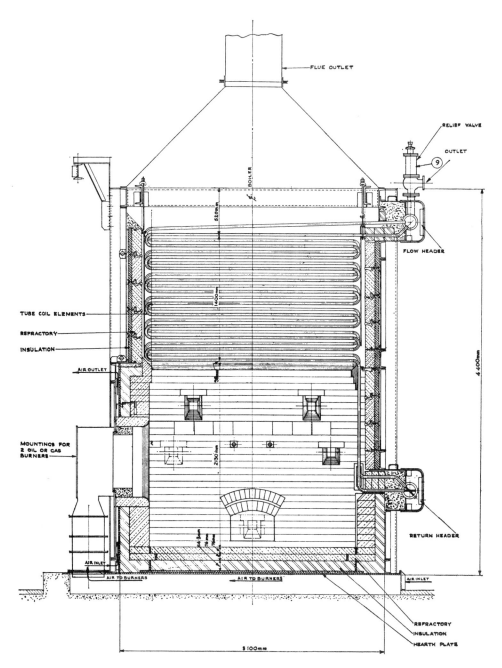

FIG. 11.4.—Section of a LaMont Boiler: 3·6 MW.

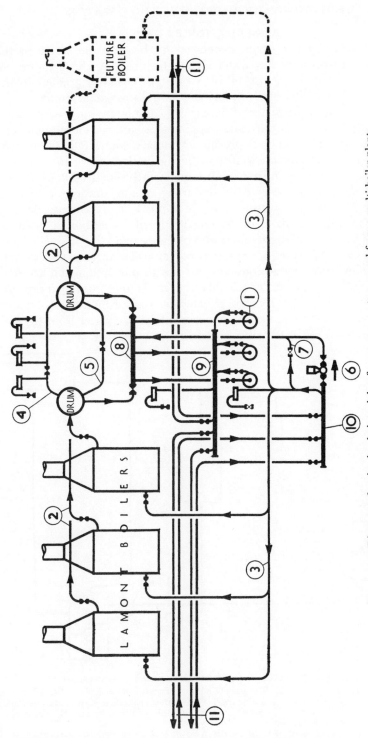

Fig. 11.5.—Diagram of main circulating piping for H.P.H.W. system served from a multi-boiler plant.

1. Circulating pumps. 2. Boiler flow mains (each run separately to drum). 3. Boiler return mains. 4. Steam balance-pipe. 5. Water balance-pipe. 6. Thermostatic mixing-valve. 7. Hand-operated bye-pass. 8. Mixed-water header (pump suction). 9. Flow header (pump discharge). 10. Return header. 11. Heating circuits to buildings. Air bottles where shown.

PRESSURIZATION BY GAS*

Instead of using the steam generated by heating the water in the boiler of a high-pressure hot-water system for creating the pressure, an alternative method is the use of an air or gas cushion, applied to the system in a cylinder. Plate XI(*a*), facing p. 225, shows an example.

With this system the boiler or boilers are completely filled with water, as in the case of a low-pressure hot-water system, and the circulation is by pump in the normal manner. A pressure cylinder is connected to some part of the system, generally the return near the boiler where the water is coolest. The pressure cylinder is maintained partly filled with water and partly with air, or gas, derived either from an air compressor or from bottled gas. In this way, pressure may be applied to any amount desired, within the pressure limits of the system and well above boiling point as determined by the working temperature of the system.

The whole of the water of expansion could be contained in the pressure cylinder if this were made large enough; but in one system the arrangement is made for surplus water to spill out of the system through a pressure control valve into a spill tank open to atmosphere. For the return of water, as the system cools down, a pump is provided to take water from the spill tank and return it to the system.

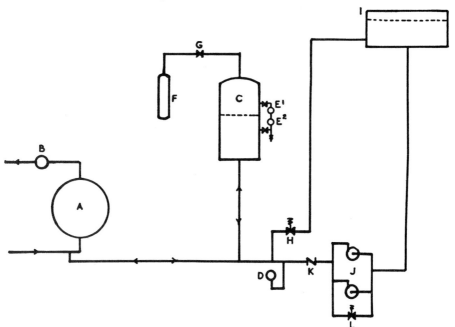

Fig 11.6.—Diagram of Gas Pressurized High Pressure Hot-Water System.
A: Boiler. B: Circulating pump C: Pressure cylinder. D: Pressure control. E^1 and E^2 High and Low level cut outs, F: Gas cylinder or air supply. G: Gas inlet regulating valve, hand operated. H: Water pressure spill valve. I: Expansion water spill tank. J: Water pumps. K: Non-return valve, L: Pump pressure relief valve.

* See also paper by J. R. Kell, *Methods of Pressurization. I.H.V.E. Journal*, Vol. 26, p. 1.

The elements of the system are shown in Fig. 11.6. An essential part of the apparatus is the pressure controller D, which regulates the admission of water from the pump J or its expulsion via the regulating valve H, level controllers EE and safety-limit controllers.

The advantages of this arrangement are that, apart from temperature being independent of pressure, no balance pipes are needed to a multi-boiler installation, boiler mountings are simplified, and the system is more stable in operation due to the absence of steam.

The reason for using gas such as nitrogen as the cushion, instead of air, is that it is less soluble in water than air. Furthermore, it is inert, so that corrosive tendencies from this source are eliminated.

In order to remove the tendency for atmospheric air to be dissolved into the spill tank water, in an alternative system this is arranged as a closed cylinder lightly pressurized with nitrogen. In another system the water of expansion is taken up by a flexible bellows. The expansion water is cooled in a heat exchanger (not shown in the diagram) before passing to the spill tank.

Gas Pressurization: Working Pressures—The calculation of the working pressure on the boiler and on the cylinder in this system starts from a decision as to the flow temperature required at the highest point in the system, such as in some high level main. Having settled this point, and the temperature, the rest follows as in the example given below.

Example of Pressure Differentials

Assume 170° C at flow to system	= 8	bar abs
15° anti-flash margin, 185°	= 11	,,
Highest point of system 10 m head above W.L. in cylinder	= 1	,,
Minimum pressure at cylinder	= 12	,,
Pressure differential on control switch	= 0·5	,,
Normal operating pressure of system	= 12·5	,,
Head pump operates between 12 and 12·5 bar		
Spill valve starts to open at +0·25 bar	= 12·75	,,
Safety valve on boiler	= 14	,,

It will be noted that an anti-flash margin of 15° C has been selected, but this might be varied according to circumstances. The object of this is to provide pressurization at the highest point of the system such that at no time can ebullition take place. The head of the circulating pump greatly augments this anti-flash margin during working conditions, but it has to be assumed that at some time the circulating pump may stop when the system is up to full temperature, and it would be unfortunate if ebullition should occur under such conditions.

Gas Pressurization: Sizing of Pressure Cylinder—It obviously simplifies control if the movement of water line for a given pressure differential is at a maximum. If no loss of heat occurs from the cylinder and associated piping during expansion of the water and compression of the gas in the cylinder, the law of compression will approach the adiabatic $PV^{1.405} = C$. If, however, the vessel and piping are left unlagged, the change becomes more nearly isothermal and it has been found by experiment that compression follows the law approximately $PV^{1.26} = C$.

It follows from the above that the movement of the water level can be calculated from the formula:

$$x = h\left[1 - \left(\frac{p_1}{p_2}\right)^{\frac{1}{1.26}} \right]$$

where x = movement of water line in metres
h = height of air space in metres before compression starts
p_1 = initial pressure in air space (bar absolute)
p_2 = final pressure in air space (bar absolute)

It should be noted that the movement of the water line is independent of the diameter of the vessel, but the latter must be considered in conjunction with the quantity to be dealt with by the head pump, which should run for about two minutes between low and high water levels.

Sizing of Head Pressure Pump—The maximum duty which this pump could be called upon to deal with at any time would be in the case of rapid contraction throughout the system due to a sudden shut down of the boilers. This is not a matter of accurate calculation, but some approximation can be made to it and an ample margin allowed. Pressure against which the pump must work will be that of the highest pressure to which the cylinder is subjected before the spill valve starts to open with the differential selected, as in the example previously given. In this case the head pump would operate between a pressure of 12 and 12·5 bar, plus a margin for friction loss in piping etc.

With regard to the various items of control equipment and fittings associated with this method of pressurization, it is not possible here to consider details, but there are a number of proprietary systems which have been developed, and reference to makers' data is suggested for those seeking further information.

Circulating Pumps—In a steam pressurized system it is preferable to place the circulating pumps in the flow main. Fig. 11.7 illustrates this problem. It will be noted that, whatever the pressure in the boiler, the additional head set up by the pump is added so that the flow main is at a pressure well above that in the boiler. Water travels through the system until, at the point of return to the boiler, the lowest pressure is reached, and therefore the whole of the rest of the system is above this pressure, excepting for the short length of main between the boiler and the pump suction.

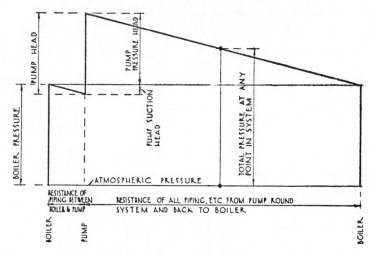

FIG. 11.7.—Pressure diagram for typical H.P.H.W. system.

In a gas pressurized system, it may be better for the pump to be in the return to the boiler with the pressure cylinder connected in the suction line. Virtually all parts of the system are then subject to pump pressure in addition to static pressure.

CONTROL OF TEMPERATURE BY MIXING

Where a boiler plant serves process heating at a constant temperature, also in the case of space heating where a variable temperature is required according to weather, use may be made of a mixing connection (controlled either by hand, or thermostatically, in such a manner that return water is by-passed into the flow from the boiler) for the heating system. Separate pumps would be provided for the process work and for the heating system. Such an arrangement is illustrated in Plate XI(*b*), facing p. 225.

In a similar manner, where it is desired to apply zoning, i.e. to arrange for different areas to be supplied at dissimilar temperatures, a series of mixing valves and pumps may be used, each delivering to its own circuit. Fig. 11.8 shows diagrammatically a gas-pressurized system which has a boiler circulating pump and three zones, together with process heating The advantages of including a boiler circulating pump are that boiler flow is uninfluenced by the variations in demand of the mixing valves. Also, where tubular boilers having high resistance are used, the necessary pressure is disposed of by the boiler circulating pump, leaving the circuit pumps only their own circuit head to deal with. Where large load variations are expected, there is a case for providing one pump per boiler.

It will be noted in Fig. 11.8 that, whilst the boiler circulating pump is in the return for the reason stated earlier, the sub-circuit pumps are in the flow—again for the reason that their pressure is added to the sub-circuit.

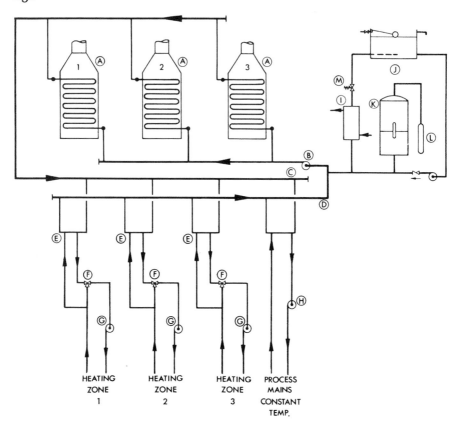

Fig. 11.8.—Example of High-Pressure Hot-Water System with Three Zones and Process Circuit. (Note that the isolating valves are omitted, for clarity).

A. Tubular boilers
B. Boiler circulating pump
C. Flow header
D. Return header

E. Accept and reject connection
F. Mixing valve
G. Zone pump
H. Process pump
I. Cooler

J. Nitrogen pressurising unit
K. Pressure cylinder
L. N₂ bottle
M. Spill valve.

EXPANSION OF WATER

The amount of expansion may be calculated from the water contents as in the following example:

Water capacity of system = 100 000 kg

Volume at 150° C (Table 1.3): $\dfrac{100\,000}{998\cdot5}$ = 100·2 m³

„   „   mean temp. 125° C (Table 11.1):

$$\frac{100\,000}{939\cdot1} = 106\cdot5 \text{ „}$$

increase = 6·3 „

Volume at mean temp. 150° C: $\dfrac{100\,000}{916 \cdot 9}$ = 109·2 m³

Increase from 125° C to 150° C   = 2·7 ,,

Increase overall 15° to 150° C   = 9·0 ,,

As already stated, in a steam pressurized system 2·7 m³ may be accommodated by variation in water level in boiler or drum: the initial warming up expansion (6·3 m³) is blown down, and on cooling restored by feed pump.

In a gas pressurized system the capacity in the cylinder is small and the whole of the water of expansion is therefore to be accommodated in the spill tank (i.e.; 9 m³).

CIRCULATING PUMP

Fig. 11.9 shows one of the types of pump available for this duty. It is of the end-suction centrifugal type direct-coupled to an electric motor.

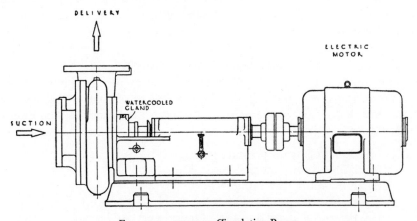

Fig. 11.9.—H.P.H.W. Circulating Pump.

Points usually taken care of in the design of circulating pumps for this purpose are:

Suction connection designed to avoid sudden changes of velocity, and consequent cavitation.

Casing generally of cast steel to withstand high pressures.

Gland cooled by water fed under pressure and with visible discharge into a tundish.

Bearings suitable for high temperature.

Pump Duty—In the type of system shown in Fig. 11.8, the boiler circulating-pump duty will be determined by the total boiler loading and design-temperature drop. The pressure of this pump will be the sum of the loss in the boiler, headers and connections, the unit being kN/m².

The zone pumps will be sized to the loading of each zone circuit, and to the overall design-temperature drop or some lesser drop—assuming there is always some return water entering the mixing valve. The total quantity handled by the zone pumps and process pump must always be exceeded by the boiler circulating pump. The pressure in kN/m^2 of the zone and process pumps will be that of their own circuit, piping, emitter and mixing-valve loss in each case, unrelated to boiler pressure loss.

In practice, standby pumps would no doubt be provided, particularly to the boiler circulating pump. The zone pumps might use a common standby with valving arrangement permitting it to be used on any circuit. **Treatment of Feed Water**—High-pressure and medium-pressure systems are prone to give trouble from scale deposit on valves, sludge deposits in apparatus such as unit heaters, and corrosion of feed-connecting pipework. Treatment of the feed water is thus highly desirable and one of the manufacturers of pressurizing equipment offers a combined pressurizing and demineralising plant. On smaller systems some form of dosing pot may be used as described in Chapter 9 (see Fig. 9.14, page 189). For feed-connecting pipework it is probably worth using cupro-nickel or a similar material.

BOILERHOUSE INSTRUMENTS

The number and type of instruments to be provided in the boilerhouse depend on the size of the installation, the degree of control required and cost. The following instruments, however, are generally considered essential:

(a) Boiler or drum pressure gauge.
(b) Boiler thermometer.
(c) Flow-main thermometer.
(d) Return-main thermometer.
(e) Pressure gauges on pump suction and delivery.
(f) Water gauges on boiler, or drum, with steam-pressurized system; water gauge on pressure cylinder with gas-pressurized system.

Other instruments usual in large systems, roughly in order of usefulness:

(g) CO_2 indicator.
(h) Boiler or drum pressure recorder.
(i) Flue-gas temperature indicator.
(k) Draught gauge.
(l) Flow and return thermometers on each outgoing circuit.
(m) Differential pressure gauge across each boiler.
(n) Outside temperature remote-thermometer.
(o) Fuel meters.

Plate XII. High Temperature Radiant Panel Installation in large constructional shop
(Ashmore, Benson, Pease and Co.) *(see p. 244)*

*Plate XIII. A workshop warmed by Continuous Strip Heating supplied
by medium pressure hot water (see p. 245)*

Most of the above may be combined with recorders to give a continuous chart of the conditions obtaining.

Another instrument which is sometimes provided, and which is in any case useful for keeping an accurate check on operating costs, is a *Heat Meter.*

SAFETY DEVICES

If the pressure in a H.P.H.W. system were suddenly reduced, as for instance by the bursting of a pipe, large quantities of steam would be generated by the boiling of the superheated water. Automatic isolating valves were at one time installed in such systems to shut off in the event of excessive flow, but these have fallen into disfavour due to the considerable hydraulic forces set up on closure. Failures of piping systems have fortunately been rare and accidents do not appear to have had the serious consequences anticipated.

Low-flow devices are sometimes provided, particularly with tubular type boilers, with the object of sounding an alarm in the event of pump failure or cessation of flow from one cause or another. Fig. 11.10 shows the principle of operation of one such device.

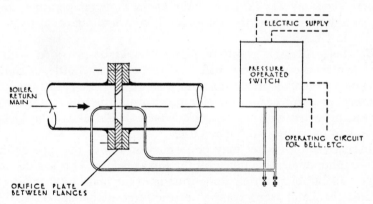

FIG. 11.10.—Diagram of Boiler Low-flow Alarm Device.

PIPING FOR H.P.H.W. SYSTEMS

Piping is invariably of mild steel, suitable for the pressure. In this connection it is to be noted that at certain points of the system the pressure is in excess of the boiler working-pressure by an amount equal to the pump head.

Joints—Screwed joints are not suitable for H.P.H.W. piping, and should not be used, as they are liable to leak. All valves should be flanged, and elsewhere the piping should be butt-welded, either by oxy-acetylene or by electric arc to an approved specification.

All bends should be of large radius (at least 3 pipe diams.), and tees should be 'swept' so as to avoid sudden drops of pressure in the piping.

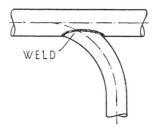

WELD

Connections—Connections to heating apparatus are best made with flanges, but to save expense the piping is sometimes welded direct to a 'tail' left on the heater battery, etc.

Headers and branches in the boiler-house may also be built up from standard tubing and bends welded together, but it is desirable to provide a certain number of flanges in the boilerhouse to facilitate erection and dismantling. Connections to pumps and boilers will be flanged.

Expansion of the Piping is a matter of the first importance, and should be carefully considered at every stage of the design. Short rigid connections between boilers, drums, headers, pumps, etc., should be avoided, and it is worth pointing out that whereas such a short connection might be practically unstrained if all the piping adjacent to it is full of hot water, cases will arise when part of the piping is out of use, and consequently full of cold water, with resulting uneven contraction, and strain.

The mains outside the boilerhouse will behave as regards expansion in exactly the same way as steam mains at the same temperature, and the basic considerations are the same as those which have already been discussed in Chapter 10.

Air Bottles—All high points of piping should be provided with vent pipes taken off the top of the main and run to an air bottle, as shown in Fig. 11.11 (a). This is made up from a short length of pipe about 6 in. (150 mm) bore, and has an air-pipe taken from the top to a convenient level and provided with a stop-cock, which should always be locked in some way to prevent unauthorized use.

Alternatively, a pocket may be formed on top of the pipe as in Fig. 11.11 (b), and an air-pipe taken off the top as before. This has the advantage that air bubbles carried along the top of the main by the water have a greater chance of being trapped in the large diameter pocket than of entering the relatively small vent pipe in (a).

Venting—Mains should be fixed to a slight fall to assist in venting, and ideally all mains should rise in the direction of flow so that the entrained air does not have to collect against the flow of water. This, however, would entail fixing the mains with falls in different directions, which is inconvenient, and is not generally necessary. The main which would be expected to contain the greater quantity of air (normally the flow main) should rise in the direction of flow, with the return main lying parallel to it.

Once the initial air has been removed there is generally little further accumulation. Due to the considerable pressure in the system, air is compressed to a much smaller volume than in the normal low-pressure system.

Valves—Trouble is often experienced in h.p.h.w. systems due to leakage past the glands of the valves. As soon as it is at atmospheric pressure, the water leaking out immediately evaporates and leaves behind any salts which it may contain, and this deposit may in time, if not regularly removed, be sufficient to prevent the valve from being turned. Trouble is particularly noticeable with small motorised valves where the operating torque is limited by the power of the motor. To overcome this difficulty many attempts have been made to design special glands for H.P.H.W., and this point should be borne in mind when specifying makes and types.

Valves used for H.P.H.W. should always be flanged, and many engineers prefer a screw-down oblique type valve as giving more satisfactory service than other types.

Regulating Valves—The amount of water passing through a H.P.H.W. system is relatively small, and the pump head available at the nearer branches may be large, so that regulation is often critical. Some designers, therefore, provide, in addition to the isolating valves on the flow and return to the branch, a third valve in the return which is used for regulating only, and once set this valve need never be disturbed. Alternatively, combined regulating and shut-off valves may be used, in which case regulation is effected by means of an auxiliary seating, which is so arranged that it cannot be tampered with after adjustment.

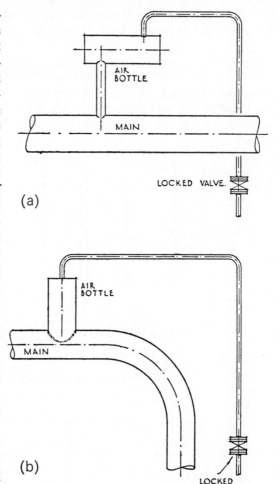

FIG. 11.11.—Air Bottles on Mains.

PIPING FOR MEDIUM-PRESSURE SYSTEMS

Medium-pressure hot water does not demand the high standards of high-pressure. Screwed joints at valves and equipment may be used and flanges are not necessary. Cast-iron valves and pumps are satisfactory, but not cast-steel. Welding of pipe runs is nevertheless preferable to screwing, in view of the expansion forces involved.

HIGH-TEMPERATURE AND MEDIUM-TEMPERATURE RADIANT HEATING, CONVECTION HEATING AND OTHER APPLICATIONS

Radiant Heating—The use of this type of heating is specially suitable for large spaces such as factories. The heating panels are of steel, having a coil welded to the plate, (see Fig. 11.12). The coil is supplied with high-

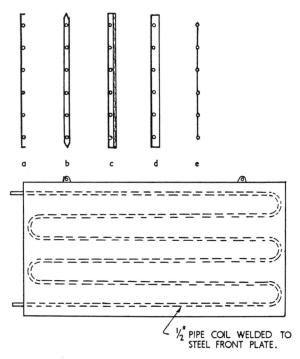

FIG. 11.12.—High Temperature Radiant Panels.

Various types of panel:
(a) exposed coil; (b) double-sided; (c) insulated back; (d) shielded back; (e) plate welded between coil. The sizes of the panels are approximately 2·5 m by 1·25 m and 2 m by 1 m. The pipe coil is welded to the steel plate at 150 mm centres.

pressure or medium-pressure hot water. Temperatures of the surface are of the order of 90° to 135° C depending on the temperature and pressure of the water. At these temperatures, flat metal plates emit powerful radiation, felt at a long distance. (See Plate XII, facing p. 240).

It is usual to suspend the panels vertically along the walls or between columns of a large shed about three to four metres above the floor. They may alternatively be fixed horizontally overhead or inclined at an angle.

Another method of achieving the same object is by the use of strip heating (see Fig. 11.13) which in one form comprises aluminium plates, clipped to heating coils, for running continuously down a workshop, overhead. (See Plates XIII and XIV, facing pages 241 and 256, the latter showing an example integrated with the structure).

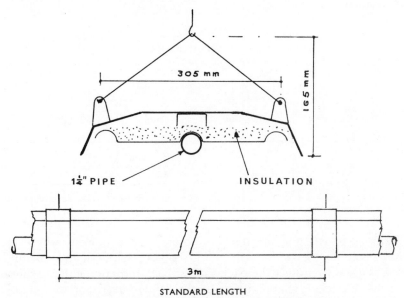

305 mm

165 mm

1¼" PIPE   INSULATION

3 m

STANDARD LENGTH

FIG. 11.13.—Strip Heating (Copperad Ltd. 'Raystrip').
Note: The Foil Insulation is omitted in the non-insulated type.

A further development, in the case of strip heating, is to combine it with provision for lighting by means of fluorescent tubes; the heating surface then also forms the lighting reflector.

The advantages of high-temperature radiant heating are:

(1) Uniform distribution of heat.
(2) Absence of temperature gradient.
(3) Objects are warmed rather than air.
(4) Reduced air change losses due to the above.
(5) No moving parts or maintenance.

The calculation of heat requirements for radiant systems in buildings is referred to in Chapter 2 (see p. 37).

Table 11.3 gives emissions for one make of panel. It will be noted that the convection and radiation components are given separately. In designing such a system it will be borne in mind that the convection component, together with convection losses from piping, serves to counteract the cooling effect of the roof. Convection currents rising to above may be regarded as supplying the heat necessary to counteract the cooling effect of the roof glass, and it is desirable to achieve a rough balance between these two quantities. It will, for instance, generally be found desirable to insulate large main piping, otherwise the convection component in the building may greatly exceed what is necessary to deal with roof losses, and such would then lead to temperature gradients which are to be avoided.

TABLE 11.3

EMISSION OF HIGH-TEMPERATURE RADIANT PANELS
kW PER PANEL

Panel size 2·5 m × 1·25 m. Freely exposed and uninsulated. Ambient air, 15° C. Water temperature drop across panel, 30° C.

Hot Water mean Temp. ° C.	Convection	Radiation	Total
100	1·9	3·1	5·0
125	3·5	4·5	8·0
150	4·7	5·9	10.6

Values are approximate for one make with plate welded between coils. For accurate emissions, consult makers' data.

Radiant panels may be freely exposed back and front where hung between columns down the centre of a shop but, where against an outside wall, it is usual to protect the back against back radiation either by insulation or by a backing plate with air space. Where fixed horizontally overhead it is also probably desirable, in most cases, to insulate the back to avoid undue wastage of heat to above.

The advantages of strip heating are that fewer connections are involved and therefore the overall cost should be less. Also in certain cases there should be less obstruction than with hot surfaces supported near walls and columns etc.

Outputs range from about 0·5 kW/m run with a single pipe at 100° C (air at 15° C) to 3kW/m run with four pipes at 150° C. Mounting heights as given by the makers are a minimum of three metres, using low-pressure hot water, and five metres using medium-pressure or high-pressure hot water. There is no upward limit to the mounting height.

High-Temperature Heating by Convection—High-pressure hot water may also be used for the heating of large spaces by means of unit heaters, convectors, and pipe coils.

Unit heaters for high-pressure hot water follow closely the designs already referred to for steam in Chapter 10, except that connections flow and return are the same size and no steam trap is required. In the design of equipment for high-pressure hot water heating an attempt is usually made to keep the velocity of water high, by arranging circuits in series rather than in parallel.

Convectors used in conjunction with H.P.H.W. also follow the designs already referred to for steam and low-pressure hot water. Their use is usually confined to the heating of small offices. Where it is a matter of serving an extensive office block adjacent to factory buildings, more advantageous results can usually be obtained by adopting low-pressure hot water.

Hot-Water Heating—For a building or group of buildings where H.P.H.W. heating is not suitable, a low-temperature radiator system may be installed, and served from a water-water calorifier using H.P.H.W. for the primary heating.

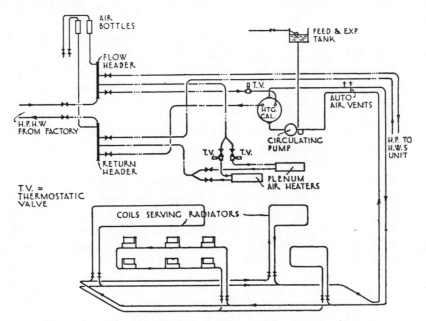

FIG. 11.14.—Diagram of Radiator System served from H.P.H.W. (see text).

Fig. 11.14 shows the piping and the apparatus for the heating and ventilation of the office-block in a factory heated by H.P.H.W. The incoming H.P.H.W. mains are connected to headers which in turn serve the heating calorifier, with its own low-pressure radiator system, pump-circulated. In addition, connections will be seen from the H.P.H.W. headers to H.W.S. units and plenum air-heaters.

The design of the low-pressure system will be exactly as given in Chapter 8. In sizing the calorifier a suitable margin should be added to the net heat losses to allow for time-lag and heating-up.

Other Applications—H.P.H.W. may be used as a source of heat for many sorts of apparatus, among which the following may be mentioned.

Steam Generators. (which are in effect water-steam calorifiers) comprise a primary coil served from the H.P.H.W. system which heats water in a cylinder above which is connected a second cylinder partly filled with water, the space above forming a steam space. Steam so produced indirectly is usually at a low pressure of about 1·7 bar absolute.

Vats and Boiling Vessels of all types. In these a coil is placed in the bottom of the vessel, or in a separate chamber at the side with mechanical circulation.

Baking and Enamelling Ovens, and similar apparatus requiring an intense radiant heat, and hot air for drying.

Vulcanizing and Plastic Presses where hot surfaces are required to promote a physical or chemical process.

Laundry Dryers, etc., in which hot dry air is used. The air is circulated over the H.P.H.W. coil by means of a fan.

The details of the design of such apparatus are beyond the scope of the present book, but it will be seen that all the processes normally served from a steam-heating system may equally well be provided by H.P.H.W. with the added advantage of greater range of control by valve operation.

PIPE SIZING FOR HIGH-PRESSURE AND MEDIUM-PRESSURE SYSTEMS

Emissions from bare and insulated piping may be estimated from Tables 10.2 and 10.3 (pages 222 and 223).

The process of pipe sizing follows that for a low-pressure two-pipe system. Loadings in kW are summated, mains losses are apportioned and provisional sizes determined on the basis of the design temperature drop, using Fig. 8.6 as already explained. Accurate sizing may then follow circuit by circuit, be duly balanced and result in the final pressure-loss calculation from which the pump head is established

CHAPTER 12

Heating by Electricity

ELECTRICITY IS AN IDEAL SOURCE of heat : it possesses every conceivable advantage:

(a) Absence of fumes or products of combustion.
(b) Avoidance of labour.
(c) Availability at any temperature at point of emission, hence great freedom in design of apparatus.
(d) Ease of thermostatic control.
(e) Transmissibility to any point regardless of levels or limitations found with other media. Portable appliances may be used.
(f) Shortness of time lag.

Two vital considerations, however, limit the free use of electricity for heating:

(1) Cost.
(2) Distribution capacity.

Cost—The generation of electricity by a conventional power station involves certain losses inherent in the steam cycle. Efficiency of generation has been steadily rising and the most modern stations achieve nearly 40 per cent. The average efficiency of all stations lies between 30 and 35 per cent. Other methods of generation are by water power which can only be on a small scale in the British Isles, and by nuclear power. It is apparently the case that this latter method will not come on a par with coal- or oil-fired stations for some years and there is no likelihood of a great abundance of very cheap current from this source, as was at one time popularly anticipated.

By whatever means electricity is generated, it then has to be distributed through the grid and through various transformer stations, and finally through local cable networks to the consumer. All these processes involve losses.

The price of electricity to be charged to the consumer must also bear its proportion of the capital charges involved in supply and distribution. In the case of commercial and industrial consumers, this is frequently in the form of a kilowatt charge based on the maximum demand during any half hour of month or quarter. In the case of domestic consumers, it is frequently on the basis of a fixed charge plus a charge for units consumed. Thus, overall, the more used the lower is the average price.

In order to supply current on demand at any time, such as for electric heating, generating plant has to be available, since there is no means of storage of electricity on any considerable scale, such as exists in the production of gas with its large gas-holders.

The peak heating load comes in the winter, when days are dark and electric lighting is required in addition to the normal industrial loads. It is the winter peak which controls the situation so far as installed plant capacity is concerned and, under extreme conditions of cold, the sudden imposition of vast extra heating loads at the worst possible time has been known to cause grave embarrassment to supply authorities in certain areas, even to the extent of loads having to be shed.

A typical daily load curve has a morning peak which usually extends from approximately 8 a.m. onwards to about noon; and it has a lesser peak in the late afternoon. It is to encourage consumption during these 'off-peak' periods that attractive tariffs are offered for off-peak supply; whereas on-peak heating may be charged at 0·85p. per unit, off-peak supply may be offered at 0·33p. per unit, or less. Some areas offer more than one off-peak rate, according to the restricted hours when the supply is taken.

The usual off-peak hours are from 7.0 p.m. to 7.0 a.m., and sometimes 10.0 p.m. to 7.0 a.m. on a more restricted period and lower price. A midday boost such as from 12.0 noon to 3.0 p.m. was available, although this advantage is tending to be withdrawn due to increasing loads generally.

Distribution Capacity—If a building is to be heated entirely by electricity, the installed load will be many times that of the supply required for normal lighting, power and other services.

It is also the case that, whereas the other supplies have some diversity as between maximum demand and connected load, it is not safe to assume a diversity where heating load is concerned, seeing that in extremely cold weather every available heater is more than likely to be full-on at the same time.

By transferring the heating load to the off-peak period, the distribution capacity for normal supplies can, in effect, be available for the heating load, but even so in a given area the total of a large number of buildings electrically heated may bring about a condition where the available capacity is fully used up. In such case, additional capital charges may become involved in order to meet the electric heating load, and the viability of the off-peak tariff is in jeopardy. Such a condition was reached between 3.0 p.m. and 3.30 p.m. in February 1970, instead of the previous peak at 5.0 p.m. to 5.30 p.m. The afternoon boost is thus no longer favoured for new consumers and, instead, new 'white meter' tariffs are being offered designed for an eight to ten hour night-charging period. In consequence, all new storage heaters are uprated as 'high-capacity' heaters.

It is as well to restate here that the watt is a *rate of energy flow*. One watt equals one joule per second. Thus the killowatthour, or unit of electricity, is equivalent to

$$\frac{J \times 3600 \text{ s}}{s} = 3600 \text{ J, or } 3 \cdot 6 \text{ MJ.}$$

In passing, however, it should be noted that the kilowatthour is not pure SI. In the SI system, the *hour* is not a preferred unit for time.

In British units, the *therm* of 100 000 Btu equals $1 \cdot 055 \times 10^8$J. The equivalent of the therm is then

$$\frac{1 \cdot 055 \times 10^8}{3600} = 29 \cdot 2 \text{ kWh.}$$

The therm, in fact, is roughly equal to $MJ \times 10^5$, which would be a convenient unit which might possibly be called the *new therm*. But, again, 10^5 is not a preferred power in SI. At the time of going to press, the new unit to replace the therm has not been declared.

Comparative costs of heat from electricity and from various fuels were given in Table 3.2 (page 65) in terms of gigajoules ($1 GJ = 1000 MJ$); but this is an inconveniently large unit. Off-peak tariffs are usually of the order of about 0·3 new pence per unit (three farthings), which equals 84 p./GJ. On-peak tariffs are about 0·8 new pence per unit, which equals 225 p./GJ.

If on-peak supply is taken, it will be seen to be an expensive way of heating; but in spite of this, in view of its many advantages, it is nowadays used to a considerable extent. This is particularly so where heating is required intermittently and heat can be taken just when it is required with no no-load losses. Some of the various methods of using electricity direct for heating in rooms were referred to in Chapter 3 and it is not proposed to consider them further.

At the lower figure above, such as may apply to off-peak supply, if the building is designed with a view to economy of heat-requirement by general weathertightness and full use of insulation, the kind of running cost involved may in many cases be acceptable, particularly bearing in mind the saving of labour, maintenance, storage space, flue, etc. It is in view of the importance of this method of heating that further attention will be devoted to it in this Chapter.

Off-Peak Electric Heating—The various methods in common use for storing heat taken electrically off-peak at night for use during the day are:

> Block storage heaters, in which the heat is stored in a solid material contained in an insulated casing and given out again continuously uncontrolled;
>
> Block storage heaters, in which the heat is retained and given out when required, known as 'storage fan heaters'—a type which is the basis of the 'Electricaire' warm-air system;

Warmed floors in which the heat is stored in the screed and floor slab;

The Thermal Storage system, in which the heat is stored in water in large cylinders.

Amount of Heat Required—The amount of heat required over a 24-hour period will depend on which of the above methods is adopted. Where the heat is emitted uncontrolled, as in the case of floor-warming and uncontrolled block storage heaters, heat is liberated from the surfaces of floor or storage casing during the whole period of warming-up at night, as well as during the period of use during the day. In effect, this amounts to continuous heating throughout the 24 hours, though with some dropping off of temperature towards the end of the discharge period, depending on the capacity of the storage medium in relation to the heat losses.

In the case of controlled storage heaters and the thermal storage water system, although there will be some losses at night, the bulk of the heat to be taken during the off-peak period will be that used for heating during the day.

The calculated heat losses per 24 hours may take into account the fact that at night-time a lower air-change rate is to be expected (see Table 12.1).

TABLE 12.1

RECOMMENDED DESIGN VALUES OF RATES OF AIR CHANGE BY NATURAL
INFILTRATION IN MULTI-STOREY OFFICE BUILDINGS*

	Infiltration rate (air change/hr.) Height of Building			
	15 m and under		15 to 30 m	
	Day	Night	Day	Night
Buildings with little or no internal partitioning (i.e. 'open plan' buildings), or with partitions not of full height, or buildings with poorly-fitting windows and internal doors.	$1\frac{1}{2}$	$\frac{3}{4}$	2	1
Buildings with internal partitions of full height, without cross ventilation, and having self-closing doors to staircases, lift-lobbies, etc.	$\frac{3}{4}$	$\frac{1}{2}$	1	$\frac{1}{2}$

Note: The basic design temperature should take into account the fact that there is no inherent reserve and a basic temperature of $-3.3°$ C for single storey buildings, and $-1.7°$ to $-1.1°$ C for multi-storey buildings is proposed.*

In the case of a floor-heated building, it is necessary to add to the net heat requirement the expected downward loss to the ground or to a room beneath the heated floor. If the total heat requirement so found is denoted by QkW, the installed load for an *uncontrolled* system will be $24Q/n$ where n is the available hours of charge.

Where the heat input is *controlled*, for an occupied period of 12 hours, and losses assessed at the equivalent of 4 hours, the installed loading required would be $16Q/n$.

* *The Heating of Buildings by Off-Peak Electricity Supplies.* H.V.R.A. report, May 1961.

It is apparent from experience that the successful application of off-peak heating whether by floor or block storage depends on adequate thermal storage in the building structure to prevent too great a fall in temperature during the day when the power is off. This temperature fall should not exceed 3·3° C; which means that if it is desired to maintain 18° C average, the temperature in an office block will be up to 20° in the morning and down to 16° in the afternoon. For domestic usage this would not be satisfactory, since it is in the evening that heat is usually most required, and hence the importance of a mid-day boost in such cases.

In the case of thermal storage systems using water, these considerations do not arise, as the available storage can be made as great as required simply by designing the capacity of the storage vessels accordingly.

BLOCK STORAGE HEATERS

Block storage heaters enable heat produced electrically to be stored during off-peak hours for warming the rooms during the rest of the day. They are direct heaters placed in the rooms and are therefore 100 per cent. efficient, and incur no losses in mains etc.

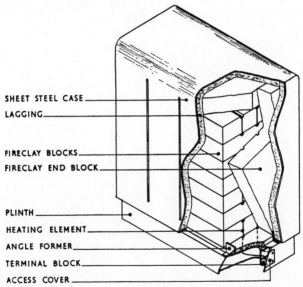

SHEET STEEL CASE

LAGGING

FIRECLAY BLOCKS
FIRECLAY END BLOCK

PLINTH
HEATING ELEMENT
ANGLE FORMER
TERMINAL BLOCK
ACCESS COVER

Fig. 12.1.—Block Storage Heater. (The controller and limit stat are not shown).

The night storage heater (as in Fig. 12.1) comprises a series of electric resistance elements enclosed in a ceramic or concrete material contained in a partially insulated steel casing. Loadings vary with different makes between 1·5 and 3·5 kW. The maximum temperature achieved on the surface of the casing after the charging period is about 85° C, and this declines to about 40° C at the end of the day. The storage material reaches a temperature of about 200° C.

This type of thermal storage system is applicable to single rooms, or a complete building, especially an existing building. The installation is simple, involving only the placing in position of the heaters and connecting them up to socket outlets wired back to a separate meter controlled by a time switch set by the Supply Authority according to the hours of off-peak supply.

Thermostatic control is rarely used— the 'charge controller' with its somewhat limited range of control is usually found adequate. A high-temperature cut-out is now included to prevent overheating.

Some makes of storage heaters include an on-peak convective element. Fig. 12.2 gives curves of heat output and room temperatures of one such make. The room temperatures are given with and without the convection boost.

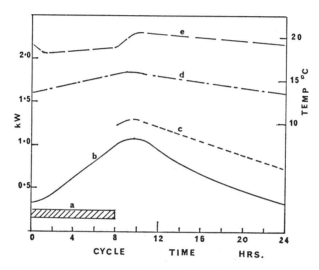

Fig. 12.2.—Output and Temperature Curves of a Block Storage Heater (E. Chidlow and Co., Ltd.).

(a) Heating element 'on'.	(d) Temperature storage only.
(b) Storage mass output.	(e) Temperature storage and direct element.
(c) Direct element output.	

During the day period, room thermostats are of course inoperative, as current is not available due to the time switch having shut off supply. The heat output is uncontrolled and about 30 to 40 per cent of the energy taken is released during the charging period.

Storage Fan Heaters—A development of the block storage heater is one in which the lagging is increased so as to reduce the amount of stray heat loss, and the unit is equipped with a fan arranged to drive air through passages in the heat storage blocks and out into the room to be heated.

So as to provide a uniform air-outlet temperature, a certain degree of mixing of air from the room with the heated air from the block takes place by means of a thermostatically controlled damper. A unit of this type can be controlled by means of a thermostat in a room, and, if desired, by a time switch, so that although the heat supply is taken only during the night, the bulk of the heat stored is released according to requirements during the day. The loss is stated to be 7 per cent. of the energy taken over 24 hours. If the day is mild, room temperature will be quickly achieved and a small amount of heat only will be withdrawn from the unit. If the day is cold, a larger amount will be withdrawn. Provided the heat storage is adequate to last throughout the day, the result should be one in which a uniform temperature is maintained internally without any problem of anticipatory control or swings of temperature, such as are inherent in other systems so far described. A unit of the controlled type is seen in Fig. 12.3.

Electric Storage Warm-Air System—This system, known as 'Electricaire', is a development of the last one in that a larger capacity block storage heater is used, capable of heating, say, various rooms of

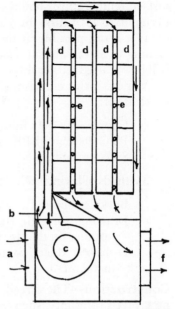

FIG. 12.3—Section through a storage Fan Heater (GSW Ltd., 'Activair').

(*a*) Return air.
(*b*) Damper set for discharge.
(*c*) Fan.
(*d*) Storage blocks.
(*e*) Elements.
(*f*) Warm air discharge.

a flat. This unit, as in the case of a gas warm-air unit, is planned-in with the rooms of the flat, so that it takes a central position. The unit is in principle similar to Fig. 12.3, but input ratings range generally from 6 to 12 kW, some being available up to 36 kW.

EMBEDDED FLOOR HEATING

The absence of a mid-day boost reduces the application of floor heating to those cases in which a decline in temperature during the afternoon and evening is unimportant, or to cases where the extra cost of an on-peak boost is acceptable. The 'white meter' tariff takes care of such a means of operation.

The heating elements are resistance wires, the total length being such as to produce the loading required, calculated as described above. This length of element is laid out in grid-iron formation in one, two, three or more circuits according to size of cable, its rating and the length required. Spacing of the coils will vary according to the loading required and to type of floor finish, bearing in mind that a hard finish emits heat more readily than say a carpet. Centres are often 150 mm, 230 mm or 300 mm. The surface temperature should not exceed 24° C, though it may be assumed to run up to 27° C first thing in the morning at the end of the charge.

If it is not possible in the area of the floor to accommodate sufficient loading to meet the heat losses over the 24 hours, supplementary heating will be required, either direct on-peak or by night storage heaters.

Various forms of resistance elements are available; proprietary systems make use of:

> Copper sheathed mineral insulated cable laid solid;
> PVC sheathed/EPR (Ethylene propylene rubber) insulated;
> Silicone rubber.

Methods of laying include:

> Burying solid in the screed;
> Drawing through steel tubes running between troughs, as for instance on the two sides of a room;
> D-shaped ducts connecting to troughs overlaid with expanded metal mesh.

Construction—The screed requires to be not less than 75 mm in thickness, of a strong mix, such as one part cement to 3½ sand and aggregate. Where the floor is at ground level, edge insulation as shown in Fig. 12.4 is necessary. Where the heating is applied to a multi-storey structure

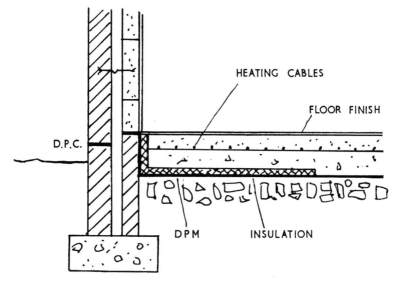

Fig. 12.4.—Edge Insulation.

and the screed is on a suspended slab, the insulation is preferably placed below the slab so that the heat-storage effect thereof assists in reducing temperature swing, though it is undesirable that the slab mass should be

Plate XIV (right). An application of Strip Heating in which the panels are integrated with the structure (see p. 245)

Plate XV (below). Electric Floor Heating: copper sheathed mineral insulated resistance cables laid ready for screeding (see p. 257)

Plate XVI. Electrode Boilers in an electrical thermal storage system (see p. 258)

too great or the system will become unduly sluggish. If the insulation is placed immediately beneath the cables, the swing will be much greater.

Most floor finishes have been used on floor-warming installations, but precautions are necessary to ensure that if adhesives are used, such as for linoleum, rubber and cork, they will be suitable under heating conditions. Residual moisture should be driven off by operating the system before final finishes are laid, particularly where a wet finish is to be applied. Various hardwoods and softwoods appear to be suitable. Where carpets are laid with an underfelt, if too thick, the temperature may rise to too high a level. Underfloor insulation is essential if a high resistance floor finish is used; otherwise there will be considerable loss of heat to the ground if on a ground floor, or to the room below in the case of an upper floor.

Heat Transmission to Below—The loss of heat from a heated floor at the ground level, depending on size and shape, may amount to about one-third of the total input in the case of small floors, or to about one-tenth for larger floors. The major part of the downward loss occurs at the edges of the floor, hence insulation at the centre has little effect, but insulation beneath the whole area is to be preferred. The downward heat loss through intermediate floors may be 25 per cent of the heat supplied if insulation is not provided, but would be reduced to perhaps 5 per cent with insulation.

The downward losses are susceptible of calculation, but it is clear that they should be taken into account when estimating loadings and consumptions.

Reference may be made again to the H. and V. R. Association publication, dated May 1961, on the *Heating of Buildings by Off-peak Electric Supplies*. Reference may also be made to *Recommendations for the Design of Floor Warming Installations* issued by the Electricity Council.

Plate XV, facing page 256, illustrates the installation of an electric floor-warming system.

Control of Floor-Warming Systems—The simplest method of controlling a floor-warming installation is by a room air-thermostat which switches the power off when the desired temperature has been reached. The clock switch installed will bring the supply on once again at the commencement of the charging period and, when the desired indoor temperature has been reached during the night, the off-peak supply will be cut off by the thermostat. Should the temperature drop, the thermostat will bring the supply on again, and so on until the off-peak supply is terminated. This may result in night-time temperatures being unnecessarily high.

Greater economy may be achieved by a control system which delays the commencement of the night-charging period as long as possible, such that in the remaining hours of supply the temperature of the floor will have been brought up to that desired by about the time that the supply is

terminated. This involves a time-temperature controller which is responsive to external weather. What is required is some form of anticipatory
control, seeing that the heat input during the night will have to last for the
next twelve hours or so and, should there be a sudden drop of temperature,
internal warmth may suffer. In a building of substantial construction with
good thermal mass, such sudden changes may not be serious. On the other
hand, a warm day following a cold one might result in discomfort in the
rooms through overheating. There is no means of cutting down the
amount of heat liberated from a warm floor.

In the case of a block of flats, time-temperature controllers for each
user may not be practicable, and resort to a simple room thermostat is
then necessary; but this may have the disadvantage that it is affected by
any direct heating appliances in use. In such case, the room thermostat is
best located in the hall, where it would not normally be affected by such
additional heating.

A thermostat may be placed in the floor itself, but this is to serve the
purpose of a limit-stat to prevent the floor temperature becoming too
high, and it will not serve for controlling the room temperature.

An application of electric floor warming not concerned with buildings
may be mentioned here: namely, the heating of road-way ramps to
prevent frost. For this application, a steel-mesh element operating at a
low voltage has been found successful.

THERMAL STORAGE—HOT-WATER SYSTEM

This system is indirect. Heat is stored in the form of hot water raised to
as high a temperature as possible, and contained in large cylinders
generally placed in the basement. The heat is generated either in an
electrode heater utilizing the resistance of the water as the element, or in
smaller systems by immersion heaters. The apparatus is automatically
controlled by thermostatic and time switches. (See Plate XVI, facing
page 257).

Operation—Fig. 12.5 shows the general arrangement of a typical
thermal storage system. The principle of operation may be described as
follows:

(a) The electrode water heater (or separate immersion heater)
warms the water in the storage cylinder, the pump accelerat-
ing the circulation. If the immersion heaters are contained in
the main storage no pumping is necessary.

(b) This heating takes place at off-peak hours, and in order to
economize in capacity of storage the temperature is raised as
high as possible without generating steam. This may be as
high as 185° C, and is determined solely by the height of the
building. An artificial head can be produced by pressurizing
by means of a separate air or gas cushion cylinder, as described in Chapter 11. The storage temperature is generally

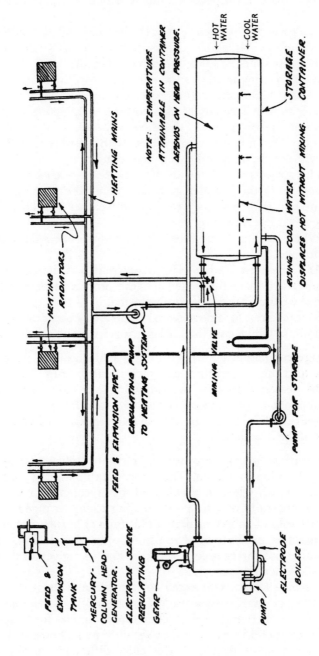

Fig. 12.5.—Diagram of Electrical Thermal Storage System.

kept about 10° C below that at which steam would be produced.

(c) At a given time the current to the heater is cut off automatically by time switch, or from the supply authority's sub-station. At the same time the storage pump is stopped.

(d) When heating in the building is called for, the heating pump is started; high-temperature water is drawn off from the cylinder and returned cooler at the bottom. A proportion of return water is mixed with the flow through a thermostatically controlled mixing valve. Thus, with possibly 140° C in the cylinder, 80° may suffice for the radiators.

(e) This process continues all day, the level of cool water at the bottom of the cylinder gradually rising, but not mixing with the hot, and by night time in cold weather probably little hot water is left in reserve.

(f) When heating is no longer required the secondary pump is stopped.

(g) At a given time, when off-peak current is available, the supply is turned on to the boiler and the process repeated. The primary pump is simultaneously started.

(h) In some areas it is possible to obtain a day boost during certain hours in which case the storage capacity may be reduced.

The storage capacity is controlled by the heat output of the plant, its period of use and the temperature at which it can be run. The electrical loading of the heaters is likewise determined by the daily total heat load and the number of hours during which current is available.

Types of Heater—The heater may be of two types according to the voltage and size of the installation:

(a) *Immersion heaters*, consisting of resistance elements inside metallic tubes or blades. These are suitable for low voltages up to 250, or, when in a balanced arrangement, up to 500, and are generally of 3 to 4 kW loading each, arranged in groups or banks of 50 kW or more. This type of heater is often placed direct in the bottom of the storage vessel as in Fig. 12.6. Installations rarely exceed 300 kW. Fig. 12.7 gives an illustration of a typical plant.

(b) *Electrode heaters*, connected to a medium voltage (up to 650 volts) or high voltage (3·3, 6·6 or 11 kV) three-phase alternating current supply. They are suitable for installations up to 5000 kW each, and one such type is shown in Fig. 12.8. They are vertical and the chief difference in the various designs is in the method adopted for load regulation. Current passes from electrode to electrode, using the resistance of the water itself as the heating element, and the load is varied by in-

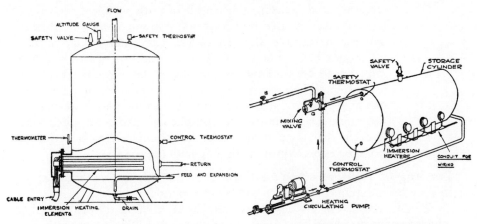

FIG. 12.6.—Immersion Type Thermal Storage Heater.

FIG. 12.7.—Diagrammatic Lay-Out of Electric Thermal Storage Plant with Immersion Heaters fitted direct to Storage Cylinder.

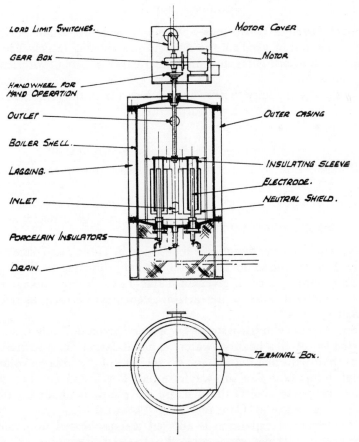

FIG. 12.8.—High Voltage Electrode Water Heater.

creasing or decreasing the length of path which the current has to take by the interposition of non-conducting shields between or around the electrodes. The conductivity of the water may have to be adjusted by the addition of soda or other salts if need be, to render it suitable for the purpose. This type of heater is kept separate from the main storage because means of load regulation involving raising and lowering or rotating gear to the sheaths complicates a direct application; and, further, one heater may be connected to a series of storage vessels to provide the capacity required.

Loading of Heaters—The total heat requirements per 24 hours must include all radiation and other losses over the period.

The hourly load in kW depends on the duration of the off-peak supply, which may be anything from eight to eighteen hours. The hourly rating of heaters in kW

$$= \frac{\text{total heat requirement (kW) over twenty-four hours}}{\text{duration of supply in hours}}.$$

Thermal Storage Cylinders—Storage vessels are usually cylindrical, either horizontal or vertical, the size and shape being largely determined by space conditions. Diameters up to 3·2 metres are usual, with lengths up to 10 metres.

The storage capacity to be provided depends on the total heat requirements per day and on the temperature range.

The capacity in kg of water

$$= \frac{\text{kJ required during the time current is off}}{\text{storage temp. °C minus minumum heating return temp. °C} \times 4\cdot2}$$

then

$$\frac{\text{kg}}{\text{density kg/m}^3 \text{ at storage temp.}} = \text{storage capacity m}^3$$

A selected range of densities will be found in Table 11.1 (page 229).

The head required to prevent ebullition, allowing 10°C excess over saturation temperature, is given in Table 12.2. Where tankage at these heights is impracticable, a pressurizing plant may be used, as referred to in Chapter 11.

Capacities of cylindrical storage vessels are given in Table 12.3.

Expansion of the heating water is much larger on thermal storage systems than with heating boilers, on account of the greater volume and the high temperature of the water stored. For instance, when water is raised from 50° to 120° C the increase in volume is about 4·5 per cent, and when raised to 150° C the increase is 8 per cent.

If the water of expansion is allowed to pass heated from the water heater, and then is allowed to cool off in the expansion pipe and tank be-

TABLE 12.2

HEAD PRESSURES FOR HOT-WATER THERMAL STORAGE

Storage Temp-erature °C	Pressure bar absolute	Pressure above Atmospheric for 10°C margin	Head of Water m
125	2·3	2·2	23
150	4·8	5·2	53
175	9·0	10·5	107
200	15·9	18·3	187

TABLE 12.3

CAPACITIES OF CYLINDRICAL VESSELS WITH FLAT ENDS
m³ per m length of cylinder

Diameter m	m³ per m Length
1·0	0·78
1·25	1·23
1·5	1·77
1·75	2·41
2·0	3·14
2·25	3·98
2·5	4·91
2·75	5·94
3·0	7·07
3·25	8·28
3·5	9·62

fore returning gradually during the day, a certain loss of heat is incurred. For this reason a sufficient space at the bottom of the cylinders should be arranged below the level of the return connection to contain the diurnal water of expansion. From the bottom of this space the feed and expansion pipe is connected as in Fig. 12.9. When the water is heated it depresses

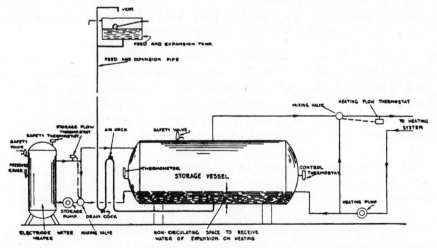

FIG. 12.9.—Diagram showing Connection of Expansion Pipe and Mixing Arrangements of Electrode Heater.

that at the bottom up into the tank and in theory only cold water should be forced up the expansion pipe. In practice, there is probably some mixing, in addition to the small heat transmission by conduction, and a small loss from this source is unavoidable.

In any case, this expansion space is only considered to accommodate the daily expansion, i.e. from about 50° C to the storage temperature, since the expansion from 10° to 50° only occurs once during the heating season and it may be accommodated in the expansion tank additionally to the daily expansion.

The height of the return pipe from the bottom of a horizontal cylinder should therefore be about one-eighth of the diameter, corresponding to an expansion allowance of 8 per cent.

Insulation—Loss of heat from the cylinders and heater is reduced as far as possible by efficient insulation, such as 125 mm of glass fibre with metal or other cladding. Cradles are also insulated with hard material, such as compressed cork or hard wood, to minimize heat loss by conduction.

Mixing Valve—The mixing valve, which is an essential part of every thermal storage plant, consists of three ports. One is the high-temperature water inlet, one the cool return water inlet, and one the mixed water outlet. The proportions of the two former are controlled by a valve or valves operated by means of water pressure, electrical solenoids, or motor, from a thermostat in the mixed outlet pipe.

Expansion Tanks—As has been said, relatively large expansion tanks are required for thermal storage systems capable of containing the water of expansion resulting from

(a) heater and storage capacity raised from 10° to maximum storage temperature;

(b) heating system capacity raised from 10° to maximum working temperature : e.g., 80° C with radiator heating.

Control—As the temperature in an electrode heater rises the resistance of the water becomes less and the load correspondingly increases. Thus a 100 kW heater at 40° C would at 150° C have an output of about 220 kW. Means to provide a constant outlet temperature, and hence constant load, are included by some installers, making use of a further thermostatically controlled mixing valve in the boiler-cylinder circulation as shown in Fig. 12.9.

Protective gear is necessary in the case of electrode heaters to prevent operation on two out of three phases, or with out-of-balance currents flowing. Such faults might cause heavy earth leakage currents since the latter are only avoided when the current in all three phases is equal.

The protective and control gear for thermal storage systems is perhaps outside the scope of the present book. It is a great advantage to have all the control instruments, relays, contactors and switchgear mounted on a common switchboard so that faults can be more easily located and proper supervision given.

Electrical Circulators—Although this chapter deals predominantly with off-peak systems, reference should be made to the direct electric water-heater generally known as a 'circulator', which may be used to replace a boiler in a conventional hot-water system. Cases exist where it is impracticable to arrange for a flue of any sort, and yet where heating by hot water is preferred for various reasons. In such cases, an electrically-heated boiler or circulator is suitable. This type of unit also finds a place in churches where the labour for stoking a boiler is non-existent, and therefore a means of automatic heating is necessary. Many Area Boards offer special tariffs for church-heating at a low rate for week-end and evening use, and the electric circulator in such case may be a suitable answer.

THE HEAT PUMP

The heat pump is a means of using energy such as is produced from electric supply to better advantage than by merely degrading it into heat in the normal way.

The heat pump makes use of the same principle as a refrigerator (see Fig. 12.10).

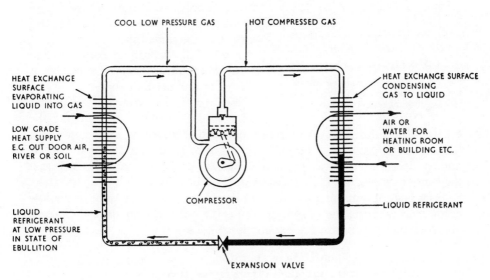

FIG. 12.10.—Diagram of Heat Pump.

A *compressor* is used to compress a gas (known as the refrigerant) and in so doing raises its pressure and temperature; that is, it puts work into the gas.

The compressed hot gas is passed on to the *condenser* where heat is removed and the gas condenses into a liquid.

The refrigerant then expands through an expansion valve where the pressure drops and the liquid enters the *evaporator* at a reduced temperature. In this state the refrigerant can absorb heat at a relatively low temperature and in so doing boils, or vaporizes, the resultant vapour returning to the suction of the compressor; and the cycle then repeats.

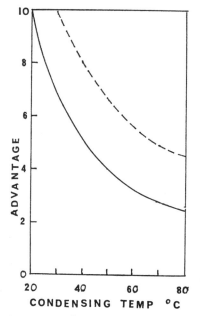

FIG. 12.11.—Heat Pump using ammonia. Evaporator temperature: 1·7° C. Theoretical advantage shown dotted; actual advantage shown solid.

The condenser and evaporator are heat-exhangers, usually of tubular form. The compressor may be driven by electric motor, or by a prime mover such as a diesel or gas engine or gas turbine.

Where this cycle is used for refrigeration, interest lies in the evaporator, where heat is absorbed, and freezing or cooling results. In the case of the heat pump, interest lies chiefly in the condenser, i.e. the heat producer, as this heat may be used for warming a building, or warming hot water.

Sometimes the same plant may perform the double function of providing 'coolth' in the summer and warmth in the winter, or coolth and warmth at the same time.

For a full theoretical consideration of the heat pump cycle the reader is referred to a paper by Faber.*

Fig. 12.11 is extracted from this paper and it will be noted that the greater the temperature difference between evaporator and condenser, the less 'advantage' is obtained. For instance with evaporator at 1·7° C and condenser at 38° C a practical advantage of 5·7 is obtained, whereas if the condenser is at 60° C the advantage is reduced to 3·5.

The heat obtainable from the condenser is equal to the sum of the low grade heat entering the evaporator, plus the heat equivalent of the power absorbed by the compressor (less any losses). Thus if

heat entering evaporator	= 3
heat equivalent of power absorbed	= 1
	—
heat from condenser	= 4
the 'advantage' is then	4 : 1.

This is sometimes referred to as a 'coefficient of performance' of 4, and sometimes inaccurately as an 'efficiency' of 400 per cent.

* *Proc. I. Mech. E.*, 1946, Vol. 154, p. 144.

The low grade heat for the evaporator may be obtained from a river, or from waste effluent, or from some cooling process. Heat from sewage has been used for heating a building by means of a heat pump.* Use has also been made of atmospheric air and the heat in the earth by burying a coil in the garden.†

The higher grade heat from the condenser may be used in radiators or panels, or for warming air direct. As explained, the lower the temperature of use, the better the coefficient of performance, and therefore embedded panel heating using water at 40 to 50° C is to be preferred to any higher temperature system such as radiators.

The following examples of practical unit type heat pumps are given by way of introduction to this subject about which a considerable literature has been built up.

Fig. 12.12. This heat pump is for room heating and derives its low grade heat supply from air outside the building by means of a fan drawing air over the external coil. The condenser forms the internal coil over which room air is drawn and delivered into the space to be heated.

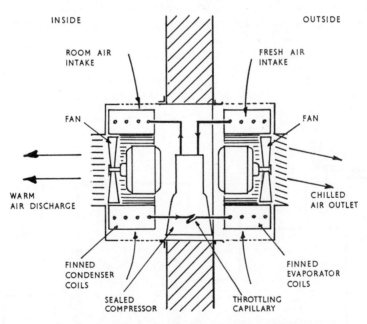

FIG. 12.12.—Diagram of Heat Pump for Room Warming.
Note.—The change-over valve for summer cooling is not shown.

The compressor is self-contained in the casing, the whole unit fitting into a window or into a hole in the external wall. For an input of 0·75 kW an output of about 3·5 kW is obtained, i.e. an advantage of about 4·7.

* Paper by Kell and Martin, *Journal I.H.V.E.*, Vol. 30, p. 353.
† Paper by J. A. Sumner, *Journal I.H.V.E.*, Vol. 23, p. 129.

The performance depends on temperature conditions as with any heat pump. Noise is a factor for consideration.

To preserve a better balance between average load and installed capacity, the heat pump may be designed for two-thirds of maximum cold weather duty, and the balance be made up by direct electric heating, or night storage heaters. The operation is reversible by means of change-over valves so that, in summer, cooling may be obtained in the room.

The importance of the heat pump is that it is or should be a fuel-saving device. The low grade heat is in effect waste heat, or normally un-recoverable heat and the power input is small compared with the heat output. The saving is however not as startling as seems possible at first sight due to the fact that if the electricity is derived from a coal-burning station the thermal efficiency of generation is no more than about 25 per cent or 4 parts of coal to 1 part of power output.

If the heat pump coefficient of performance is 4, then 1 part of power produces 4 parts of heat, so that the overall conversion is 100 per cent efficient. This is higher than the efficiency of a normal heating boiler, but it is at the expense of a good deal of plant and complication.

Fig. 12.13: A more impressive case can be made out if the heat pump

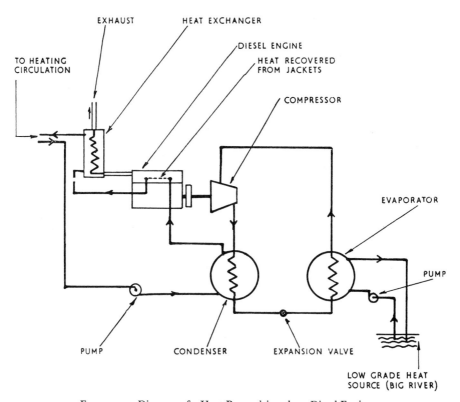

FIG. 12.13.—Diagram of a Heat Pump driven by a Diesel Engine.

is driven by a prime mover, and the waste heat from the latter used to augment the heat from the condenser as in the diagram. If a diesel engine is used, the thermal efficiency is of the following order:

Thermal efficiency of diesel engine:

Power output	30 per cent
Heat output	40 per cent

Assume a heat pump having a coefficient of performance of 4:

Power input from diesel is increased to 120 per cent.
Total heat output = 120 + 40 = 160 per cent.

The disadvantage of such an arrangement is the running of the prime moving plant and the maintenance associated therewith.

The economic case for the heat pump is often found to be disappointing owing to the high cost of plant involved and it remains to be seen whether future developments bring about an improvement.

ADDITIONAL NOTES AND CALCULATIONS

In conclusion, by way of an appendix to this chapter, some notes concerning electricity generally and its application to heating systems are appended here.

Units—Current is measured in *amperes*. Pressure, or potential difference, in *volts*. Resistance in *ohms*. The current I passed through any resistance R, when the potential difference is E, is given by Ohm's law:

$$I = \frac{E}{R},$$

so that
$$E = IR \text{ and } R = \frac{E}{I}.$$

Power supply P is measured in *watts*.

$$1 \text{ watt} = 1 \text{ volt} \times 1 \text{ amp.}$$

or $P = EI$. Thus power P, which is equivalent to the heating effect, is given by substitution:

$$P = I^2 R.$$

Types of Supply—Electricity is supplied direct and alternating: direct current is almost obsolete as a public supply.

With *direct current* one pole is maintained constantly at positive and the other at negative potential, so that the simplest supply is two-wire, one positive and one negative.

In a three-wire D.C. system a neutral or middle wire is introduced at earth potential, one 'outer' being maintained positive to this and the other negative. The difference of pressure between the neutral and each outer is half that between the outers. With a 480-volt D.C. three-wire supply, the connection to the mid-wire would give 240 volts, thus:

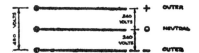

When a three-wire D.C. system is 'balanced', the current taken by the positive-to-neutral exactly equals that from neutral-to-negative, and no current flows in the neutral. As soon as the load on either side is altered with respect to the other, the system becomes out of balance and current flows in the neutral either in one direction or the other.

With *alternating current* the polarity of the supply is reversed in regular cycles. The unit of frequency in SI being the hertz (Hz), 1 Hz = 1 cycle per second. Thus in a 50 Hz supply (standard in this country) the reversal takes place 50 times per second. The cycle of a single-phase supply is of sine wave from, thus:

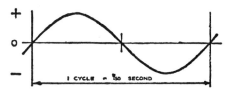

A.C. is usually either single-phase, or three-phase (two-phase supplies are obsolete).

Single-phase current may be treated as D.C. supply for the purpose of heating loads which are all non-inductive and for which Ohm's law still holds good. The question of '*power-factor*'* does not enter on this account, i.e. power-factor is taken as 'unity'.

Three-phase three-wire may be visualized as three single-phase supplies 120° out of phase, represented thus:

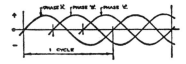

The three wires may be connected to apparatus in star or delta formation:

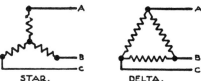

STAR.   DELTA.

* Where the load is inductive, as in a motor, the voltage and current get out of step with one another. The apparent watts as shown by a voltmeter and ammeter are greater than the true watts as shown by a watt meter. The power factor is the ratio $\dfrac{\text{true watts}}{\text{apparent watts}}$.

The voltage between points *A* and *B*, *B* and *C*, *C* and A is the declared voltage; 415 volts is now standardized.

Three-phase four-wire is the same as the three-wire, but a neutral line is introduced at the centre of the star formation, and this is earthed, thus:

The voltage between *A* and *N*, *B* and *N*, *C* and *N*, is then the declared three-phase voltage divided by $\sqrt{3}$.

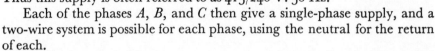

With the standard voltage of 415, it is $\dfrac{415}{\sqrt{3}} = 240$ volts.

Thus this supply is often referred to as 415/240 V. 50 Hz.

Each of the phases *A*, *B*, and *C* then give a single-phase supply, and a two-wire system is possible for each phase, using the neutral for the return of each.

When a three-phase four-wire system is balanced, the load on each phase is the same and no current flows in the neutral. If any one of the loads is varied, current flows in the neutral wire.

This system is the standard now adopted in Great Britain.

Application to Heating Systems—Direct electric heating elements up to 3 kW (13 amp. at 240 volts) are served by a two-wire circuit. Heavier duty elements are usually arranged in banks and the load divided over three phases of a three-phase supply if available.

For thermal storage with immersion heaters the usual arrangement is balanced on a three-phase three- or four-wire supply.

Thermal storage with electrode heaters is invariably served from three-phase three-wire supply at medium or high voltage.

The current passing in the conductors may be calculated from the load and voltage. Taking as an example a heating load of 500 kW, the current passing through the circuit will be as follows:

for two-wire D.C. *or single-phase* A.C. at 240 volts,

$$\frac{500 \times 1000}{240} = 2080 \text{ amperes};$$

for three-wire A.C., *three-phase* at 415 volts between phases, load balanced,

$$\frac{500 \times 1000}{\sqrt{3} \times 415} = 700 \text{ amperes per phase};$$

for four-wire A.C. *three-phase* at 415 volts between phases (240 volts phases to neutral), load balanced, the current will be the same as with the three-wire system, no current flowing in the neutral.

In the case of a number of small heaters, each will be served from a separate way on the distribution boards. Having established the current flowing in the circuit supplying one heater, these are totalled up to give the current supplied by the main to the distribution board. From these currents the electrical distribution system is then designed. For a more complete study, reference should be made to one of the many text-books on electrical engineering.

kW and kVA—The difference between kilowatts (kW) and kilo-volt-amperes (kVA) should be understood. The latter term is used in connection with alternating currents and refers to 1000 volt-amperes, or 1000 'apparent' watts. The kilowatt refers to 1000 'true' watts. At unity power-factor (see page 270) the kVA and the kW are the same, but, if the load is inductive, then kVA × power factor = kW. Normal heating elements are non-inductive and the terms are then synonymous, but such is not the case where motors or power transmission lines are concerned.

CHAPTER 13

Hot-Water Supply

LOCAL AND CENTRAL SYSTEMS

THE HEATING OF HOT WATER for baths, basins, showers and sinks, may be by a local or a central system.

Local hot-water supply systems derive their heat either from gas or electricity, and the heater is placed near to the point of consumption. A central system is one in which the hot water is heated at some point remote from the fittings to be served, and is conveyed thereto through a system of piping. The heat supply for a central system may be produced from the burning of solid fuel, oil or gas, or by the consumption of electricity.

Choice of System—The type of system will depend on the type of building. Small houses and flats may very often be conveniently served by local systems which require no attendance, cleaning or maintenance. A small domestic boiler system in a house, provides some general warmth to the kitchen and, where fired with solid fuel, a means of disposal of a certain amount of refuse. In addition, one or two small radiators may be added to heat the hall, etc.

Large Block of Flats—The tendency of late has been to provide each flat with its own local hot-water heater, heated either by gas or electricity. The tenant then pays for just what he or she consumes through the gas or electricity meter. The alternative of a large central system for the complete block of flats or a number of blocks is favoured by certain Authorities and in the better class of property, in which case the cost of hot water is included in the rent.

Office Blocks, Factories, etc.—If the lavatories are few and widely separated, the advantage will be with the small separate local hot-water supply units, electrically or gas heated. If, however, the washing facilities are arranged in a compact manner the central system will be more economical and will give greater reserve for sudden peak load draw-off.

Hospitals, Hotels, Institutions—In this class of building the central system is almost exclusively adopted for the reason that it alone can give a great reserve for heavy demands, and the maintenance of a large number of small units is avoided. In building complexes of this type the heat is often derived from a central boiler plant which provides all the heat requirements, burning a cheap fuel, and therefore a greater economy in running cost is achieved with such a system.

Losses—The determining factor of local versus central system may, in a doubtful case, be considered in terms of losses versus heat usefully supplied.

In a central system there are mains losses from circulating piping continuing during the hours of running per day whether or not any water is drawn off. Where large quantities of hot water are drawn off daily, these losses may bear a small ratio to the total heat supplied. If, however, draw-off is small in quantity and spasmodic, it may well be found that the mains losses far exceed the actual amount of heat in the water drawn off. In the former case a central system would be well justified; in the latter there would be much in favour of a local system.

Local Gas Hot-Water Supply—The instantaneous gas water heater of the type shown in Fig. 13.1 is a well-known example of this type of unit.

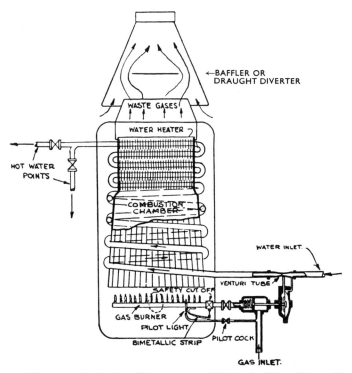

FIG. 13.1.—Diagrammatic Section of Instantaneous Multi-point Water Heater ('Ascot').

There is no storage of hot water, but it is heated as in the older type of geyser, as the water flows through. The type illustrated is a multi-point type which can serve a number of draw-off points provided they are within the limit of the length of draw-off (see p. 310). Or it may be a single-point type serving only one basin. A pilot is continuously burning and when a hot-water tap is opened water begins to flow and, by the pressure difference across the venturi, the main gas valve is opened, whereby the gas burners are ignited. A safety device cuts off the gas should the pilot light be extinguished. The cold water supply is usually connected from a tank, and a head of at least 3·5 m is required. The products of combustion are conveyed to outside by means of a flue with a suitable terminal, or in another model of balanced-flue type there is an inlet and outlet duct, both communicating with the outside.

The rate of flow from this type of heater is restricted, and it is not to be assumed that a good supply may be obtained from more than one tap at a time.

Local Electric Hot-Water Supply—For single-point draw-off, these take the form shown in Fig. 13.2 of non-pressure type. The hot water is stored

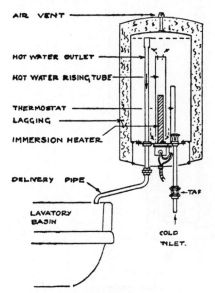

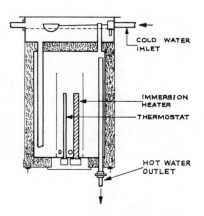

FIG. 13.2.—Electric Hot-Water Storage Heater; Closed, Non-Pressure Type.

FIG. 13.3.—Open Tank Electric Hot-Water Heater.

in a cylinder heavily insulated and containing in the centre an electric immersion heater with thermostatic control. When hot water is required the cock in the cold-water inlet is opened and this, by displacement, causes hot water from the top of the storage cylinder to be discharged through the open delivery pipe. It is therefore of non-pressure type.

The second type shown in Fig. 13.3 comprises in the one unit a feed tank in the top fed by cold water through a ball valve connected by an internal feed pipe to the base of the storage vessel. Upon opening one of the taps, hot water is displaced by the cold falling from the tank. The head being small, the flow rate is apt to be limited.

Units of the two types illustrated vary in size from 6 litres to about 130 litres, with loadings from 0·5 to 3 kW. Sizes of less than 30 litres are not adequate where a bath is to be supplied.

The pressure type of electric storage heater illustrated in Fig. 13.4 is more akin to a central system. It is fed by means of a tank, hot water being displaced upwards and out of the delivery pipe to the taps. The cylinder is thus under pressure. The length of draw-off, seeing that no circulation is possible, is limited by that permitted by the Water Authority.

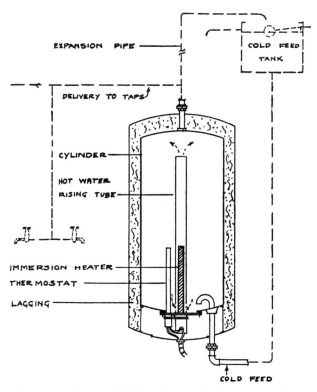

FIG. 13.4.—Electric Hot-Water Storage Heater, Pressure Type.

For use with off-peak supply under the 'white meter' tariff (see page 250), both the tank feed and the pressure type models are available in capacities up to 225 litres with loadings up to 6 kW. They are specially shaped to be tall and smaller than normal in diameter in order to preserve stratification of the hot and cold water. The Authors are aware of cases where families of up to five persons have been satisfactorily served by 200 litre tank-feed units.

CENTRAL SYSTEMS

A central system comprises a boiler or water heater of some form, coupled by circulating piping to a storage vessel, the combination of the two being so proportioned as to permit of all the normal demands for hot water being met upon the opening of the taps according to the type of usage. For instance, in a hospital there may be a continuous demand for hot water all day, and a comparatively small storage but a high heating-up rate may be suitable. In a sports pavilion, however, there may be only one sudden demand at the end of games, in which case a large storage and a small heating-up rate may be adequate.

The system may be *direct*, in which case the water drawn off is the same water as circulates through the boiler as shown in Fig. 13.5. If the

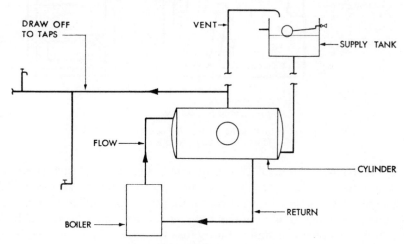

FIG. 13.5.—Direct Hot-Water Supply System.

water is hard, scale deposit may occur in the boiler and hence boilers for direct hot-water supply have to be of a simple type with large waterways and big clean-out openings. Fig. 13.6 is a sectional type and Fig. 13.7 a

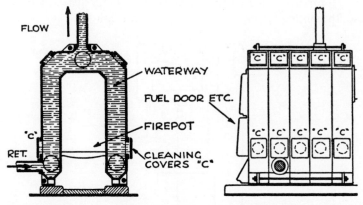

FIG. 13.6.—Typical Cast-Iron Sectional Hot-Water Supply Boiler. (Ratings up to 120 kW).

welded steel type. If, alternatively, the water is soft, direct boilers of iron or steel may cause discoloration of the water through rusting, and hence in the past such boilers have sometimes been made of copper or treated with an anti-corrosion treatment such as Bower-barffing, or one of the vitreous coatings.

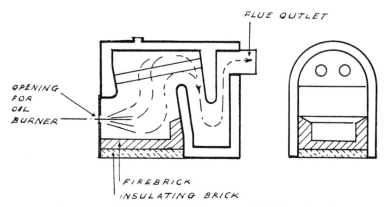

FIG. 13.7.—Oil-fired Boiler for Heating or Hot-Water Supply.
(Hartley and Sugden 'Oilex').

In the *indirect system* all this is avoided. Fig. 13.8 illustrates an in-direct system, in which any type of heating boiler may be used. The storage cylinder becomes indirect containing some form of heating surface, a common form being of annular type as shown, and within which the water from the boiler circulates. The water to be heated is contained in the vessel outside the indirect surface, and hence never comes in contact with the boiler. An indirect cylinder is alternatively termed a *Calorifier*.

In the indirect system scale formation is generally negligible as the surface temperature is not usually high enough to cause scale deposit. Although the use of copper for indirect cylinders was at one time confined to soft-water areas, this material is now frequently used in hard-water areas also, the piping system being in copper throughout.

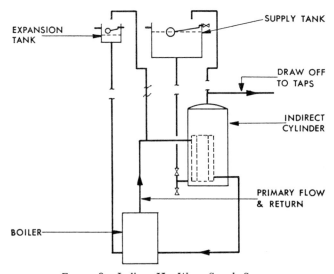

FIG. 13.8.—Indirect Hot-Water Supply System.

The boiler in such a system may be of much more efficient type than is possible with a direct type of boiler. It will be noted that the boiler, or primary system, requires a separate feed tank or cistern as in the case of any ordinary heating system.

Combined Systems—It has become common practice to combine the duties of heating and hot-water supply served from one boiler or group of boilers. This can only be done by using the indirect system, the boiler or boilers being run at a constant temperature with the heating served through a mixing valve at varying temperature, as discussed on page 180 (and see Fig. 9.7).

During summer when the heating is off, the boiler is of course over-sized—which, in the days when hand-fired solid fuel was usual, was considered unsatisfactory due to the likelihood of occasional overheating or boiling. With automatically-fired boilers burning gas or oil, this disadvantage does not arise and it is quite common to find a wide disparity of load successfully handled from one boiler, though there is clearly a limit. A large boiler running intermittently on a light load may involve corrosion troubles if oil-fired, and hence this practice is best confined to plant burning class D oil and to duties where the combined load is under 300 kW. With gas firing the question of corrosion does not arise, but a large gas boiler on light load tends to low efficiency.

Thus, where combined system loadings are above the limit suggested above, it is preferable to provide a small boiler for summer hot-water supply duty, but so arranged that, when the main boiler or boilers are in use during the heating season, they carry the hot-water supply load and the small boiler is out of commission. Such an arrangement is shown in Fig. 13.9.

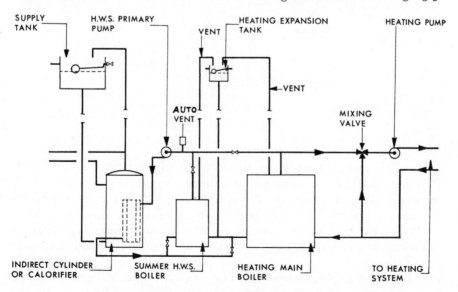

FIG. 13.9.—Combined Heating and Hot-Water Supply System.

The circulation from boiler to indirect cylinder may be by gravity, but it is worth considering the use of a pump, as shown in the Figure. Not only does this enable the size of piping connections to be reduced but, by thermostatic switching of the pump on and off, a simple means of temperature regulation of the hot-water supply is achieved.

Heat Supply—The heat supply to a central system may then be derived from the boiler of a central-heating system fired in any of the various known ways.

If using gas, the unit may be quite independent, as in Fig. 13.10. If using electricity on-peak, an immersion heater will be used, as in Fig.

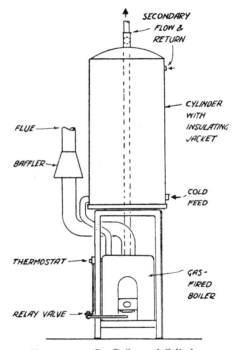

FIG. 13.10.—Gas Boiler and Cylinder.

13.11. This shows the cylinder served alternatively from a direct boiler. If using electricity off-peak either a large storage vessel will be used or, where hot-water thermal storage supplies all the heating needs, the hot-water supply will derive its heat through a calorifier, as in Fig. 13.12.

Where the heat source is steam or medium-pressure or high-pressure hot water, a calorifier will be required, types of which are discussed later in this chapter.

Hot-Water Supply Draw-off—The draw-off piping from a central system may comprise a series of dead legs as shown in Fig. 13.13, but the

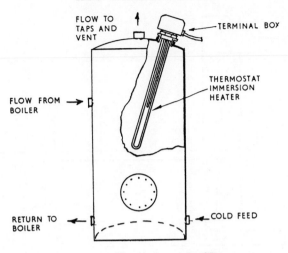

FIG. 13.11.—Electric Immersion Heater.

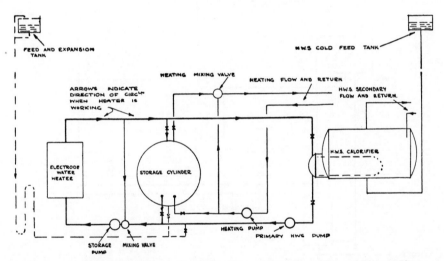

FIG. 13.12.—Diagram of Connection of Hot-Water System from Thermal Storage Plant.

runs are strictly limited by virtue of the fact than an undue time would have to pass before hot water came from the tap, in the case of a very long run. This causes wastage of water and Water Authorities commonly limit the length of dead-leg draw-off to about 8 metres. On anything other than small systems, therefore, it is necessary to form a secondary circulating system as illustrated in Fig. 13.14, such that the length of branches to any draw-off points is kept to a few feet and hot water is available almost instantly. Naturally, such circulations are insulated as there is a continuous loss of heat from them summer and winter. The secondary circulation may also serve towel rails, as indicated, and linen cupboard coils.

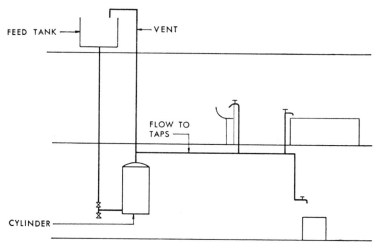

FIG. 13.13.—The Dead-Leg System.

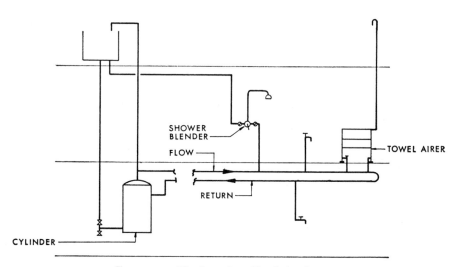

FIG. 13.14.—The Secondary-Circulation System.

In more extensive systems still, where a gravity secondary circulation is not practicable, a circulating pump is introduced as shown in Fig. 13.15. Alternatively, the pump may be in the secondary return.

Head Tank System—Fig. 13.16 introduces the old type of hot-water supply system, which is mentioned only with the object of drawing attention to its many defects, because it is still in existence.

In this system, the boiler is at the bottom and the hot-water tank at the top, and when no draw-off occurs, the water gradually rises from the

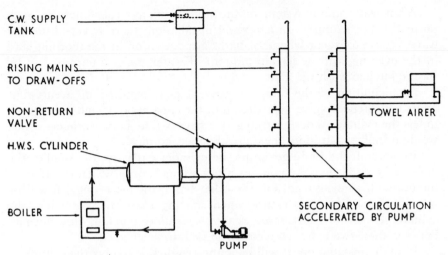

C.W. SUPPLY TANK

RISING MAINS TO DRAW-OFFS

NON-RETURN VALVE

H.W.S. CYLINDER

BOILER

TOWEL AIRER

SECONDARY CIRCULATION ACCELERATED BY PUMP

PUMP

FIG. 13.15.—Hot-Water System with Pumped Secondary Circulation.

boiler to the head-tank, which then acts in the same way as a cylinder, the level between the hot and the cold water gradually falling in the normal manner.

The circulating pipes between the two are, however, extremely long, so that the circulation is sluggish, and it is frequently necessary for the boiler to get up to boiling point before there is much circulation.

When a draw-off occurs, it is largely a matter of chance whether the water will be taken from the boiler or from the top tank, with the consequence that, even though the boiler has been alight for a considerable time, it may happen that a mixture of hot and cold water is delivered to any individual tap, since the cold water runs down the return pipe and passes the boiler so quickly as to benefit very little by such passage before reaching the draw-off point. It should be regarded as obsolete.

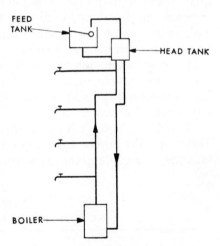

FEED TANK

HEAD TANK

BOILER

FIG. 13.16.—Old-type-Tank System (not recommended).

Capacity of Storage Cylinders—The decision as to the size of cylinder and boiler necessary depends on considerations quite different from those which apply in a heating system where, as a rule, a fairly constant quantity of heat has to be supplied, and where this quantity is susceptible of fairly accurate calculation.

A hot-water supply system, on the other hand, as a rule functions intermittently. For example, in a normal installation, there is very little hot water required except when hot baths are drawn off in the morning and in the evening, with a certain amount of water taken intermittently by basins, kitchen wash-up, etc.

A calculation of the total quantity of heat required in twenty-four hours will therefore give no criterion of the capacity of cylinder required, unless the cylinder is designed to give a twenty-four hour storage—which would usually result in its being grossly uneconomical.

In general, the more generous the cylinder capacity, the smaller the boiler power that may be used, as it has a long time in which to catch up the draw-off at peak load. On the other hand, the more sluggish will be the raising of the temperature when starting from cold, or when the cylinder temperature has for any reason been allowed to fall below normal. Between these two extremes a compromise is to be effected.

In many installations it will be found reasonable to give the cylinder a capacity equal to the maximum draw-off of hot water in any one hour at peak-load conditions, and the boiler may generally then be sized on a basis of heating this quantity of water up to the desired temperature in some longer time, depending on the installation. In many cases it will be adequate if this heating takes 2 to 3 hours, but where there is little draw-off between the peak-load conditions, this period may be further extended, and, on the contrary, where the supply approximates more to a continuous one, it may need to be shortened.

In most installations hot baths constitute the peak load for the hot-water supply system, especially in hotels and similar buildings. In any case, it is necessary to consider whether during such peak loads a supply is also required for kitchen, basins, etc.

Table 13.1 gives the capacities of various standard fittings, and Table 13.2 the approximate figures of consumption for various types of building, as a guide to the cylinder capacity.

TABLE 13.1

CAPACITY OF VARIOUS STANDARD FITTINGS

	Capacity in Litres	Temperature usually required, ° C.
Lavatory basin, normal filling - -	4	38°
,, ,, full - - - -	9	
Sink, normal - - - - -	20	60° to 65°
,, large - - - - - -	45	
Bath, average - - - - -	130	38°
Shower Bath, spray type - - -	0·1 litre/s	32°
,, ,, 6″–7″ rose type - - -	0·5 ,,	

TABLE 13.2

HOT-WATER CONSUMPTION IN VARIOUS TYPES OF BUILDING

Type of Building	Consumption per Day per Occupant in Litres (Water at 65° C.)	Peak Consumption per Hour per Occupant in Litres
School (Boarding) - - - -	90	18
Block of Flats - - - - -	110 to 160	45
Hotel - - - - - - -	90 to 130	45
Factory (excluding process work) - -	20	9
Block of Offices (including cleaning) -	22	9
Hospital (Infectious) - - - -	230	45
Hospital (Sick) - - - - -	160	30
Hospital (Mental) - - - -	110	22

As an example, in a hotel with 100 rooms, each having its own bathroom, taking the peak demand at 45 litres per hour from Table 13.2 it would be desirable to allow

$$100 \times 45 = 4500 \text{ litre at } 65° \text{ C.}$$

This is equivalent, when mixed with cold water at 10° C, to 9000 litres at 37·5° C or the equivalent of 66 baths taken during one hour.

Here it should be noted that the 4500 litres is the hot water at 65° actually required, but as there is bound to be some mixing with the entering cold supply, something should be added to arrive at a satisfactory storage capacity. It is usual to take the effective storage at 90 per cent of the actual to allow for this incomplete stratification. Thus in the example, the actual storage to be provided is

$$4500 \times \tfrac{100}{90} = 5000 \text{ litre.}$$

Public bathrooms not attached to a particular bedroom may, of course be used much more frequently in one hour, particularly when the number of bathrooms is small compared with the number of bedrooms. Special consideration is necessary in such cases.

This sort of assessment of storage capacity probably overstates the case. Records of hot-water demands taken in university halls of residence[*] show that the actual demands are much lower than those forecast by conventional methods. Thus a considerable diversity factor on storage capacity appears justifiable, and can best be determined by a careful study of possible load curves and records of comparable installations where they exist. There must, however, always remain some element of intelligent guesswork in arriving at storage capacity.

* See *I.H.V.E. Journal* (1964), Vol. 32, p. 353. The same no doubt applies to hotels, etc., generally.

Table 13.3 gives the capacities for flat-ended cylinders of various dimensions.

TABLE 13.3

CAPACITIES OF HOT-WATER SUPPLY CYLINDERS IN LITRES (ASSUMING FLAT ENDS)

Diameter metres	Length: metres					
	1	1·5	2	2·5	3	4
0·4	125	188				
0·5	196	296				
0·6	282	423	564			
0·7	385	580	770			
0·8	503	755	1006	1258		
0·9	636	956	1272	1592		
1·0	785	1180	1570	1965	2355	
1·2		2240	2262	3371	3393	
1·4		2300	3080	3850	4620	
1·6			4022	5040	6033	8044
1·8			5090	6360	7635	10 180
2·0			6286	7850	9429	12 572

Boiler Power Required—The boiler power is arrived at, as already stated, by assessing the allowable re-heating period. In addition to the heat required for raising the hot water to the required temperature, the losses due to radiation from boiler, mains, cylinder and fittings must be properly allowed for.

Taking the above example, and assuming a re-heating period of two hours, and radiation losses from the system of 10 kW the boiler power is estimated as follows:

Heat to raise water temperature in two hours

$$4500 \text{ litre} = 4500 \text{ kg (approx.)}$$

$$= \frac{4500 \times (65 - 10^\circ \text{ C.}) \times 4\cdot2}{2 \times 3600} = 144 \text{ kW}$$

Radiation   10 ,,

Net boiler power   154 ,,

HOT-WATER SUPPLY CALORIFIERS

Calorifiers for hot-water supply differ from those for heating in having a much larger shell so as to provide storage. The capacity of the storage is determined as in the case of cylinders already discussed, making due allowance for the space occupied by the heating surface. With any type of heat exchanging surface the element is best kept near the bottom of the shell so as to promote convection over as great a volume as possible. A vertical arrangement gives better stratification than a horizontal one.

Water-to-Water—For low-pressure hot water, where the storage capacity is below about 900 litres, an annular type element is commonly used, as shown in Fig. 13.17, being economical in cost. Calorifiers of this type are generally referred to as 'indirect cylinders'.

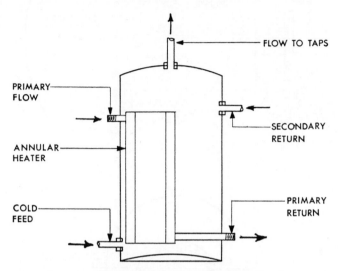

FIG. 13.17—Indirect Cylinder for Hot-Water Supply.

For larger capacities a tubular battery is used, as shown in Fig. 13.18. Means for withdrawal are provided for cleaning purposes.

A new form of indirect heater for water-heater heat transfer is shown in Fig. 13.19 and comprises an immersion heater having waterways so arranged as to give a high rate of transmission. This is specifically designed for domestic use in conjunction with a micro-bore system.

For medium-pressure and high-pressure hot water, the calorifier frequently takes the form shown in Fig. 13.20, in which the primary water follows a single path in order to maintain a high velocity which is necessary due to the relatively small amount of water in circulation. Tube withdrawal in this case necessitates sufficient height for the coil to be pulled out vertically.

Control of water-water calorifiers can be simply achieved by a direct-acting valve actuated by an expansion element located in the shell about one-third up from the bottom, or by electrical on-off control as shown in Fig. 13.21. Or control may be by on-off switching of the primary pump if such is included, as referred to earlier. In very small systems the primary is often left uncontrolled.

Steam-to-Water—The heating surface generally adopted is a tubular battery, as in Fig. 13.22. The steam, on condensing, collects in the lower part of the steam chest and passes thence to the steam trap. The battery is withdrawable for cleaning—hence the horizontal arrangement is pre-

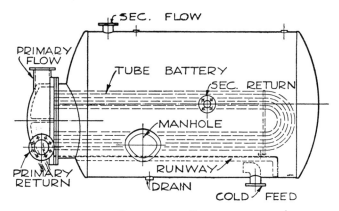

LONGITUDINAL SECTION.

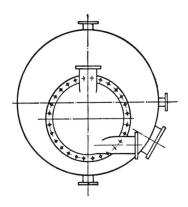

END ELEVATION.
FIG. 13.18.—Calorifier for Large Water-Water System.

ferred. Control may be by direct-acting 'Horne's' valve, as shown, or by various other means.

Centralised Distribution of Hot-Water Supply—Where a number of buildings are to be served from a central boilerhouse, or where a number of flats are to be served there are two methods of distribution which may be employed: *Centralised Secondary* and *Centralised Primary*. *Centralised Secondary* means that water is heated at the central point and is distributed through a system of circulating pipework serving all points directly—i.e., the water circulating is the water drawn off. This is simply an enlargement of the system shown in Fig. 13.14. *Centralised Primary* means that the water circulated or the steam supplied from the central heat station is passed into a number of calorifiers from which the hot-water supply is derived locally. Fig. 13.21 shows diagrammatically, such a system served from high-pressure hot water.

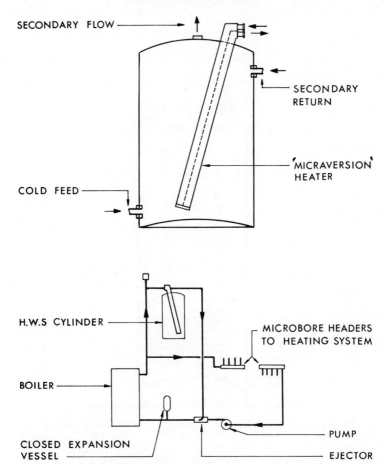

SECONDARY FLOW

SECONDARY RETURN

'MICRAVERSION' HEATER

COLD FEED

H.W.S CYLINDER

MICROBORE HEADERS TO HEATING SYSTEM

BOILER

CLOSED EXPANSION VESSEL

PUMP

EJECTOR

Fig. 13.19.—Above: Hot-Water Immersion Heater. Below: as used with a Micro-bore System (Wednesbury 'Micraversion').

The choice as to which is the best system in a certain set of circumstances depends on many factors such as size, water characteristics, space maintenance and cost. Further consideration to this matter will be given in Chapter 23 on *District Heating*.

Rating of Calorifiers—The heating surface required depends on the temperature difference between the primary water or steam and the secondary water, and on the transmission rate per degree difference.

Water-to-Water Calorifiers—Considering Fig. 13.17, for simplicity in calculation it is generally assumed that the primary temperatures of flow and return are suitably fixed at say 80° C flow and 70° C return. It is then also assumed that cold feed is entering continuously at some temperature such as 10° C and hot water is leaving also continuously at say 60° C. The mean temperatures are then: primary 75°, secondary 35°, which equals 40° C difference.

u

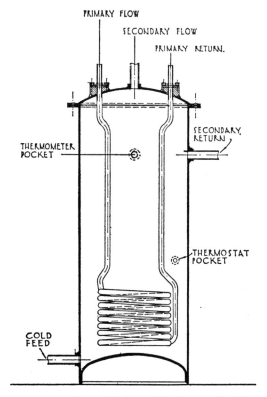

FIG. 13.20.—H.P.H.W. Vertical Storage Calorifier for Hot-Water Supply.

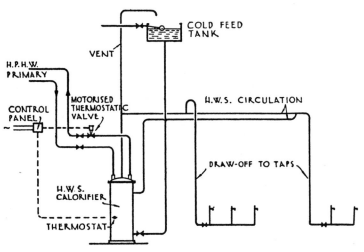

FIG. 13.21.—Typical arrangement of Local Hot-Water Supply System served from H.P.H.W.

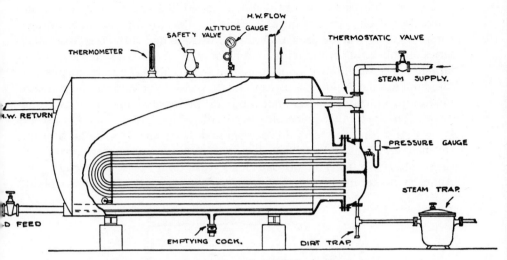

FIG. 13.22.—Hot-Water Supply Calorifier for Steam-Water System.

In practice, conditions of stability such as this may never exist. In periods of heavy draw-off, the whole cylinder may be emptied of hot water when the temperature difference will then be nearer to 65° C. As the cylinder warms up the difference will become less, until when fully heated the difference may be only about 15° C. In the former case considerable boiler power would be necessary to maintain the full rate of transmission at the increased temperature difference, and in the second case when the cylinder is warm, the heat requirement is much less. This shows that some means of control of the primary heat source is needed.

Steam-Water Calorifier—Where steam is used as the heating medium, the temperature of the steam will depend on its pressure (see Chapter 10 on *Steam Heating*, p. 220). The temperature of the water may be taken at the arithmetic mean of the secondary. The temperature difference is then multiplied by the coefficient per unit area of heating surface. This transfer rate, divided into the estimated duty, gives the area of surface required. Certain data for heating calorifiers were given in Table 10.4, page 223, based on the *IHVE Guide*, but these are not directly applicable to storage calorifiers where scaling has to be allowed for; also, water velocities on the secondary side are those due to natural convection only.

The *Guide* coefficients for heat transfer in storage calorifiers using 1 inch (25 mm) copper tube range from 400 to 1170 W/m²° C for water-to-water and 640 to 1180 W/m²° C for steam-to-water, depending on whether clean or fouled, on water velocity and temperature differences. But, due to variables involved, the size and spacing of tubes and types of design, it is best to consult the makers for ratings and not rely on theory only.

The Angelery Hot-Water Generator—This equipment (illustrated in Fig. 13.23), suitable for steam or hot-water as the primary medium, tackles the hot-water supply problem in a different manner from the conventional one. In certain cases it is possible to dispense with storage completely. It will be seen that the heat-exchange surface consists of a series of interlacing coils giving a high capacity in a small space and being largely self-de-scaling. It is not an instantaneous heater, but, by means of a 'fluid operated computer' and a special flow-regulator valve in the primary medium, the heater relates heat input to 'periodic' demand, so avoiding peaks. To apply this principle successfully, the form of load must be known and the size of unit determined from maker's data. Storage (termed an 'accumulator') is required: for instance where batch or surge loads occur. For normal load variations without storage, the Angelery heater in effect draws on the surplus capacity of the boiler plant.

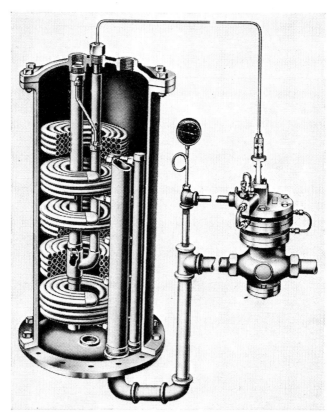

FIG. 13.23.—The Angelery Hot-Water Generator. (*By courtesy of British Steam Specialties Ltd.*)

MATERIALS FOR CYLINDERS, CALORIFIERS AND PIPING

Galvanized steel being cheaper than other materials was at one time most in use; however, copper piping is now preferred for secondary systems regardless of water characteristics. As mixed metals are undesirable in such a system due to probable electrolytic corrosion, the storage vessel must in this case also be of copper.

A galvanized cylinder with galvanized pipework is suitable with waters having a pH value as follows:

pH value	temporary hardness
7·3 - - - -	greater than 210 ppm
7·4 - - - -	150
7·5 - - - -	140
7·6 - - - -	110
7·7 - - - -	90
7·8 - - - -	80
7·9 and over - -	70

Copper for cylinders and pipework is necessary with waters having a pH value of 7·2 and under.

In hard water districts there is often a good case for water softening for the feed to the hot-water supply system in order to reduce soap consumption, remove calcium scale in baths, etc., avoid furring of heating surfaces and improve the washing characteristics of the water generally. This is particularly the case with laundries. Materials for the hot-water supply system will then be selected according to the resultant water condition.

FEED CISTERNS

Feed cisterns (often referred to as tanks) are required on hot-water supply systems, to supply the water drawn off from the taps, and in addition to allow for expansion as in the case of a heating system.

The sizing of the cistern depends to some extent on the supply of water available. If the pressure and flow are good, a one-hour storage under peak load conditions will probably suffice, but if poor, sufficient for two or three hours would be advisable.

The feed pipe from the cistern has to be connected to the cylinder so as to prevent mixing with hot water, thus preserving stratification. This may be achieved by finishing with a tee or bend inside the cylinder when entering through the bottom of the shell (see Fig. 13.18).

Indirect systems require a feed cistern to both boiler and calorifier circuits, and these have been shown in the diagrams already illustrated. In order to avoid the need for two cisterns, special indirect cylinders have been devised for domestic use in which the primary is fed from the secondary side. Obviously, with the inevitable expansion and contraction, this would lead to continual change of water in the boiler and heating system with great risk of furring-up or other hazards. Hence these units contain means to prevent such mixing—as, for instance, by an air seal. Within the limits they are intended to cover, they appear to be successful.

CHAPTER 14

Piping for Hot-Water Supply Systems

PIPE RUNS FOR HOT-WATER SUPPLY systems fall into three categories, which call for separate consideration. These are:

(a) Primary pipes, which have to circulate water steadily from the boiler to the cylinder or calorifier.

(b) Secondary pipes and feed pipe which are required to pass water spasmodically and in relatively large quantities when a tap is opened.

(c) Secondary circulating pipes, which in addition to (b), should circulate enough to make good their own heat-loss, and so keep hot water constantly available at the draw-off points.

Primary Circulation (Direct System)—The sizing of the pipes connecting boiler and cylinder on the direct system should be ample in order, first to allow for the furring which will inevitably take place in them if the water is hard, and, secondly, so that the frictional resistance may be so low that a brisk circulation takes place.

The sizes may be calculated in exactly the same manner as already described for a heating system operating by gravity circulation, taking the circulating head as the difference of pressure between the hot rising and cold falling columns from centre of boiler to centre of cylinder, with a temperature drop of, say, 30° C. This head, divided by the length of travel of flow and return, including bends, etc., will give the circulating pressure per metre run, from which the pipe size necessary may be determined from Fig. 8.6 (page 158). Alternatively, the sizes based on return 10° C, flow 40° C, height from centre of boiler to centre of cylinder 1·5 m, travel 6 m plus 4 bends, may be taken from the following Table:

TABLE 14.1
SIZE OF PRIMARY FLOW AND RETURN MAINS FOR 'DIRECT' SYSTEM

Size	kW flow
1¼ in. (32 mm)* - - - -	10
1½ ,, (40 ,,) - - - -	15
2 ,, (50 ,,) - - - -	30
2½ ,, (65 ,,) - - - -	50
3 ,, (80 ,,) - - - -	80

* These equivalents apply to steel. Copper differs slightly.

A size of 1 in. (25 mm) should be a minimum for small systems.

The maximum output of a hot-water boiler is required at periods of heavy draw-off—i.e., when the bottom of the cylinder is certain to be cold.

294

As the heating of water in the cylinder proceeds and the return to the boiler rises in temperature, the flow temperature will also rise, though not by the same amount since the circulating head at higher temperatures will be increased and more water will be passed. Thus, when the water is returning to the boiler at 60° C, an equal heat transmission will be taking place with the same pipe sizes with a flow at 80° C—that is to say, a 20° C rise as against 30° C. At this point, however, the output from the boiler should be reduced so as to meet the radiation losses only, otherwise the temperature will continue to mount, which, in most cases, would be unnecessary and wasteful.

This adjustment of the boiler output is to a small extent automatic, since, as the temperature of the system rises, the radiation losses will increase.

It will be seen from the above that the primary circulation is a constantly varying one, and as the maximum duty is required at the poorest circulating temperatures, great care should be exercised in seeing that the piping is kept as short as possible with easy radius bends to avoid undue resistance.

Primary Circulation (Indirect System)—With the indirect system, no allowance need be made in the sizing of the primary circulation for furring since the same water is constantly re-used as in a heating system.

A rapid circulation is essential, as the transmission of heat from the primary to the secondary water through the walls of the heat exchanger depends on a difference of temperature being maintained right through to the outlet.

The flow and return temperatures may be taken at higher figures than with the direct system; for example, with a combined heating and hot-water supply apparatus the primary water circulation will be the same as that supplied to the radiators in cold weather at, say, 80° C flow and 60° C return.

Where the indirect cylinder or calorifier is served from a boiler whose sole duty is to provide hot-water supply, it is of advantage to run this at as high a temperature as possible so as to economize in heat-exchange surface.

Based on a flow and return temperature of 80° C and 60° C respectively, a height from centre of boiler to centre of calorifier coils of 1·5 m,

TABLE 14.2
SIZE OF PRIMARY FLOW AND RETURN MAINS FOR 'INDIRECT' SYSTEM
Flow 80° C, Return 60° C

Size					kW flow
1 in. (25 mm)*	-	-	-	-	8
1¼ ,, (32 ,,)	-	-	-	-	12
1½ ,, (40 ,,)	-	-	-	-	20
2 ,, (50 ,,)	-	-	-	-	40
2½ ,, (65 ,,)	-	-	-	-	60
3 ,, (80 ,,)	-	-	-	-	100

* See footnote to Table 14.1.

and a travel of 6 m, plus 4 bends, plus equivalent of 3 m of pipe for resistance of indirect coil, the heat transmitted by the various pipe sizes is given in Table 14.2 (see p. 295).

SECONDARY SUPPLY

Outflow from Taps—The sizes of the pipes from the cold water storage cistern to the cylinder and from the cylinder to the taps depend on the outflow from the latter, and not in any way on the boiler load. The pressure available for delivering the water is the gravity head or distance between the lowest water level in the supply cistern and the topmost tap. Obviously the topmost must be taken as this is bound to be the worst case with the up-feed system.

This pressure must deliver the calculated quantity of water through the pipes, and as the length of the latter can be measured from the plans, making due allowance for bends, etc., as for a heating system, the available pressure per metre run of pipe may be calculated.

Sizing Secondary Flows—The rate of flow from any tap will depend on its size and on the pressure or head available at that point, and thus, in a building with various floors fed from a cistern on the roof, the flow rate will progressively increase going down the building. It would be an added complication to allow for this in calculating pipe sizes; furthermore, at high discharge rates there is liable to be considerable splashing, and throttling-down may be expected.

The *Guide* gives recommended rates of flow from fittings from which Table 14.3 gives adapted values. Rates for cold water supply are also given for convenience.

TABLE 14.3

APPROXIMATE DISCHARGE RATES FROM HOT AND COLD-WATER FITTINGS, AND SIZES

Fitting	Rate of flow litre/s		Usual Size	
	Hot	Cold	in	mm
Bath (private) - -	0·3	0·3	$\frac{3}{4}$	20
(public) - -	0·6	0·5	$\frac{3}{4}$ or 1	20/ 25
Basin - - - -	0·1	0·1	$\frac{1}{2}$	15
Basin with spray tap -	0·2	—	$\frac{3}{8}$	10
Shower nozzle - -	0·1	0·1	$\frac{1}{2}$	15
Shower 100 mm rose -	0·3	0·3	$\frac{1}{2}$	15
150 mm rose -	0·6	0·6	$\frac{1}{2}$	15
Sink - - - -	0·3	0·2	$\frac{3}{4}$	20

Where there are only a few taps to be served, as in a private house, it may be assumed that at some time all the taps may be open at one time, although it is unlikely.

With any sizeable installation there will be increasing diversity of demand the greater the number of taps. For instance, it may take three minutes to fill a bath, but at least twenty minutes before the next filling may occur, during which time other baths may be filling. A hot tap to a basin may be open for thirty seconds and at least another sixty seconds will elapse before the tap is opened again. In a row of ten basins this would mean a maximum of about three taps open simultaneously even if people were queuing up to use them.

The type of usage is important. A sports pavilion when play ends may have nearly all showers running simultaneously, but, in a hospital, it is probable that only a small proportion of baths would be filling at precisely the same time.

Various attempts have been made to establish a basis for evaluating simultaneous demand.* The *I.H.V.E. Guide*, 1965, postulates a method based on the theory of probability and adopts the conception of a 'demand unit'. This unit is taken as unity for a lavatory basin tap. The type of application is taken into account by weighting the units according to an assumed interval of use varying from five minutes to twenty minutes for a basin and from twenty minutes to eighty minutes for a bath. From this theory, a simplified comparative table (Table 14.4) is derived under three categories of application:

'congested' (where times of draw off are regulated or co-ordinated)
'public' (normal random usage)
'private' (infrequent or spasmodic)

TABLE 14.4
COMPARATIVE DEMAND UNITS (APPLIED)

Fitting		Service	Category of application		
			Congested	Public	Private
Lavatory Basin	-	Hot	1·0	0·5	0·3
		Cold	0·7	0·5	0·2
Bath - - -		Hot	5·0	2·0	1·1
		Cold	3·0	1·2	0·6
Shower nozzle	-	Hot	1·0		
		Cold	1·0		
rose -	-	Hot	3 to 6		
		Cold	3 to 6		
Sink - - -		Hot	6·0	3·0	1·7
		Cold	2·5	1·3	0·7

Having established progressive totals of 'demand units' throughout the system to be sized, the flow required may be read from Fig. 14.1 which is based on probability of simultaneous use.

* e.g. Burberry and Griffiths: *The Architects Journal*, 1962, pp. 1185–91.

Fittings such as wash fountains and washing-up machines requiring a continuous flow may be included additionally according to type and maker's flow requirement.

No method such as the above can be more than a guide—a certain amount of common-sense must be exercised in applying it. For instance, in

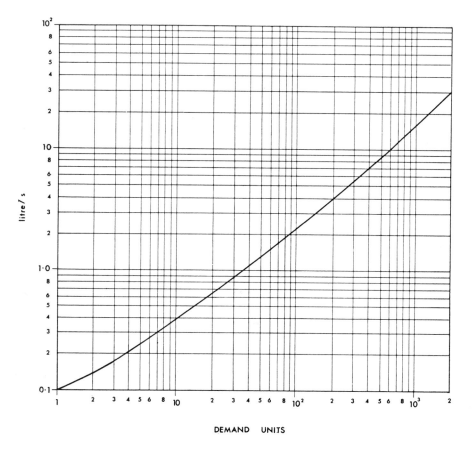

DEMAND   UNITS

Fig. 14.1—Flow Probability and Demand Units.

an hotel, cleaner's sinks on various floors are unlikely to be in use simultaneously with baths, but sinks in the kitchen may well be.

Example of Flow Pipe Sizing

To illustrate the method of sizing, consider Fig. 14.2 which is a diagram of a typical riser system serving four floors. From this the number of fittings and hence the demand units may be set down in tabular form, as shown below the Figure, finally arriving at the flow required in litre/s for each section of pipework.

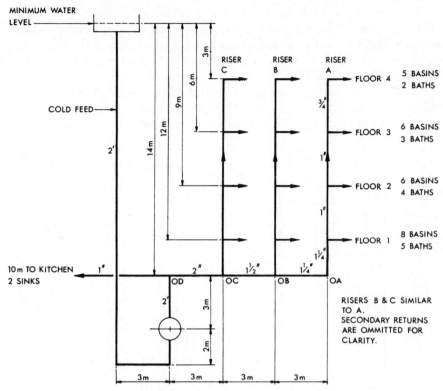

FIG. 14.2—Example of Pipe Sizing for Up-Feed System.

From Fig. 14.2 : *Summation of Demand Units* (*taken as 'public' category*)

Floor	Fittings	Demand Unit		Running total		Litre/s from Fig. 14.1
Riser A						
4	5 basins	0·5	2·5			
	2 baths	2·0	4·0	6·5		0·3
3	6 basins	0·5	3·0			
	3 baths	2·0	6·0	9·0		
			—		15·5	0·5
2	6 basins	0·5	3·0			
	4 baths	2·0	8·0	11·0		
			—		26·5	0·7
1	8 basins	0·5	4·0			
	5 baths	2·0	10·0	14·0		
			—		40·5	1·1
Main OA to OB				40·5		
Riser B as riser A				40·5		
OB to OC					81·0	2·0
Riser C as riser A				40·5		
OC to OD					121·5	2·6
2 kitchen sinks (assumed running simultaneously) @ 0·3 =						0·6
Cold feed and cylinder connection to OD						3·2

To proceed with the sizing: use may be made of the pipe-sizing charts (Fig. 8.6, page 158, or Fig. 14.4, page 306), but for greater convenience pressure drops should be based on the head of water, as it is the head from cistern to tap which causes flow. Also, flows expressed in volume per unit time, not kW, are required. Table 14.5 has been compiled in this form from I.H.V.E. data for water at 75° C in galvanised heavy-weight steel pipes, and Table 14.6 for copper pipes to B.S. 659.* The differences with medium-weight steel and

TABLE 14.5

FLOW OF WATER AT 75° C, LITRE/S, IN GALVANIZED STEEL PIPES TO B.S. 1387
(HEAVYWEIGHT)

Head loss m water per m run	Nominal Bore										
in.	½	¾	1	1¼	1½	2	2½	3	4	5	6
mm	15	20	25	32	40	50	65	80	100	125	150
0.01			0·1	0·4	0·6	1·2	2·4	3·8	7·8	14	22
0·015		0·1	·2	·5	0·8	1·5	3·0	4·7	9·6	17	27
0·02		·1	·2	·6	0·9	1·7	3·5	5·4	11·1	20	32
0·025		·1	·3	·6	1·0	1·9	3·9	6·1	12·5	22	36
0·03		·2	·3	·7	1·1	2·1	4·3	6·7	13·7	24	39
0·035		·2	·3	·8	1·2	2·3	4·6	7·2	14·8	26	42
0·04		·2	·3	·8	1·3	2·4	5·0	7·8	15·8	28	45
0·045		·2	·4	9	1·3	2·6	5·3	8·2	16·8	30	48
0·05		·2	·4	·9	1·4	2·7	5·6	8·7	17·7	32	51
0·06	0·1	·2	·4	1·0	1·5	3·0	6·1	9·5	19·4	34	56
0·07	·1	·3	·5	1·1	1·7	3·2	6·6	10·3	21·0	37	60
0·08	·1	·3	·5	1·2	1·8	3·5	7·1	11·1	22·4	40	64
0·09	·1	·3	·5	1·2	1·9	3·7	7·5	11·7	23·8	42	68
0·10	·1	·3	·6	1·3	2·0	3·9	7·9	12·4	25·1	45	72
0·15	·2	·4	·7	1·6	2·5	4·8	9·7	15·2	30·8	55	89
0·20	·2	·4	·8	1·9	2·9	5·5	11·3	17·5	35·6	63	103
0·25	·2	·5	·9	2·1	3·2	6·2	12·6	19·6	39·8	71	
0·3	·2	·5	1·0	2·3	3·5	6·8	13·8	21·5	43·6		
0·4	·3	·6	1·2	2·6	4·1	7·8	16·0	24·8			
0·5	·3	·7	1·3	3·0	4·6	8·8	17·9	27·8			
0·6	·3	·8	1·5	3·2	5·0	9·6	19·6				
0·7	·3	·8	1·6	3·5	5·4	10·4					
0·8	·4	·9	1·7	3·8	5·8	11·1					
0·9	·4	·9	1·8	4·0	6·1	11·8					
1·0	·4	1·0	1·9	4·2	6·5						
Equivalent lengths for K = 1 m (average)	0·4	0·6	0·8	1·1	1·4	1·9	2·7	3·4	4·7	6·2	7·8

Pipe fittings, K =
Bends, tees, reducers, enlargements—allow = 1·0
Screw down valve or tap   „   10·0
Connections to cylinder   „   1·0

* Since the density of water at 75° C is 975 kg/m³ and standard gravity is 9·808 m/s², the force exerted by a column of water one metre high (at that temperature) is 975 × 9·807 = 9560 kg/ms² (N/m²). By taking this as being approximately 10 000 (as in Tables 14.5 and 14.6), an error of only five per cent arises, which is insignificant in the context of the other variables existing.

TABLE 14.6

FLOW OF WATER AT 75° C, LITRE/S, IN COPPER PIPES TO B.S. 659

Head loss m water per m run	Nominal Bore										
in. mm	½ 13	¾ 19	1 25	1¼ 32	1½ 38	2 51	2½ 63	3 76	4 102	5 127	6 152
0·01		0·1	0·2	0·4	0·7	1·5	2·8	4·5	9·8	17	28
0·02		·1	·3	0·6	1·0	2·3	4·1	6·6	14·3	25	41
0·03		·2	·4	0.8	1·3	2·8	5·1	8·2	17·9	32	51
0·04	0·1	·2	·5	0·9	1·5	3·3	6·0	9·7	20·9	37	60
0·05	·1	·3	·6	1·0	1·7	3·8	6·8	10·9	23·6	42	68
0·06	·1	·3	·7	1·2	1·9	4·2	7·5	12·1	26·1	46	75
0·07	·1	·3	·7	1·3	2·1	4·5	8·1	13·1	28·4	50	81
0·08	·1	·3	·8	1·4	2·2	4·9	8·7	14·1	30·5	54	87
0·09	·1	·4	·8	1·5	2·4	5·2	9·3	15·1	32·5	58	93
0·10	·1	·4	·9	1·6	2·5	5·5	9·9	16·0	34·4	61	99
0·2		·2	·6	1·3	2·3	3·7	8·0	14·4	23·2		
0·3		·3	·7	1·7	2·9	4·7	10·1	18·0			
0·4		·3	·9	2·0	3·4	5·5	11·7				
0·5		·4	1·0	2·2	3·8	6·2					
Equivalent length for $K=1$ m (average)	0·6	1·0	1·5	2·0	2·6	3·5	4·6	6	8	11	14

Pipe Fittings
Values of K may be taken from Table 14.5

with copper to B.S. 3931 respectively are slight, but the accurate data are in the *Guide* if required.

Next, consider the pressure due to the head of water at the taps. The worst case is the top-floor tap on riser A which is subject to only 3 m head and has the longest travel.

Travel $T =$ cold feed $= 24$
allow for
fittings $\quad$ 5 $\qquad$ 29 m

main $\qquad$ 9
fittings $\quad$ 2 $\qquad$ 11 m

riser $\qquad$ 11
fittings $\quad$ 3 $\qquad$ 14 m

$\qquad$ 54 m

$$\frac{H}{T} = \frac{3}{54} = 0.055 \text{ m/m run}$$

Thus, taking the litre/s flow rates from the tabulated example, it is possible to set down the nearest pipe size for each section, reading from Tables 14.5 or 14.6 according to the material used for piping.

	Pipe Size. Inches	
Riser A	*Galvd.*	*Copper*
Floor 4		
	1	$\frac{3}{4}$
3		
	1	1
2		
	$1\frac{1}{4}$	$1\frac{1}{4}$
1		
	$1\frac{1}{2}$	$1\frac{1}{2}$
Main OB – OC	2	$1\frac{1}{2}$
OC – OD	2	2
Cold Feed and cylinder		
connection to OD	$2\frac{1}{2}$	2

It will be noted there is a slight saving in pipe size in this example by using copper.

The branches at each floor next require attention: floors 3, 2 and 1 on riser *A* each have a greater pressure head available. The loss on the index run to the point of branch may be taken out in each case and deducted from the head available. The surplus may then be used with a new H/T to size the branch. The same applies to risers *B* and *C* which have a shorter travel, even to the top floor. However, if the plant is of modest size, the time spent in further refinement of calculation may not result in much saving in cost and the designer may content himself with using one value of H/T throughout. On an extensive installation this could not be regarded as economical design.

Nevertheless, in this example the kitchen would be too glaringly over-sized if taken at the index H/T. It will in fact be as follows:

$$
\begin{array}{ll}
H & 14\,\text{m} \\[4pt]
T \quad 19+3+5+10= & 37\,\text{m} \\
\quad\quad \text{fittings} & 11 \text{ ,,} \\
\hline
& 48\,\text{m} \\[6pt]
\dfrac{H}{T} \quad\quad = & 0.29
\end{array}
$$

A flow of 0.6 litre/s requires $\frac{3}{4}$ in. galvanized pipe or 1 in. copper pipe. Had it been sized on the index pressure loss, the size would have been $1\frac{1}{4}$ in. in both cases.

With regard to showers, it should be borne in mind that a head of about 1·8 m is required for the spray to function. This should be allowed for by deduction from the total available.

Sizing Secondary Returns—Having established the sizes of the secondary flow pipes, it is a simple matter to calculate the heat emission from the circulating portions of these from Tables 7.2 and 7.3 (page 145), but for convenience Table 14.7 may be used. In this Table it is assumed that mean water temperature is 57° C and air is at 17° C—a difference of 40° C. Insulation has been taken as 25 mm thick.

TABLE 14.7

EMISSION FROM HOT-WATER SUPPLY SECONDARY
CIRCULATING PIPING

Based on 40° C difference mean water to air W/m run

Galvanised Steel				Copper to B.S. 659		
Nominal Bore		Bare	Insulated	Bore mm	Bare	Insulated
in	mm					
½	15	40	12	13	28	8
¾	20	48	14	19	39	11
1	25	58	16	25	48	13
1¼	32	71	19	32	58	16
1½	40	78	20	38	68	17
2	50	96	22	51	88	20
2½	65	120	28	63	106	24
3	80	140	31	76	120	26
4	100	170	37	102	160	35
5	125	200	44	127	190	42
6	150	230	50	152	220	47

Notes

Insulation is taken as 25 mm thick magnesia, or equivalent.
Emission from towel airer: average = 250 W.
Emission from linen cupboard coil: average = 250 to 500 W.

To this must be added the emission from linen cupboard coils and towel airers, average values for some examples being also given in the Table.

The secondary returns must also carry sufficient water for their own heat emission so that the temperature drop back to the cylinder does not exceed a predetermined maximum such as 10° C. Emissions from these returns must therefore be added to each section, and since their size is not yet known they may be assumed at, say, one or two pipe sizes less than the flow in each case, or taken at a rough approximation of two-thirds of the flow.

With the above totals marked on the plans for each branch or section of main it is possible to arrive at totals working back to the cylinder exactly as for a heating system.

The circulating pressure if by gravity may be obtained as described on p. 294, by determining the average height from the centre of cylinder to average point of heat emission. The appropriate circulating head per metre of height may be taken from Table 8.1 (page 169), allowing flow at say 65° C and return at say 55° C: the value in this case will be found to be 47·36 N/m² per metre height. Alternatively if a pump is necessary owing to long runs or mains below the cylinder level, the pump pressure may be taken at 20 000 N/m² overall for most systems, and 40 000 to 60 000 N/m² for large buildings or institutions with considerable distances between the blocks. The pump pressure necessary may be estimated by allowing a maximum of 10 N/m² per metre of travel, allowing the total run of return pipe, plus one-third of the run of flow pipe (on account of its larger size) plus 25 per cent allowance for bends and resistances.

Having determined the pressure, whether for gravity or pump circulation, the resistance of the flow mains (the sizes of which have already been determined) to the furthest point should be calculated as for a heating circuit. This figure, deducted from the total circulating pressure, will give the available pressure for the return mains. The latter divided by the metres run of return main, including allowance for bends and single resistances, will give the available pressure per metre.

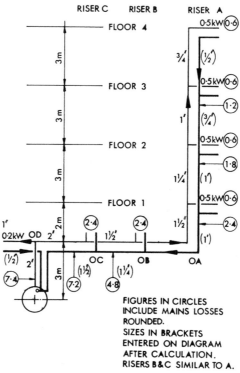

FIGURES IN CIRCLES
INCLUDE MAINS LOSSES
ROUNDED.
SIZES IN BRACKETS
ENTERED ON DIAGRAM
AFTER CALCULATION.
RISERS B & C SIMILAR TO A.

FIG. 14.3—Up-feed System: Return Sizing

Example of Sizing of Secondary Returns

Consider Fig. 14.3, which is the same system as Fig. 14.2 but with returns added.

The emission of heat from the flow piping, which has already been sized and is assumed as insulated copper, is as follows:

Riser			
$\frac{3}{4}''$	3 m	11 W/m	33 W
$1''$	3 ,,	13 ,,	39 ,,
$1\frac{1}{4}''$	3 ,,	16 ,,	48 ,,
$1\frac{1}{2}''$	2 ,,	17 ,,	34 ,,
			154 ,,

Three risers thus   =   462 W

Main $1\frac{1}{2}''$ 3 m   17 W/m   51 W

          $2''$ 9 ,,   20 ,,   180 ,,   231 ,,

                                                                        693 ,,

Taking emission of returns at 2/3rds of flow =   462 ,,

                                                                        1155 ,,

The branches serve one coil and one towel airer per floor and amount to 2 kW per riser, i.e. a total of 6 kW. Mains losses thus equal nearly 20 per cent of the total heat emitted, and it will be sufficiently near to apportion these losses equally. Thus we may now proceed to add the loads progressively, starting from the top with the 20 per cent included and entering the figures on the diagram as in the circles.

Assuming gravity circulation, the circulating pressure to floor 1 from the centre of the cylinder

$$= 5\text{ m} \times 47{\cdot}36\text{ N/m}^3 = 236{\cdot}80\text{ N/m}^2$$

and to floor 4

$$= 20\text{ m} \times 47{\cdot}36\text{ N/m}^3 = 947{\cdot}20\text{ N/m}^2$$

The pressure loss in the flow pipework to floor 1 may then be set out as below, dividing kW by the temperature drop of 10° C to give kW/°C—for which operation Fig. 14.4 should be used.   Fig. 14·4 is a pipe-sizing chart for copper tube which has been prepared by computer, like Fig. 8.6 (page 158). For galvanised steel, Fig. 8.6 may be used even though drawn for black steel, there being little practical difference.

Section	Size inches	Length including fittings inches	kW per °C	Pressure loss N/m²/m from Fig. 14.4	Pressure loss N/m²
Cylinder to D	2	5	0·74	2	10
D–C	2	5	0·72	2	10
C–B	$1\frac{1}{2}$	5	0·48	4	20
B—floor 1	$1\frac{1}{2}$	7	0·24	1	7
At floor 1	$\frac{3}{4}$	3	0·06	3	9
		—25			—56
Floor 1–2	$1\frac{1}{4}$	4	0·18	2	8
Floor 2–3	1	4	0·12	2	8
Floor 3–4	$\frac{3}{4}$	4	0·06	3	12
	$\frac{3}{4}$	3	0·06	3	9
		—15			—37
Total		40			93

Note:

    In arriving at the equivalent length for fittings, an arbitrary basis has been used for this example. In a practical case, equivalent lengths would be assessed as explained in Chapter 8.

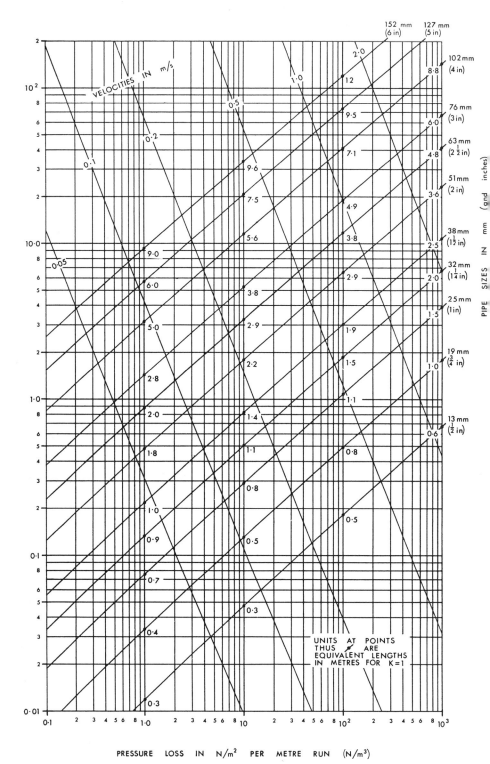

PRESSURE LOSS IN N/m² PER METRE RUN (N/m³)

FIG. 14.4—Pipe Sizing Chart for Copper to B.S. 659. (*Note:* this chart was prepared in advance of the issue of a British Standard for metric copper pipes—still pending.)

The pressure available to overcome friction loss in the return is (omitting decimals):

At floor 1	236 N/m²
less loss in flow	56 ,,

CP available	180 ,,
taking travel T for	
return as for flow = 25 m	

$$\frac{CP}{T} = \qquad 7 \text{ N/m}^2$$

at floor 4	947 N/m²
less loss in flow	93 ,,

CP available	854 ,,
travel	= 40 m

$$\frac{CP}{T} = \qquad 21 \text{ N/m}^2$$

It will be seen that the index circuit is that to floor 1. Reference to Fig. 14.4, using the same values of kW/° C as for the parallel sections of flow piping, now enables return sizes to be entered, taking $CP/T = 7$, as marked on the diagram in brackets. Fig. 8.6 would be used for galvanized steel.

For the remainder of the riser a separate calculation may be made for each floor, but, taking them as for floor 4 for purposes of this example, we use $CP/T = 21$ and again mark in the sizes.

Connections at each floor and the sub-division of the branches remain to be sized using the same method. The kitchen is again a special case having

$$CP \text{ available} = 3 \times 47 \cdot 36 \qquad = 142 \cdot 08 \text{ N/m}^2$$

loss in flow (0·02 kW/°C in			
a 1″ pipe)	= 0·1 N/m² × 12 =	1·2	N/m²
		140·88 N/m²	

travel	=	12 m

$$\text{thus } \frac{CP}{T} \qquad = 11 \text{ N/m}^2.$$

This results in a size less than half an inch which is, however, the minimum usable on account of possible deposits. In fact, this circuit will run at a lesser temperature drop, lower CP and a greater flow to balance. Some designers prefer three-quarters of an inch as a minimum size for reasons of mechanical strength and to allow for possible furring-up of the pipe bore.

Were this a pumped system, regulating valves in the returns would be desirable to enable any short branches such as this to be checked down.

Draw-off through returns—No account has been taken in the example

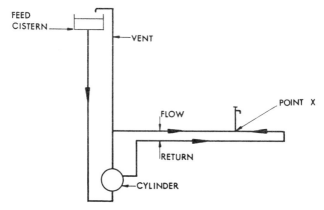

FIG. 14.5—Showing Flow of Water in Secondary Circulation
when tap is opened.

of water flowing to the taps from the return pipe. In Fig. 14.5, when a tap
is opened at X, water will obviously flow to it in both directions from the
cylinder, the water in the return pipe flowing in the opposite direction to
the usual owing to the circulation being completely overcome by the much
greater head from the tank to the tap.

The reduced frictional resistance brought about in the out-flowing
system, due to the double pipe, will vary inversely as the distance of any
tap from the cylinder. This effect is usually ignored when sizing pipes for
small systems, but on a large layout it may be taken into account.

DROP SYSTEM

A second method of secondary circulation is as shown in Fig. 14.6. In
this case the flow riser delivers direct to a hot-water head tank above the

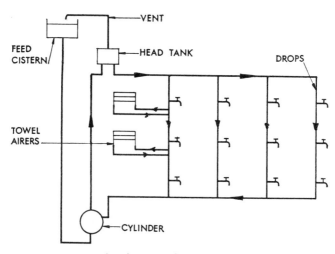

FIG. 14.6—Hot-Water Supply Drop System.

level of the topmost tap, and the supplies to the taps are taken from a system of drops connected from the bottom of the tank.

The advantage of this method over the up-feed in the case of high buildings, is that the pipe sizes are generally less. The upper floors are fed downwards from the top tank or partly downwards and partly upwards.

As a result a greater height (H) may be taken from the cold water supply cistern to the top tap than in the case of the up-feed system. An assumption that the top one-third of the drops are fed downwards is generally reasonable. The remaining two-thirds of the height is fed from the returns working in the reverse direction

The disadvantages of the drop system are:

(i) That it involves ranges of large piping at or near roof level, often difficult to accommodate.

(ii) That towel airers and linen cupboard coils connected to the drops operate by local gravity circulation only, so that there is not the same possibility of connecting them when at a distance from the drop.

UP-FEED HEAD TANK SYSTEM

In order to avoid mains at or near roof level necessary with the drop system, the up-feed method may be used with a separate hot-water head tank to each riser as shown in Fig. 14.7.

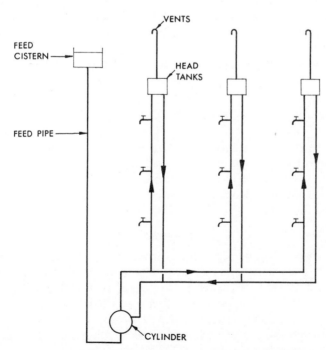

FIG. 14.7—Hot-Water Supply System with separate Head Tanks.

At periods of heavy draw-off on the lower floors of any particular riser, when the supply to the top floor might tend to be reduced, this tank will supply the riser downwards by drawing on its storage, so ensuring a good supply at all levels. As it is unlikely that the heavy draw-off continues for more than a minute or so, the tank will not have emptied before the demand ceases, when the level in the system will balance up again by water flowing up the riser in the normal way.

The sizing of this system may be simplified by ignoring the top one or two floors and sizing flow mains and risers for the discharge up to the next floor down. The Height H is thus greater than with a plain up-feed system and smaller pipes will result.

Dead Legs—These are the non-circulating branches to taps and it is desirable to keep them as short as possible. *Code of Practice* no. 342 (Centralized Domestic Hot-Water Supply) gives 7·6 m as the maximum permissible. Spray taps are a special case where the circulation should be taken almost up to the tap—a dead leg of no more than 0·6 m if possible—on account of the small flow.

Sizing of Head Tanks—The size of the head tank can be determined by allowing two or three minutes' capacity for the discharge which is assumed to come from it. Thus if 1 litre per second comes downwards in a drop system, and 2 litres upwards, a head tank of $1 \times 3 \times 60 = 180$ litres would be adequate. This storage should not be taken as reducing the main cylinder capacity.

Pump Sizing—In the case of an accelerated secondary circulation, the size of the pump may be determined from the emission in kW as for a heating system.

For example, if the total emission is 50 kW, and a 10° C drop has been assumed, the capacity of the pump will be

$$\frac{50}{10 \times 4·2} = 1·2 \text{ litre/s}$$

Allowing a margin, a suitable size pump would be 1·5 litre/s. The pressure of the pump will be arrived at as discussed on page 177.

As for a heating system, pumps of centrifugal type are most suited for the purpose. The pump body should be of a type readily opened for inspection and for removal of scale, and is preferably of gun-metal or bronze.

GENERAL

Vent and Feed Pipes—Vent pipes from the top of each riser carried above the feed cistern are necessary unless it can be so arranged that the topmost tap acts as a vent when it is opened. Frequently it is possible to accommodate only one open vent, other risers being vented by the taps. In the case of the head tank system, particular care must be taken to ensure that each tank is vented; otherwise a collapse could occur when water is drawn off.

It should be noted that the cold feed pipe has to furnish the whole supply, and an adequate size is essential. Even in the smallest installation 1 in. diameter (25 mm) should be the minimum size for this pipe, and is so called for in *Code of Practice* No. 342.

Insulation—With a hot-water supply system it is most desirable that every metre of circulating pipe should be insulated, as the losses are going on twenty-four hours a day summer and winter, and may mount to a considerable figure. The heat emitted from such pipes if not insulated is also objectionable from the point of view of the temperature in the building in hot weather.

Such insulation may take the form of any of the insulating coverings already referred to for heating systems. It is needless to say, of course, that the boiler and cylinder should also be adequately lagged, unless the latter is used for the warming of a linen cupboard in a small domestic system.

Valves—Valves on hot-water supply systems often give trouble due to fur or scale collecting on the faces, which in time renders them useless for shutting off the water. For this reason plug cocks are much to be preferred, especially if of the lubricated type. These can be relied upon to shut off tight even with the most severe internal encrustation, though approval may be required by certain Water Authorities.

Showers—Blending valves for showers are either hand-operated or thermostatic. It is a great advantage in both cases for the pressures of cold and hot to be roughly equal. This is achieved by arranging for the cold feed to the blender to be fed from the same tank as the hot-water supply, preferably from a separate down-feed so that the pressure is not affected by other draw-off points.

CHAPTER 15

Running Costs of Heating Systems

THE TYPE OF FUEL FOR a heating system will be determined by cost, cleanliness, convenience and availability (see page 65).

Cost depends on quantity of heat required and price per unit quantity. The amount of heat required for any given plant and conditions can be arrived at with a fair degree of accuracy, but will naturally vary from year to year according to the weather. Efficiency and ratio of period of utilization to total hours will affect the comparison as between one fuel and another. The price per unit quantity of heat will generally be the determining factor.

Cost includes other items, and is made up of

(1) Fuel consumption (referred to above).
(2) Running of auxiliaries.
(3) Labour.
(4) Maintenance.
(5) Interest and depreciation.
(6) Insurance.

These will be dealt with in detail later.

Cleanliness and convenience are equally important, yet it is often difficult to give a money value to them. Conditions of cleanliness which might be quite satisfactory, for instance, in a factory or workshop would probably be intolerable in a bank or office building.

The fuels and methods of firing come in the following order, the least cleanly and convenient being placed first:

Hand-fired solid fuel.
Automatic coal stokers.
Magazine-fired boilers.
Oil firing.
Gas firing.
Electric thermal storage.
Electric direct heating.

FUEL CONSUMPTION

The quantity of fuel required for the warming of a building depends on

(a) heat losses;
(b) heat content per unit of measurement;
(c) average working efficiency over the season;

312

(*d*) period of use;

(*e*) proportion of full load operation.

(*a*) **The Heat Losses** as calculated for design purposes will be in excess of those to be used as a basis for an estimate of consumption. This is due to the fact that if one considers any building, heating design must be such that on *each* aspect sufficient warmth can be provided to maintain a satisfactory temperature. In practice, air infiltration resulting from wind will occur only on the windward side; other aspects, i.e. the leeward side will ex-filtrate.

Furthermore, adventitious heat gains from occupants, lighting, solar heat and sundry other sources all contribute to the sum total of heat supply.

Based on a variety of evidence, a fair assessment of heat input requirement may be made on the basis of:

(1) Heat losses as normally calculated in kW but deducting 20 per cent for air change not applying in all parts simultaneously.

(2) A deduction of 3° C for incidental heat gains, equivalent to 5 per cent averaged over 24 hours.

(3) The sum of the above, being only a rough approximation, may be taken at a deduction of 25 per cent from the calculated heat losses.

Any losses from mains in a central system must be taken additionally to the building heat losses.

(*b*) **Heat Content** is a clearly established figure for each fuel per unit of weight or volume and has already been discussed. Solid fuels are sold by weight, which often includes a quantity of free moisture. Oil is sold by volume, gas by heat content and electricity by energy content which is equivalent to heat.

(*c*) **Average Working Efficiency over the Period**—This is the most controversial point of all. Test figures can be produced which, while probably true for the conditions under which the test was made, are misleading for ordinary working—due to boiler heating surfaces not being always clean, excess air owing to lack of adjustment, periods of light load working, intermittent working and so on. It is to be expected that large boilers will be run more efficiently than small. Table 15.1 (p. 314) is an attempt to postulate the kind of efficiencies which may be expected, assuming modern plant and controls. 'Large' installations for this purpose are taken as 500 kW and over. The Table also gives the derived fuel consumption.

(*d*) **Period of Use**—Here we have to assess the occupancy of the building and the length of time the heating system must be at work. The necessity for reheating the air, before occupation, is allowed for by the preheating period mentioned below. For a light construction some drop in temperature of the fabric may occur at night-time, but less heat is required to restore the temperature in the morning.

TABLE 15.1

CENTRAL HEATING SYSTEMS
PROBABLE EFFICIENCIES WITH VARIOUS FUELS

	Test	Allow	
		large	small
	per cent	per cent	per cent
(a) Solid fuel boiler hand-fired - -	70	55	50
(b) Solid automatic feed - - -	75	60	55
(c) Gas boiler - - - - -	85	70	65
(d) Oil-fired boiler - - - -	80	65	60
(e) Electric thermal storage - - -	—	95	—

Fuel	Heat content per unit of measurement	Derived heat sent out per unit of measurement	
		Large	Small
(a) Solid fuel, hand-fired - - -	30 000 kJ/kg	16 500 kJ/kg	15 000 kJ/kg
(b) Solid fuel, automatic feed - -	30 000 kJ/kg	18 000 kJ/kg	16 500 kJ/kg
(c) Gas boiler - - - - -	1 MJ/MJ	0·7 MJ/MJ	0·65 MJ/MJ
(d) Oil, Class D - - - - -	45 000 kJ/kg × 0·835 s.g. = 37 500 kJ/kg	24 200 kJ/kg	22 500 kJ/kg
(e) Electric thermal storage - - -	—	0·95 kWh/kW	—

The usage may be resolved into periods per year, per week, and per day.* For the yearly use a figure of thirty weeks is commonly assumed, i.e. from the end of September to mid-May.

The weekly use depends on the kind of building, whether seven days a week, five (when week-ends are omitted) or only one or two days.

The daily use depends on the type of building. It will be appreciated that very few are heated continuously for twenty-four hours per day. The shorter the period of heating each day the greater will be the no-load losses.

In other words, with solid fuel (hand stoked or magazine fed) the boilers will be banked a part of the day and all night with little result in the building. On opening up in the morning, however, less preheating is necessary after night banking, as the system will all be warm and the building will not have cooled off so much. The whole of the heat at night is not therefore wasted. With automatic stokers with kindling control, a somewhat similar condition exists.

On the other hand, if the system is oil firing, gas or electricity, no night running may be necessary (except in severe weather), but the preheating in the morning will be slightly longer. The assumption will be that the method of firing may be classed in two groups:

Continuous Firing, using solid fuel by hand, from magazines or automatic stokers.

Intermittent Firing, electricity, gas or oil fuel.

* In pure SI, the time scale in seconds should be as follows: one hour = $3·6 \times 10^3$s, one day = $8·7 \times 10^4$s, one week = $6·1 \times 10^5$s and one year = $3·18 \times 10^7$s. These being impossible to understand or visualise, this book prefers to stick to the time scale of history which everyone knows.

In addition, those buildings which are not occupied at week-ends will call for week-end banking, or for complete relighting early on Monday mornings, which will often take as much fuel as banking, with less desirable results. Whichever method is adopted there will be a longer period of preheating necessary on Monday mornings than on other days for these cases.

It is suggested that the 'period of use' may be divided into about six categories as follows:

(1) *Continuous Heating.* Hospitals, three-shift factories, police stations.

(2) *Continuous with Reduction at Night-Time.* Houses, flats, hotels, boarding schools, universities.

(3) *Daytime Heating.* Offices, public buildings, shops, factories (one shift).

(4) *Part Daytime Heating.* Day schools, cinemas, factories, clubs.

(5) *Occasional*—two days per week. Churches with mid-week meetings. Public halls used two days per week.

(6) *Occasional*—one day per week. Churches, Sunday schools, sports accommodation.

Many more intermediate divisions are no doubt possible, but these cover the majority of cases.

The allowance for preheating for the various categories, in terms of hours equivalent full use, may be taken as in Table 15.2.

(*e*) **Proportion of Full-Load Operation**—This will be a variable factor depending on the weather. Obviously no system will be called upon to operate at 100 per cent output (based on $-1°$ C outside) during the whole season. What proportion of this full load can be assumed for the purpose of calculation, and how does it vary for different parts of the country?

A basis of comparison is afforded by the degree-day method. The British degree-day assumes that in a building maintained at $18°$ C no

TABLE 15.2

PREHEATING AND BANKING PERIODS

Type of Heat Requirement (See above)	Daily Preheating, Hours	Continuous Firing			Intermittent Firing. Additional Preheating in Morning, Hours
		Hours Banked at Night, at 15 Per Cent	Extra Monday Morning Pre-heating, Hours	Hours Banked at Week-end, at 15 Per Cent	
1	—	—	—	—	—
2	1	24 hours, less occupation period, less preheating	—	Total hours unoccupied, less preheating	
3	2		2		1 hour per day of use
4	3		2		
5	6		—	—	
6	8		—	—	

heating is required when the external temperature is 15° or over. The difference between the daily mean temperature and 15° is then taken as the number of degree days for the day in question. Thus, for an outside daily mean temperature of 5° C there are 10 degree days. The total degree days for the year forms a basis of comparison between the heat requirements of one place and another and the fuel consumption of one year with another.

It should not be regarded as an absolute unit, nor should it be pressed beyond its usefulness for comparative purposes. For instance, the effect of low night temperatures may give exaggerated degree-day readings even though the buildings may not be heated at night.

TABLE 15.3

DEGREE DAYS AND WEATHER FACTORS FOR GREAT BRITAIN

District	Degree days C 20 year average 1947–67	Weather Factor
Thames Valley - - -	1970	0·58
South Western - - -	2260	0·67
Southern - - - -	1970	0·58
Western - - - -	1750	0·52
Severn Valley - - -	2090	0·62
Midland - - - -	2220	0·66
Lancashire - - - -	2240	0·67
North Western - - -	2260	0·68
North Eastern - - -	2420	0·72
Yorkshire - - - -	2060	0·61
Eastern - - - -	2190	0·65
South East Scotland - -	2350	0·70
West Scotland - - -	2490	0·74
East Scotland - - -	2470	0·73

If it is assumed that full load on a heating system takes place when the outside is at $-1°$ C, the maximum degree days may be taken as 16 in any one day. For a heating season of 30 weeks, i.e. 210 days, there are possible 3360 degree-days.

The Gas Council (Coke Department) publish monthly and annual degree-day records for various parts of the country based on Meteorological Office records, and from these may be deduced a factor, taking 3360 as unity.

This factor we herein term the *weather factor*. It may not be an accurate figure, for the reason that the degree-day basis itself is not entirely trustworthy;* also the published figures may be for a longer period than 210 days. The factor may however be used as a guide to proportion of full-load operation and for comparison of one part of the country with another.

Table 15.3 quotes the 20 year average degree-days as published by the Gas Council, and from these are derived the weather factors as stated.

Meteorological Office records may be obtained in respect of any particular town, if desired, and the degree-days calculated therefrom.

* N. S. Billington. H.V.R.A. Report No. 35: Oct., 1966. *Some Aspects of Heating Design.*

TABLE 15.4

EQUIVALENT HOURS PER ANNUM FULL USE FOR VARIOUS HEAT REQUIREMENTS

Type of Heat Requirement	Hours of Use	Number of Hours per Day	Daily Preheating in Hours	Days per Week	Total Hours per Week A	Hours Banked at Night at 15%	Days per Week	Equivalent Total Hours per week B	Extra Monday Morning Preheating in Hours C	Week-end Banking at 15% in Hours	Equivalent Total Hours per week D
1. Continuous	—	24	—	7	168	—	—	—	—	—	—
2. Continuous except night-time	8 a.m.–10 p.m.	14	1	7	105	9	7	9½	—	—	—
3. Daytime, excluding week-ends	9 a.m.–6 p.m.	9	2	5½*	60½	13	5	9¾	2	40	6
4. Part daytime	9 a.m.–4 p.m.	7	3	5½*	55	14	5	10½	2	39	6
5. Occasional, 2 days per week	10 a.m.–8 p.m.	10	6	2	32	14†	2	4¼	—	—	—
6. Occasional, 1 day per week	10 a.m.–8 p.m.	10	8	1	18	14‡	1	2	—	—	—

* The half-day is presumed to be until 1 p.m.
† For use at 7 a.m. on Sunday, preheating would commence at 2 p.m. on Saturday and would go on until 6 p.m., followed by 14 hours banking, with a further 2 hours preheating before occupation commenced. This complete cycle occurs twice a week.
‡ A similar cycle to No. 5, but starting at noon on Saturday and occurring once a week.

TABLE 15.4—*continued*

Type of Heat Requirement	Continuous Firing							Intermittent Firing			
	Hours per Week Useful Heat. Columns A+C E	Weather Factor	Column E × Weather Factor	Equivalent Hours Banking per Week —Columns B+D F	Total Equivalent Hours per Week	Number of Weeks' Use per Annum	Equivalent Hours per Annum	Total as Column E	Add for Extra Preheating due to Absence of Banking, 1 Hour per Day of Use	Total Hours per Week G	Column G × Weather Factor (0·6) × 30 Weeks = Equivalent Hours per Annum
1. Continuous	168	0·6	101	—	101	30	3030	168	—	168	3030
2. Continuous except night-time	105	0·6	63	9½	72½	30	2175	105	7	112	2016
3. Daytime, excluding week-ends	62½	0·6	37½	15¾	53¼	30	1597	62½	6	68½	1233
4. Part daytime	57	0·6	34	16½	50½	30	1515	57	6	63	1134
5. Occasional, 2 days per week	32	0·6	19	4¼	23¼	30	697	32	2	34	612
6. Occasional, 1 day per week	18	0·6	11	2	13	30	390	18	1	19	342

It is proposed in the calculations which follow to take a factor of 0·6 as the proportion of full load use or 'weather factor'. For other parts of the country this would be increased or reduced in direct proportion to the number of degree-days.

Hours of Use per Annum—Table 15.4 attempts to combine all the foregoing conclusions in a simple form.

The first part of the Table shows the equivalent hours firing per week which would be required for useful heating (i.e. during the occupation period and for preheating) and for banking, assuming that a boiler can be kept alight at 15 per cent of its normal consumption. These figures are all on the basis of ' − 1° C' weather conditions, i.e. weather factor = unity.

The second part of the Table gives the hours firing per annum (30 weeks' heating season) with a mean weather factor of 0·6. It will be noted that this factor has not been applied to the figures for night banking, since this loss is constant at the 15 per cent given, no matter what the weather may be.

It is admitted that there is, in a Table like this, plenty of room for argument as to what allowances should properly be made, and it is put forward solely for what is believed to be average conditions. There can be nothing final or dogmatic about it, especially when divorced from any particular job in hand. It does, however, give a rational method for estimating the various factors.

Annual Heat Consumption—The heat requirement per annum is now simply a matter of arithmetic:

if heat loss of system in kW, as adjusted
(see page 313), plus mains losses, if any   = H
if equivalent hours of corrected full load use
per annum   = E
if heat sent out per unit quantity of fuel
measurement (Table 15.1)   = F
if fuel consumption per annum   = A

$$\text{Then} \quad A = \frac{H \times E \times 3600}{F}$$

Example: An office block (5½ day week)
Heat loss for − 1° C adjusted   = 600 kW
Equivalent hours of corrected full load
continuous firing   = 1597 h.p.a.
intermittent firing   = 1233 h.p.a.
Heat contents, taken from Table 15.1, for *large* installations:

(*b*) Solid fuel auto-fired
$$\frac{600 \times 1597 \times 3600}{18\,000} = 192\,000 \text{ kg} \qquad = 192 \text{ tonnes}$$

(*c*) Gas boiler
$$\frac{600 \times 1233 \times 3600}{1000 \times 0\cdot7} = 3\,800\,000 \text{ MJ} \qquad = 3800 \text{ GJ}$$

(*d*) Oil-fired
$$\frac{600 \times 1233 \times 3600}{24\,200} \qquad = 110\,000 \text{ litres}$$

(e) Electric hot-water thermal storage

$$\frac{600 \times 1233}{0\cdot95} = 770\ 000\ \text{units}$$

If these are priced at the figures shown below, the comparison is as follows:

Solid fuel @£8/tonne =£1536
Gas   @   50p/GJ =£1900
Oil   @ 1·4p/litre =£1540
Electricity @0·33p/unit =£2570

Electric Floor Heating—Heating with this system must be assumed to be virtually continuous. Allowing for the weather factor from Table 15.3 there should thus be 3030 hours of operation. But temperatures are not constant and published consumptions vary from about 1500 kWh/kW of heat loss to about 3000 for office blocks. Taking say 2500 as a mean, the annual consumption for the example above would be

$$600 \times 2500 = 1\ 500\ 000\ \text{units}.$$

If it is necessary to compare this with the central systems previously dealt with, it will be noted that electrical thermal storage therein considered amounted to a consumption of 770 000 units, or just over half of the floor-heating system, the difference being due to the night losses of the latter.

Block Heaters Uncontrolled—This system avoids any question of floor loss and in practice consumptions appear to be lower than for floor warmed systems by 10 to 20 per cent. Results may not, however, be considered comparable.

Storage Fan Heaters—Consumption here is similar to a continuously-fired central system. A small amount of heat is emitted throughout the 24 hours and, when required by thermostat or clock switch, heat is liberated which in turn has to be made good during the night. Fundamentally, the estimation of consumption is no different from that for a central system and the same processes may be followed.

Fuel Consumption for Hot-Water Supply—The consumption of heat for a central system may best be assessed by estimating:

(a) heat losses from storage vessel(s), circulating piping, towel airers, linen cupboard coils, drying coils, etc.,

(b) actual hot water drawn off per day by the occupants.

It is worthy of mention that components (a) and (b) are often equal. A knowledge of the type of building will decide whether (a) is continuous for 24 hours per day, as in a hospital, or for some lesser period such as 8 hours per day in a school. Similarly the days per annum will vary.

The heat consumption of (b) taken from cold at say 10° C to hot at say 60° C, will be derived from data of water consumptions of comparable buildings.

The total of (a) and (b) will then form the basis of a sum in which calorific value of fuel and efficiency are taken into account as for heating.

RUNNING OF AUXILIARIES

A further item in running costs is the electric current consumed by motors driving blowers, and other auxiliaries, for automatically fired boilers. This applies to oil firing, automatic coal stokers and to gas boilers with forced air.

As a rule, for the lower range of boiler sizes up to say 1000 kW, the cost of the current is not a significant item and is ignored.

For an oil-fired boiler of 1000 kW burning oil requiring preheating,

the consumption is about 0.03 kg/s of oil;

the temperature rise (say 10° to 80° C) is 70° C;

the specific heat capacity of oil class E or F is approximately 2kJ/kg.

Thus

the heat required $= 0.03 \times 70 \times 2$	$= 4.2$ kW
the current needed for the boiler	$= 2.0$ kW, approx.
the current for the induced draught fan (if any) $=$	3.0 kW „
the current for the controls	$= 1.0$ kW
Total per hour	$= 10.2$ kW

If the running hours are 1500 per annum, the consumption involves 15 300 units which, at 1p per unit, will cost £153 per annum. This will be found to be about five per cent of the cost of the fuel.

A forced-air gas-fired boiler of the same size will involve only the burner power and the controls, i.e. about one third of the above.

Similarly, the cost of current for the circulating pumps is relatively of small magnitude and indeed most of the energy paid for comes out as heat somewhere in the system and so is not lost. A system of 1000 kW boiler load, as above, might involve a pumping power of 2 kW, the pump running continuously for the heating season (say 5000 hours). The consumption would thus be

$$5000 \times 2 = 10\ 000 \text{ units,}$$

which, at 1p per unit, amounts to £100 per annum, or of the order of three per cent of the cost of the fuel.

The above examples may serve to illustrate a method of estimating these costs; but, for any particular case, the makers lists will provide actual power requirements from which consumptions may be calculated.

LABOUR

The cost and scarcity of labour for manual work being such as it is, all new heating plant is assumed to be fully automatic. Thus the old estimates based on hand-stoking, or even the filling of hoppers and the clearing of ash, are no longer relevant.

Below a certain size, the occasional attention required for oil-fired, gas-fired or electric systems is so small as to be no more than a very part-time occupation, and it cannot be given a value. What that size is it is difficult to say, as there comes a point, such as in a hospital or a factory, where boilermen must be employed anyway, even though their duties are more in the nature of shift engineers or technicians than involving actual labour. The cleaning of boilers, keeping the log, turning pumps off and on and minor maintenance jobs, such as oiling and greasing, are still required. This may be worked on a three-shift basis which with weekends and holidays involves four men with a possible fifth as stand-in.

Costs for this scale of staff are then bound up with the whole staff structure of the enterprise and become but a tiny fraction of the overall running expenses. In other words, the old methods of evaluating the labour required to run such and such a system as against another one are no longer the criteria that matters. Other broader aspects of management enter in which are beyond our present scope.

However, if costing must be applied to heating apparatus, an assessment can be made at current rates of remuneration for the kind of attention mentioned above.

MAINTENANCE

Average maintenance of small installations is often covered by a maintenance contract with the installer or fuel supplier, involving two or three visits a year and costing a nominal sum per visit.

In the medium range of installations it is now common practice to employ one of the firms specialising in the maintenance of heating plant— which covers the cleaning of boilers, adjustments of controls, lubrication, checking burner equipment and operation, the cleaning of calorifiers and fan convectors (if any) and general overhaul. Annual charges will vary according to the size of the installation. In the larger scale plants such as are dealt with by hospital authorities, estate companies and industrial concerns, maintenance staff is part of the organisation, probably operating on a planned maintenance basis.

Taken by and large, maintenance of heating plant may be expected to be covered by an allowance of the order of ten per cent of the fuel cost per annum.

Interest and Depreciation—In making comparisons between one system and another, it is often necessary to take depreciation into account. In certain plant, such as boilers, the combustion equipment may be given a life of fifteen to twenty years, rotating equipment such as pumps twenty to thirty years, and static equipment such as pipes and radiators fifty to sixty years. A sinking fund may be established over these periods to cover replacement.

Interest on capital expended on the basis of a diminishing principle should take into account builders' work in addition to engineering work, such as for a flue where an alternative system would require no flue. The same applies to space for plant and fuel storage. The rate of interest in the comparison will of course depend on the current market rate or Government loan sanction rates where these apply.

Insurance—Insurance and annual inspection of steam boilers and other pressure vessels is mandatory, but this does not apply to low-pressure hot-water apparatus. Nevertheless, in any sizeable system it is customary for the owner to insure as a matter of self-protection, and also to cover against fracture, burn-outs and accidents of all kinds.

DIRECT HEATING SYSTEMS

If these are of convective type, the same method of calculation of the fuel consumption may be followed as for central systems—inserting the appropriate efficiencies, running hours and weather factor, assuming thermostatic control. There will, of course, be no mains loss for inclusion.

If the system is to be of luminous radiant type, a true comparison with a convection system is hardly possible unless based on globe thermometer readings. An assessment of running cost can best be made by estimating the time when the heaters are likely to be on and multiplying by their rated consumption. There is usually no true control of temperature with such systems. Data from comparable installations is probably the best guide.

CHAPTER 16

Ventilation

THE WORD *Ventilation* MEANS LITERALLY the causing of air movement or wind (L. *Ventus*), but has acquired the meaning of a system which gives a regulated supply of air to an enclosed space so as to make the conditions better for human habitation.

The purpose of ventilation is fundamentally to supply the untainted air necessary for human existence, because life depends on a constant supply of oxygen. The supply of this air involves the removal of a corresponding volume of expired or vitiated air: with the smells and noxious gases which are associated with concentrations of people.

The normal adult at rest inhales about 0.5 m³ of air per hour (0.14 litre/s). Of this about five per cent is absorbed as oxygen by the lungs. The exhaled breath gains from three to four per cent of carbon dioxide (CO_2), equal to about 0.02 m³ per hour (0.005 litre/s). Thus, if it is desired to keep the concentration of CO_2 down to say 10 parts per 10 000, with external air containing say 3.5 parts per 10 000, it will be necessary to supply

$$0.02 \times \frac{10\ 000}{10 - 3.5} = 30 \text{ m}^3 \text{ air per hour (8.3 litre/s) per occupant.}$$

In addition, the human body at normal temperatures at rest gives out about 50 grammes of water vapour per hour and 100 W (0.1 kW)* of sensible heat.

The 30 m³ per occupant, as above, will become warmed and humidified as follows:

$$\text{Temperature rise} = \frac{0.1 \times 3600}{30 \times 1.22\dagger} = 10° \text{ C assuming no heat loss from room.}$$

$$\text{Increase in humidity} = \frac{50}{30} = 1.7 \text{ grammes moisture per m}^3.$$

It has long been accepted that the CO_2 content is not a satisfactory criterion of good ventilation. The CO_2 content may be as much as 50 parts per 10 000 and yet the air feel fresh and pleasant, or it may be under 10 and yet feel stuffy owing to other more important effects.

To state what is a satisfactory criterion is more difficult. Freshness of air in a room may be judged by the sense of smell, an analytical method

---

* This varies greatly with temperature.   See Fig. 17.17.
† Volumetric specific heat capacity of air at 20° C.   See Table 1.3.

more delicate than any laboratory one. Freshness or stagnation are in addition judged by the rate of loss of heat and moisture from the skin, on air-speed, and by dust content.

The interplay of air temperature, air movement, and radiation have been referred to under Equivalent Temperature (p. 34). Where heating to a room is from a radiant source, the air temperature requires to be lower and the air movement greater than in a room with the same equivalent temperature, but with air and surfaces all at a uniform temperature, for the same degree of comfort.

The inter-relation of all these factors, as depicted in Fig. 16.1, is very complicated, and nothing short of an exhaustive investigation of each one

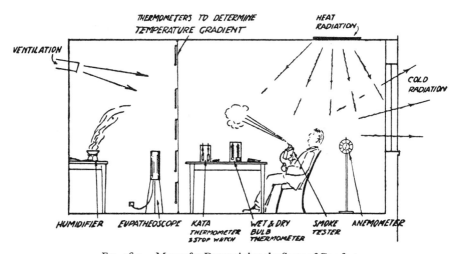

FIG. 16.1.—Means for Determining the State of Comfort.

will suffice to determine the precise conditions, and even with such knowledge it may be difficult to say whether a room is well ventilated or not. Yet, as is well known, a person (particularly of the female sex) entering the same room will be able to say at once, 'This room is stuffy' or 'This room is draughty'. After a time, due to acclimatization, the same room may appear to have become more tolerable.

The practice of ventilation is made somewhat more indefinite by the lack of an easy method of detecting slight air currents. A speed of between 0·2 and 0·5 m per second is generally regarded as the maximum which can be allowed without a feeling of draught but a lower level will be necessary if the air stream impinges on the back of the neck.

Such relatively small velocities can be measured either by the Kata Thermometer or by the Hot Wire Anemometer, in which air flow is measured by means of the cooling effect on a heated wire. These do not, however, indicate direction, for which some visual detector is necessary, such as a cold smoke produced by the mixture of ammonia and hydrochloric acid fumes, or titanium tetrachloride. Smoke, however, tends to diffuse rapidly and is useful only as a local index.

Good ventilation cannot, therefore, be defined in simple terms, and reference has to be made to conditions which have been found in practice to give reasonably satisfactory results. These may be discussed under the following headings:

> Volume of air necessary.
>
> Distribution and air movement.
>
> Temperature.
>
> Humidity.
>
> Purity.

VOLUME OF AIR NECESSARY

(a) **Temperature Rise Basis**—Where the occupancy is known the temperature rise can be estimated as with *Air-Conditioning* (see Chapter 17). If a rise of, say, 8° C is accepted as reasonable and 0·1 kW is assumed to be emitted per occupant, the volume of air required will be

$$\frac{0\cdot1 \times 3600}{8 \times 1\cdot22} = 36 \text{ m}^3 \text{ per hour (10 litre/s) per occupant.}$$

This assumes no building heat losses or gains other than from occupants.

It frequently happens that in crowded rooms much heat is also released from electric lamps, motors, cooking appliances, plant, and other apparatus. The heat from these can generally be calculated, and an increased air-change is needed to carry this extra heat away (see p. 37).

In such cases the air per person per hour may greatly exceed 36 m³ per hour (10 litre/s).

For summer a higher rate may be necessary, as there may be heat gains from the sun and a higher inlet air temperature; but it will be seen from a consideration of Chapter 17 that the removal of this heat gain without cooled air calls for very large ventilation volumes, which may be undesirable from other points of view and unnecessarily high for winter.

(b) **Legislation Basis**—For theatres and music halls, including all places licensed for music and dancing, boxing, etc., a minimum of 28 m³ of fresh air per hour (7·8 litre/s) per occupant is required by most licensing authorities.

Factories where manual labour is employed for gain are governed by the current Factories Act, 1961. This does not, however, lay down numerical standards. H.M. Factories Inspectors interpret these requirements according to each separate case. Six changes per hour may be considered an average for normal conditions, but there are many exceptions such as where noxious vapours, steam, dust, heat and fumes occur. Orders made under the Act lay down requirements for special industries. In the case of textile factories special requirements cover the humidity permissible. Offices, shops and railway premises are subject to the Act of this name of 1963. A recent *Department of Trade and Industry Note* (No. 19) recommends a minimum fresh-air rate of 4·7 litre/s per head.

(c) **Air-change Basis**—Where the occupancy is unknown or variable, an arbitrary basis must be taken. Table 16.1 may be used as a guide.

TABLE 16.1

VENTILATION ON AIR-CHANGE BASIS

Type of Room	Air-changes per Hour
Offices above ground - - - - - -	2–6
Offices below ground - - - - - -	10–20
Factories, large open type - - - - -	1–4
Factories and workrooms closely occupied - -	6–8
Workshops with unhealthy fumes - - - -	20–30
Laundries, dye-houses, spinning mills - - -	10–20
Kitchens above ground - - - - -	20–40
Kitchens below ground - - - - -	40–60
Lavatories - - - - - - - -	6–12
Boilerhouses and engine rooms - - - -	10–15
Foundries, with exhaust plant; rolling mills - -	8–10
Foundries, without separate exhaust plant - -	10–20
Laboratories - - - - - - -	10–12
Hospital operating rooms - - - - -	20
Hospital treatment rooms - - - - -	10
Restaurants - - - - - - -	10–15
Smoking rooms - - - - - - -	10–15
Stores, strong-rooms - - - - - -	1–2
Assembly halls - - - - - - -	3–6
School class rooms - - - - - -	3–4
Living rooms - - - - - - -	1–2
Sleeping rooms - - - - - - -	1
Entrance halls and corridors - - - -	3–4
Libraries - - - - - - - -	2–4

(d) **Effect of Cubic Content on Ventilation Rate**—The cubic content per occupant and length of time occupied obviously have an important effect on the rate of air supply necessary. Thus, in a large hall of 15 000 m³ capacity, seating 500 persons, each will have 30 m³ of air already stored in the building. If occupied for three hours and the desired rate of supply is say 30 m³ per hour per occupant, the actual fresh air needed is only

$$500 \times 30 - \frac{15\ 000}{3} = 10\ 000 \text{ m}^3 \text{ per hour.}$$

= 20 m³ per hour per occupant instead of 30.

(e) **Odours**—One of the essentials of good ventilation is the removal of odours arising from human occupation. The problem only becomes serious in crowded places. A supply of fresh air at the rate of 17 m³ per hour (5 litre/s) per person is found to be the minimum to obviate trouble from this source. With workpeople in their dirty working-clothes, the figure may need to be 40 to 50 m³ per hour (10 to 15 litre/s).

(*f*) **Condensation**—In certain special circumstances the need to combat condensation can be the criterion for determination of the ventilation rate. One such instance is that of the swimming-pool hall where, of the requirements to supply air to occupants, to dissipate the characteristic 'chemical smell' and to combat condensation on glazing, the last is by far the most important factor. It has been shown* that, with double glazing, the fresh-air supply rate should be not less than 0·02 m³/s (20 litre/s) per m² of wetted surface. In this context 'wetted surface' is defined as 1·2 times the water surface of the pool or pools in order to allow for splash-over etc. onto the pool surrounds.

DISTRIBUTION AND AIR MOVEMENT

The admission of fresh air to a room should be such that:

(*a*) It is evenly diffused over the whole area at the breathing level.
(*b*) It should not strike directly on the occupants.
(*c*) It should give a feeling of air movement and prevent stagnant pockets.

Various methods of approaching this end are discussed in Chapter 18.

The extraction of vitiated air from the room has little directional effect apart from a general upward, downward or crosswise motion.

Distribution often determines the air volume to be circulated. Though a small quantity of fresh air may suffice in the case of a large room sparsely populated, it may be impossible to diffuse this evenly throughout and avoid stagnation. The volume must then be increased, either all as fresh air, or by re-circulating room air mixed with the small quantity of fresh air.

Similarly, in a small crowded room, distribution difficulties may prevent admission of the necessary air-change rate without draught, and the volume then has to be reduced with unavoidable rise of temperature unless full air-conditioning with cooling is employed.

Distribution limits, consistent with the maintenance of comfort, appear to be about 3 air-changes per hour minimum, and 20 air-changes per hour maximum where mechanical means are employed. With natural inlet, lower rates than 3 will, of course, be possible and often sufficient, but distribution can then be only a matter of chance.

TEMPERATURE

The temperature of the air admitted must be not too much below that of the room, or the air will fall to the floor without proper mixing and cause cold draughts. The air must not on the other hand be too warm, or it will rise to the ceiling without adequate distribution.

In a 'straight' ventilation system, which is here under discussion, it is

---

* See Doe, Gura and Martin, *I.H.V.E. Journal*, Dec. 1967.

assumed that the warming of the building is accomplished by a separate radiator or other direct heating system to supply the heat necessary to balance the heat losses in winter, say to 20° C. Under these conditions the air supply will preferably be about 3° C lower than the desired room temperature and, when occupied, the leaving air will be of room temperature.

Under summer conditions without provision for air-cooling, no control can be exercised over the inlet temperature except by the small reduction possible by the process of evaporation, referred to later.

HUMIDITY

The fresh ventilation air admitted from outside will have the same moisture content as outside. In cold weather this may result, as has been shown earlier, in an unduly low relative humidity after heating. This may be adjusted, where mechanical inlet is provided, by humidification by means of water sprays or other device for adding water vapour, and the amount of moisture added may be controlled to give the desirable relative humidity of about 40 to 70 per cent at room temperature.

In normal mild weather the outside humidity in Great Britain will generally be satisfactory without alteration. The human body is not critical of humidity variations at normal temperatures.

There is, further, a reservoir effect caused by the hygroscopic contents of buildings, such as fabric, wood, paper, etc., which steadies up humidity changes, tending to equalize them over long periods.

In summer, when the outside humidity is high, there can be no control of the inside humidity without full air-conditioning which includes means for de-humidification. An air washer or other humidifying device, unless supplied with chilled water, can only *add* moisture, but due to evaporation may cool the air to a temperature approaching the wet-bulb.

Thus, with outside air at, say, 24° C dry bulb and 18° C wet bulb (relative humidity 55 per cent), a single bank air washer may be able to reduce the temperature to 19° C, at the same time increasing the relative humidity to 90 per cent. This final condition may be more oppressive than the initial, and it is indeed often found that in hot weather, air washers, where provided, are shut off for this reason. Reference to Fig. 2.8 will show that the above initial condition is just inside the summer comfort zone, but the latter condition is well outside it.

AIR PURITY

Most buildings requiring special ventilation are in the centre of towns where atmospheric pollution is highest. Such pollution is produced partly by the burning of coal and other fuels, and consists of soot, tar, ash and sulphur dioxide and partly by road traffic causing dust and fumes.

When air is introduced into a building for ventilation a proportion of solids will be brought in also. If the inlet is natural, nothing satisfactory can be done to prevent this, but where mechanical inlet is provided, filters and washers may be employed to remove the greater part. The various devices available are discussed in Chapter 18.

Vitiated air may be re-vitalized by removal of deleterious products in exhaled breaths by using activated carbon filters, the same also serving to reduce the content of sulphur dioxide and other noxious gases. But such filters are costly and would be used only in special cases.

METHODS OF VENTILATION

Ventilation may be:

Natural, in which air movement is induced by effect of temperature difference or wind.

Mechanical, in which air movement is caused by power-driven fan or fans.

Inlet and extract have to be considered separately.

There are thus four possibilities.

	Inlet	Extract
(1)	Natural.	Natural.
(2)	Natural.	Mechanical.
(3)	Mechanical.	Natural.
(4)	Mechanical.	Mechanical.

(1) **Natural Inlet and Extract**—This applies to most rooms and buildings with low occupancy, such as living rooms, bedrooms, small offices, schools, hospitals, shops, etc., and even then is dependent on clean outside air and conditions which permit open windows.

If an open fireplace or gas fire with flue is provided, the flue serves the double purpose of carrying away the products of combustion and of exhausting air from the room (Fig. 16.2). This air is replaced by cold air

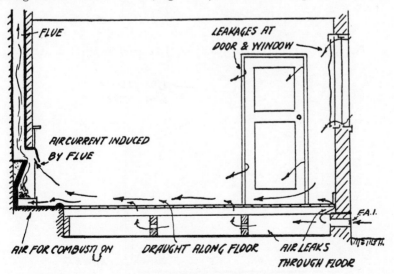

FIG. 16.2.—Ventilation by Open Fire.

drawn through cracks around doors and windows, between floor-boards, etc., thus causing ventilation of the room, though often at an unnecessarily high rate causing draughts. A restriction of flue throat is advised in order to reduce this loss.

Various methods of achieving natural ventilation without relying on opening of windows have been devised, one of which is shown in Fig. 16.3.

For larger rooms, assembly halls, factories, etc., roof ventilators combined with fresh-air inlets behind radiators (Fig. 16.4) provide a cheap solution, but the operation is spasmodic and unreliable, depending partly on temperature difference and partly on wind. Thus, in hot still weather

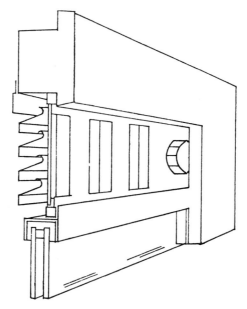

Fig. 16.3.—Natural Window Ventilation (J. Gerrard and Sons Ltd. M/C).

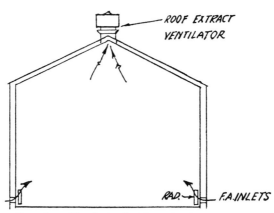

Fig. 16.4.—Ventilation by Natural Inlet and Extract.

when ventilation is perhaps needed most, no appreciable air flow occurs. In freezing weather with high winds the fresh-air inlets cause draughts, and are frequently found permanently closed.

Consider the warm air inside a flue with cold air outside as forming two legs of a U-tube.

Outside temperature (absolute) T_0

Inside   ,,   ,, T_1

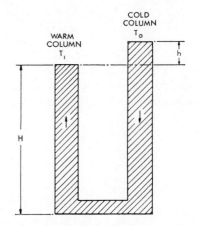

The density of air at the two temperatures is proportional to the absolute temperatures.

If the height of column T_1 is H, then the cold column, being denser, will in terms of the density of T_1 be equivalent to H plus some increased height h such that

$$h = H \cdot \frac{T_1 - T_0}{T_1} .$$

This difference in the height of the two columns causing flow may be substituted in the usual formula

$$P = \tfrac{1}{2}\rho v^2$$

where P is a pressure difference in N/m^2 and ρ is the specific mass in kg/m^3. By transposition,

$$\frac{P}{\rho} = \frac{N}{m^2} \times \frac{m^3}{kg} = \frac{kg\,m}{m^2\,s^2} \times \frac{m^3}{kg} = \frac{m^2}{s^2}$$

and thus

$$h = \frac{v^2}{2g}$$

where g is acceleration due to gravity ($9 \cdot 81 m/s^2$), v is in m/s and h is in metres.

Then $$v = \sqrt{2 \cdot g \cdot H \cdot \frac{T_1 - T_0}{T_0}} .$$

This applies to any flue in which it is assumed atmospheric pressure is exerted at the top. In the case of a building with inlet at floor and outlet at ceiling, it is generally considered that a neutral zone at atmospheric pressure exists about halfway between floor and ceiling; the lower half is at an increasingly negative pressure from the neutral zone downwards, thus causing inward flow, and the upper half at an increasingly positive pressure from the neutral zone upwards to the ceiling, causing outward flow.

The top of the imaginary U-tube must therefore in this case be at the zone of atmospheric pressure, which is at a level of $\dfrac{H}{2}$.

Then

$$v = \sqrt{2g \cdot \frac{H}{2} \cdot \frac{T_1 - T_0}{T_0}}$$

$$= \sqrt{g \cdot H \cdot \frac{T_1 - T_0}{T_0}} .$$

Table 16.2, calculated for an outside temperature of 5° C., gives the theoretical velocity in metres per second for various heights between inlet and outlet and inside temperatures.

TABLE 16.2

Indoor Temp. °C	Theoretical velocity in m/s for following heights between inlet and outlet								
	2 m	4 m	6 m	8 m	10 m	15 m	20 m	25 m	30 m
10·0	0·60	0·84	1·03	1·19	1·33	1·63	1·88	2·10	2·30
12·5	0·73	1·03	1·26	1·46	1·63	1·99	2·30	2·57	2·82
15·0	0·84	1·19	1·46	1·68	1·88	2·30	2·66	2·97	3·26
17·5	0·94	1·33	1·63	1·88	2·10	2·57	2·97	3·32	3·63
20·0	1·03	1·46	1·78	2·06	2·30	2·82	3·26	3·63	3·99
22·5	1·11	1·57	1·93	2·22	2·49	3·04	3·52	3·92	4·30
25·0	1·18	1·68	2·06	2·38	2·66	3·26	3·76	4·21	4·60
27·5	1·26	1·78	2·18	2·52	2·82	3·45	3·99	4·46	4·88
30·0	1·32	1·88	2·30	2·66	2·97	3·64	4·20	4·70	5·15

The theoretical velocity is not obtained in practice, due to resistance to air flow. The practical flow is commonly taken at ½ to ⅔ of the theoretical. The effect of wind is to increase the velocity, but this depends on the type of roof outlet. Most makers give factors of performance for the various forms of roof-ventilating appliances.

Natural Ventilation (Industrial)—There are a great many heavy industries where natural ventilation can be supplied successfully to single-storey shed-type constructions. These industries are those in which there is a considerable release of heat or steam, such as foundries, rolling mills, welding shops, plastics processing shops and dye-houses. The quantity of heat to be removed in such cases can usually be derived from the known input of energy per unit of time in the form of gas, electricity, oil, coal or steam as well as that from the sun. A temperature rise of the order of 2° C per metre of height of building may be assumed (taking as a maximum 25° C rise).

On this basis, the air quantity required to be moved may be assessed, which, divided by velocity, will give the area of opening necessary. The velocity in this case will be the practical velocity—i.e. the theoretical velocity from Table 16.2 reduced by the factor of performance of the ventilating unit. It will be noted that this will hold good for one outside temperature only (namely 5° C) and that for other outside temperatures the velocity should be recalculated. It will be obvious that the higher the outside temperature the lower the velocity obtainable by thermal motive force, unless the inside temperature is allowed to rise in proportion.

One form of roof ventilator for industrial use is the well-known Robertson's, as shown in Fig. 16.5.

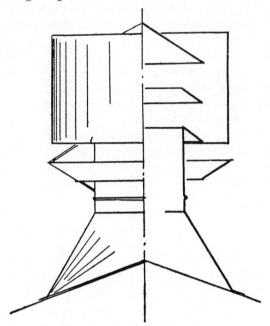

FIG. 16.5.—Robertson's Ventilator.

A more versatile form is the Colt M.F. (multi-function) shown in Fig. 16.6. This is rectangular, giving an opening from about 350 to 600 mm in width and from about 1250 to 2200 mm in length. The side dampers may be moveable so that they may be closed when extraction is not required. The weather cap also may be openable, giving a completely unobstructed outlet as shown in the right-hand diagram.

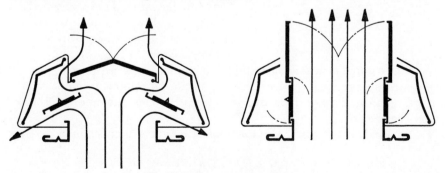

FIG. 16.6.—Colt M.F. Type Roof Ventilator.

Another form of Colt roof ventilator consists of a series of openable louvres (as shown in Plate XVII, facing p. 354) giving a very large area of aperture and a view of the sky which may have a psychological benefit. Both this and the M.F. type, when open, may of course admit rain or snow but, where there is a considerable up-draught of hot air, such may not reach the shop floor or it may be evaporated before it does so. Controls are available, however, to close the louvres almost completely, acting through a rain-sensing device.

The layout of roof vents is largely a matter of common sense—for instance, where a high concentration of heat or fumes is likely to occur, the vents should be closely grouped over this area.

Fire Venting—A development of free-opening roof vents of the types just described is in their use as fire vents. These may be made to open automatically in the event of fire, so allowing smoke to be cleared rapidly and thus preventing it from spreading to other areas. This is a valuable aid to fire-fighting where smoke is generally the principal hazard. The principles underlying this application are covered in research papers prepared by the Joint Fire Research Organisation.*

Air Inlets—Air inlets behind radiators have been referred to earlier (page 330). The free area of inlet gratings should be the same as the extract, or based on a velocity of 1·5 m/s. Convectors may similarly be fitted with fresh-air inlets, but the heater elements are liable to become quickly clogged with airborne dirt. Fan convectors are better for the purpose as they may be fitted with air filters.

Natural air inlets on an industrial scale often present a serious problem. In heavy industries, particularly where there is much radiant heat, wall openings at low level fitted with louvres may be allowable. These may be controllable and will match in size the area of roof outlet. One form of panel louvre is shown in Fig. 16.7.

Fig. 16.7.—Inlet Panel Louvre.

Where some means of warming the incoming air is required to prevent cold air impinging on the working population, mechanical inlet becomes necessary to overcome the resistance of heating elements and is discussed later (page 341).

(2) **Natural Inlet and Mechanical Extract**—A mechanical-extract

* *Investigations into the Flow of Hot Gases in Roof Venting* (Fire Research Technical Paper No. 7); also *Design of Roof Venting Systems* (Fire Research Technical Paper No. 10). Both are published by H.M.S.O.

system will function irrespective of wind and temperature differences, and is positive in action. Owing to the negative pressure set up in the space there is a tendency to inward leakage rather than outward, thus preventing escape of steam and smells, etc., to other parts of the building, and it should be particularly suitable for laundries, kitchens, lavatories, laboratories, and rooms where fumes or noxious vapours are given off from process work and where natural extract would be too unreliable.

However, the difficulty arises, as in the previous section, in achieving a satisfactory means of admitting air to balance the extract and warming it in winter. Fresh-air inlets behind radiators (Fig. 16.8) have only a limited

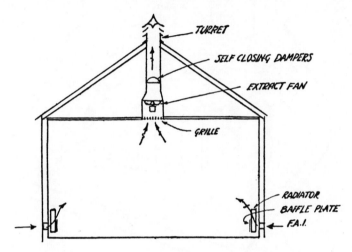

FIG. 16.8.—Ventilation by Natural Inlet and Mechanical Extract.

application. On a small-scale building, leakage may suffice. Sometimes replacement air may be drawn from another part of the building, as in a kitchen where air may come from the restaurant or the canteen. In general, however, mechanical extract goes with mechanical inlet. Nevertheless, in this section we will deal with the means for mechanical extraction alone.

A typical industrial application is as in Fig. 16.9 where a series of

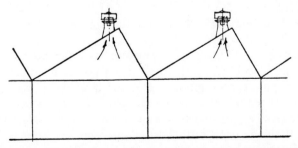

FIG. 16.9.—Mechanical Extraction as applied to North-light Factory.

extractor fans are mounted on the roof. In Fig. 16.10 the fan is centrifugal, and in Fig. 16.11 the fan is of the propellor type.

The air discharged from such units tends to hug the roof; so, where fumes or other pollutants are carried in the air, it is better to use a vertical-

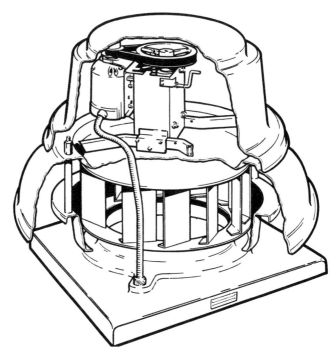

Fig. 16.10.—Cut-away view of belt-driven centrifugal Roof Extract Unit (Brooks Ventilation Units Ltd., Croydon).

discharge unit, as in Fig. 16.12. This unit contains vanes to exclude rain when not running but, when in operation, these open and the velocity is such that rain cannot enter.

For use in a vertical or horizontal duct, the arrangement may also be as in Fig. 16.12.

Where exhausting through a wall, the fan would be fixed as in Fig. 16.13.

The effect of wind blowing in opposition to a fan discharging through a wall opening is to hold up the delivery of air, or to make it hunt, producing a surging gusty action, often with some noise, and with a considerable reduction in volume exhausted. If it is possible to turn the discharge vertically upwards this is preferable, as the wind then aids the extraction. If this is not possible a baffle placed in front of the fan discharge will improve the operation.

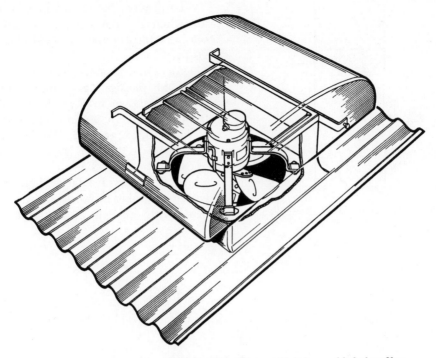

FIG. 16.11.—Cut-away view of Pitched Roof Extract Unit in moulded glass fibre
(Brooks Ventilation Units Ltd., Croydon).

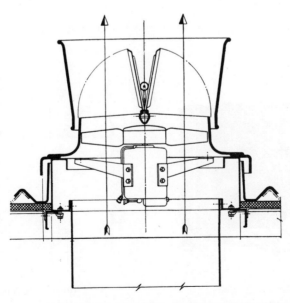

FIG. 16.12.—Vertical Jet Discharge Unit (Brooks Ventilation Units Ltd.).

K.H.A.C.

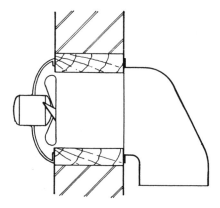

To prevent the escape of warm air when the fan is shut off, self-closing dampers may be fitted (Fig. 16.14).

For small rooms one of the moulded plastic type window or wall fans shown in Fig. 16.15 will frequently give a useful solution to a ventilating problem. These are also made to provide inlet as well as extract.

In the case of a larger hall, or where a number of rooms has to be dealt with, an extract duct system

FIG. 16.13.—Wall Extract Fan.

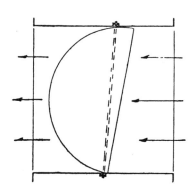

FIG. 16.14.—Butterfly Self-closing Dampers.

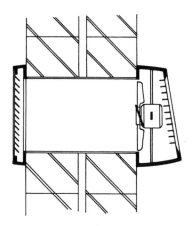

Mounted in Window.   Mounted in Wall.

FIG. 16.15.—Small Extract Fans (Vent Axia or Expelair).

becomes necessary (Fig. 16.16). The fan must then be capable of over-coming the resistance of the ducts, and the ordinary propeller fan then tends to become noisy. Use may then be made of a cased fan (p. 432)or of an 'axial flow' fan (p. 433).

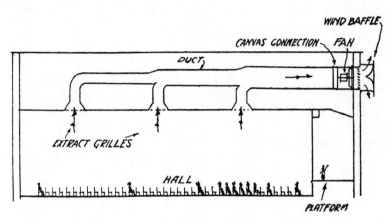

FIG. 16.16.—Ducted Extract System to Hall.

Sizes of fans can only be selected from makers' data. It is not possible to give a summary here as the range for each size and type is so wide.

The arrangement of ducting may be as simple or as extensive as is necessary to meet the case. A typical arrangement for a kitchen is shown in Fig. 16.17. Here it will be noted that hoods are provided over the main fume-producing items of equipment.

Internal Lavatories—A review of previous practice and existing Local Authorities' requirements carried out by the B.R.S. has revealed many in-consistencies. As a result a recommended minimum basis and a method of ensuring absence of interference of one floor with another has been evolved.*

This enquiry and research was primarily directed to blocks of flats, but the principles apply generally.

Briefly the recommendations may be summarised as follows:

Per W.C. compartment for housing a minimum of 20 m³/hour (5·5 litre/s)
Per Bathroom without W.C.   „   „   „ 20 m³/hour (5·5 litre/s)
Per Bathroom with W.C.   „   „   „ 40 m³/hour (11 litre/s)

In order to overcome stack effect, and other chance effects such as doors open on one floor and not on others, a design of duct system as in Fig. 16.18 is advised. It will be noted that there is to each floor a leg with sizeable resistance, thus tending to inherent regulation.

* Paper by Wise and Curtis, *I.H.V.E. Journal*, Vol. 32, p. 180.

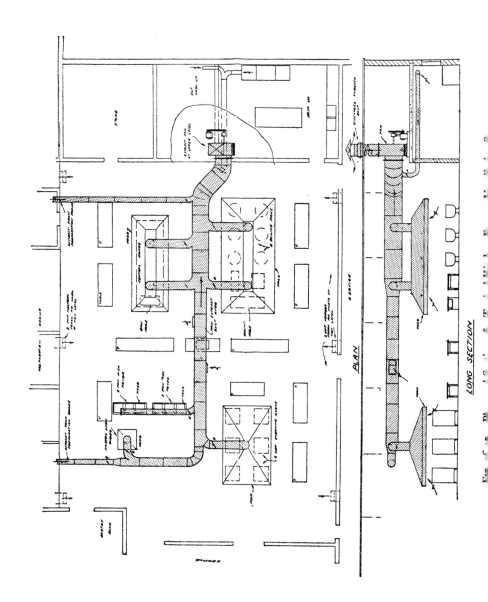

Local Authority regulations require that exhaust fans must be in duplicate. This involves an arrangement such as shown in Fig. 16.19.

Provision must be made for inlet air by suitably placed fresh-air inlets. In a block of flats, for instance, where lavatories are vertically above one another, these inlets may connect to a rising shaft communicating with outside air at the bottom, where a heater may be placed. The shaft may form the duct for the various pipes serving the lavatory fittings. Alternatively air may be drawn from a hallway or corridor, subject to approval of the local Officer of Health. In office blocks and other large buildings an extract rate of 10–12 changes per hour is desirable. Some authorities require a separate air supply to the lobby.

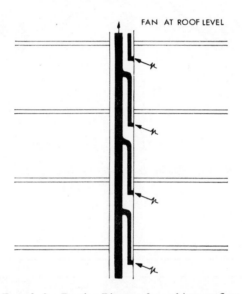

FAN AT ROOF LEVEL

FIG. 16.18.—Ducting Diagram for multi-storey flats.

(3) **Mechanical Inlet and Natural Extract**—This method is suitable for certain types of factory, offices, boilerhouses, etc.

In the case of a factory the inlet may take the form of a series of fresh air unit heaters with extract by natural roof ventilators, as in Fig. 10.13 (page 210).

In the case of offices and similar rooms a ducted system for the inlet may be used, delivering the air into the room at high level, and with louvred openings at low level, allowing the extract air to pass out into the corridor (Fig. 16.20). Provision must be made for the corridor to connect to a space of free escape, such as a staircase connecting to the street, or to a natural vent shaft open at the top, subject to the requirements of the Fire Authority.

Single offices, canteens and the like, may be served by a cabinet type

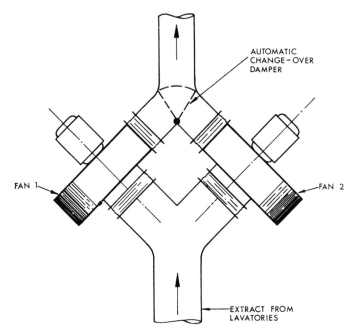

FIG. 16.19.—Duplicate Fans for Internal Lavatories.

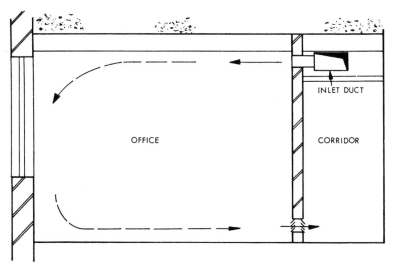

FIG. 16.20.—Ventilation by Mechanical Inlet and Natural Extract.

fan convector, one type of which is shown in Fig. 16.21. This may have provision for filtering, as well as heating, and may be set to introduce fresh air or recirculate. If provided with cooling coils it becomes a unit air-conditioner, referred to on page 354.

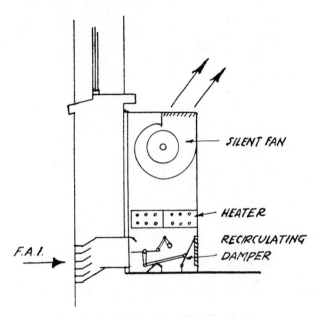

FIG. 16.21.—Cabinet Type Unit Ventilator.

Ventilation of Boilerhouses—Boilerhouses are a special case owing to the air consumed in combustion. In any boilerhouse provision must be made for this air to enter. The volume may be estimated in the manner given in Chapter 4 (page 79). Where a boilerhouse is above ground this air often provides adequate ventilation without augmentation. In the case of a confined basement boilerhouse, additional air supply may be necessary to keep the temperature down to a reasonable limit. The amount of heat liberated and volume of air necessary can be calculated.

It is then preferable to arrange for this air to be provided by mechanical inlet rather than by mechanical extract, since the latter may tend to create a negative pressure counteracting the effect of the flue. With mechanical inlet, provision must be made for the surplus extract air to escape by natural means such as by a shaft adjacent to the chimney.

(4) **Mechanical Inlet and Extract**—This method of ventilation is capable of the widest application, as distribution and pressure are both under control, as is also temperature. Air filters may be included for the cleaning of the air. It is particularly applicable to theatres, cinemas, restaurants, dance halls, banqueting halls, smoking rooms, libraries, offices, canteens, kitchens, etc. Fig. 16.22 shows a diagrammatic arrangement.

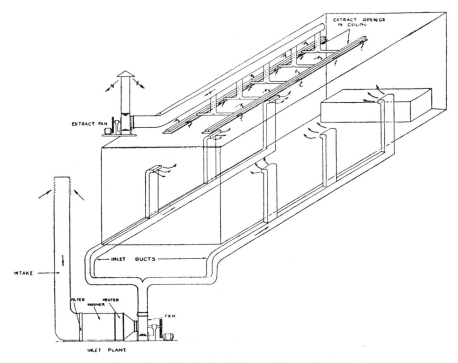

FIG. 16.22.—Mechanical Inlet and Extract System.

In all these cases both inlet and extract would most probably require to be served from fans at a distance by means of ducts.

In normal living rooms, offices, etc., where no fumes or smells are generated, the extract should be less than the inlet, so that air movement is outward not inward; it is then usually   -   -   75 to 90 per cent of inlet.

Where fumes arise, and a negative presure in the room is necessary, as in laboratories, kitchens, shops, etc., the extract should exceed the inlet, usually taken at about -   -   -   -   -   110 to 120 per cent of inlet.

There are a great number of combinations of inlets and extracts with different kinds of fans, with or without ducts, etc., to suit various purposes, but enough has no doubt been said to indicate the general principles and possibilities.

From what has been said earlier it will be apparent that, where mechanical extract is employed for industrial buildings, the problem of introducing replacement air can as a rule only be met by mechanical inlet. All too often the fact is ignored that if air is removed it must be replaced, and this involves warming in winter. If no provision for inlet is made in the design, all sorts of subterfuges have to be made later, often with highly unsatisfactory results. One solution to the problem is to provide fresh-air unit heaters, as already mentioned. These may serve also as the means of warming the building, being changed over to recirculation for warming-up.

Where local relief on the working zone necessitates inlet air being directed to specific areas, ducted units may be employed, as shown in Fig. 10.13 (page 210).

Another method of warming inlet ventilation air is that of the direct oil-fired or gas-fired heater, one form of which is shown in Fig. 3.1 (page 47). Air distribution from large output units of this kind relies on a high-velocity discharge at high level and return at low level. If the unit is handling 100 per cent fresh air, there is no low-level return and temperature gradients will be high. Thus a substantial proportion of return air—say 50 per cent—is best allowed for, fresh air making up the other 50 per cent. The pattern of air flow will be similar to that in Fig. 2.3 (page 29).

Replacement air for certain heavy industries, such as foundries where contamination already exists, may be warmed by gas burners in the air stream. The volumes handled and the dilution are such that no hazard to health arises. This is a cheap method of dealing with vast air quantities.

DESIGN OF INLET SYSTEM

The design of the inlet ventilation system with ducts is dealt with in Chapter 18. Fans, heaters, filters, air-washers, controls, ducts, may all be similar to those described in the subsequent chapters.

The method of distributing the air will follow one of the systems mentioned in Chapter 18. With straight ventilation the upward system is more general than the downward.

Heat for Air—When inlet air is introduced at about room temperature, the heating system is relieved of the duty of warming infiltration air because leakage tends to be outward. Radiators, heating panels, etc., should therefore be sized for the fabric losses only. This leads to obvious economy and avoids the risk of over-heating.

In any system where air is the sole heating medium, there are two quantities of heat requirement involved: first, the heat necessary to raise the fresh-air supply from say $-1°$ C to a room temperature of say $20°$ C, and second, that necessary to off-set building heat losses through the structure. The mass flow of air will usually be determined by the ventilation requirement and thus the air-outlet temperature from the heater will need to be considerably higher than that required in the room. For instance, consider a space requiring a fresh-air supply of 1800 m³ per hour (500 litre/s) which has a building-fabric heat loss of 15 kW to maintain an internal temperature of 20° C when outside conditions are at − 1° C. The air supplied will require to be

$$\frac{15 \times 3600}{1800 \times 1 \cdot 22} = 24 \cdot 6° \text{ C above room temperature.}$$

That is to say the duty of the heater is to raise the air temperature

$$(21 + 24 \cdot 6) = 45 \cdot 6° \text{ C.}$$

Should the requirement lead to an unduly high leaving temperature from the heater, the addition of some recirculating units is called for. Thus, if unit heaters are being used, some would introduce fresh air and some would not: discharge temperatures not exceeding 50° C to 55° C should be aimed at.

PLENUM SYSTEM

Reference to this system is retained here more for historical interest than as indicating current practice. It was primarily a system of heating and only secondarily one of ventilation. Its application was mainly

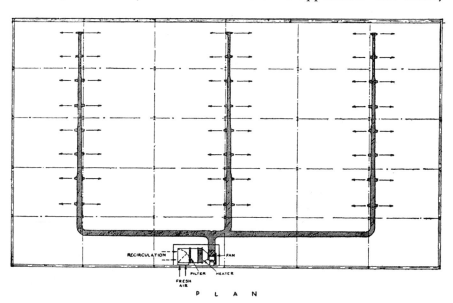

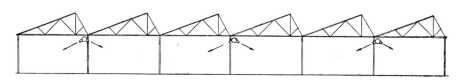

FIG. 16.23.—Typical Lay-out of Plenum System.

industrial but it lacked the flexibility generally demanded to-day. It was a fixed system which once installed could only be altered with difficulty. Furthermore, it was more costly and space-consuming than unit heaters and the like which have largely replaced it.

The system draws air from outside and, after heating, distributes it through a series of ducts to the space to be heated, as in Fig. 16.23. A slight pressure or plenum is supposed to be created within the building so that leakages should be outward rather than inward. The plant comprises an intake chamber with change-over damper—fresh to recirculating—an air filter, a heater battery (fed by steam or hot water for raising the air temperature to 60° C or so) and a fan and motor. The ducts in the Figure are shown overhead, but in some cases underground ducts have been used. Main trunks require insulating.

It will be clear that, when using fresh air, the heat put into the air to raise it to room temperature is lost; only the additional heat above this to meet fabric losses is useful. On the other hand, if recirculating, the supposed advantage of plenum no longer exists. The system was in fact susceptible to the vagaries of wind, making for patchy heating—under-heating on the windward side and overheating on the leeward side.

As mentioned, for a variety of reasons the plenum system has fallen out of use and it is not proposed to discuss it further.

PUNKAH AND DESK TYPE FANS

Fans of this type merely stir round the air of the room, and in so doing cause a rapid movement over the skin with resultant evaporation and cooling, which accounted for their widespread use in the tropics before the

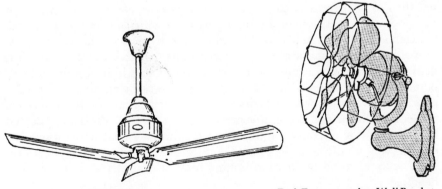

Punkah Fan.   Desk Fan mounted on Wall Bracket.

FIG. 16.24.

days of air-conditioning (Fig. 16.24). They should not be considered as providing ventilation but, if a ventilating system lacks cooling, they can give some relief.

CHAPTER 17

Air-Conditioning

THE SCIENCE OF AIR-CONDITIONING may be defined as that of supplying and maintaining a desirable internal atmospheric condition irrespective of external conditions. As a rule 'ventilation' involves the delivery of air which may be warmed, while 'air-conditioning' involves delivery of air which can be warmed or cooled and have its humidity raised or lowered.

The desired atmospheric condition usually involves a temperature of 18° to 22° C in winter and 21° to 24° in summer; a relative humidity of about 40 per cent to 60 per cent; and a high degree of air purity. This requires different treatments according to climate, latitude, and season, but in temperate zones such as England it involves:

In Winter—A supply of air which has been cleaned and warmed. As the warming lowers the relative humidity, some form of humidifying plant, such as a spray washer with preheater and main heater whereby the humidity is under control, is generally necessary.

In Summer—A supply of air which has been cleaned and cooled. As the cooling increases the relative humidity, some form of dehumidifying plant is essential. This dehumidifying is generally accomplished by exposing the air to cold surfaces or cold spray, whereby the excess moisture is condensed and the air is left saturated at lower temperatures. The temperature of the air has then to be increased, to give a more agreeable relative humidity. This can be done by warming or by mixing with air which has not been cooled.

Dehumidifying can also be brought about by passing the air over certain substances which absorb moisture. Thus, in laboratories, a vessel is kept dry by keeping a bowl of strong sulphuric acid in it or a dish of calcium chloride, both of which have a strong affinity for moisture. Silica-gel, a form of silica in a fine state of division exposing a great absorbing surface, is used also for drying air on this principle, but this process is complicated by the need for re-conditioning of the medium by heat and subsequent cooling, and is not generally used in comfort air-conditioning applications.

The application of air-conditioning may be considered under various headings:

1. Where crowds of people congregate such as in restaurants, cinemas, theatres and the like.
2. Where work has to be carried on in a confined space, the work being of a high precision and intensive character, such as in operating theatres, instrument assembly shops and the like.

3. Where the exclusion of air-borne dust is essential.

4. Where the process to be carried out can only be efficiently done within strictly controlled limits of temperature and humidity.

5. Where the type of building and usage thereof involves considerable heat gains such as in multi-storey office blocks with large glass areas subject to solar gain, and including heat-producing office machinery, computers, intensive electric lighting, etc.

6. In a great variety of conference rooms, lecture theatres, laboratories and animal houses.

7. The core areas of modern buildings planned in depth, where the accommodation in the core is remote from natural ventilation and windows and is subject to internal heat gains from occupants, lighting, etc.

In tropical and sub-tropical countries, air-conditioning is primarily required to reduce the high ambient temperature to one in which working and living conditions can be tolerable. In the temperate climate of the British Isles, and in similar parts of the world, long spells of warm weather are the exception rather than the rule, but modern forms of building and modern modes of living and working have produced conditions in which, to produce some tolerable state of comfort, air-conditioning is the best answer. Thus we find modern buildings incorporating to a greater or lesser extent, almost as a common rule, some form of air-conditioning. This great variety of applications has produced an almost equally great variety of systems, although all are fundamentally the same in basic principle: that is, to achieve a controlled atmospheric condition both in summer and winter, as referred to earlier, using air as the medium of circulation.

The installation of complete air-conditioning in a building as a rule eliminates the necessity of heating by direct radiation, and it naturally incorporates the function of ventilation, thus eliminating the need for opening windows or reliance on other means of ventilation.

All air-conditioning systems involve the handling of air as a means for cooling or warming, dehumidifying or humidifying. If the space to be air-conditioned has no occupancy, no supply of outside fresh air is needed, that inside the room being continually recirculated. In most practical cases, ventilation air for occupancy has to be included and, in the design for maximum economy of heating and cooling, this quantity is usually kept to a minimum depending on the number of people to be served. Thus, in most instances it will be found that the total air in circulation in an air-conditioning system greatly exceeds the amount of fresh air brought in and exhausted. Where, however, it is a matter of contamination of the air, such as in a hospital operating theatre, or where some chemical process or dust-producing plant is involved, one hundred per cent fresh air may be needed and no recirculation is then possible.

With a certain design of plant it is possible to arrange for 100 per cent fresh air to be handled during periods of medium weather, such as in spring and autumn, when neither cooling nor heating is required, or when it can do useful cooling.

The basic elements of air-conditioning systems of whatever form are:

> *Fans* for moving air;
>
> *Filters* for cleaning air, either fresh, recirculated, or both;
>
> *Refrigerating plant* connected to heat exchange surface, such as finned coils or chilled water sprays;
>
> *Means for warming* the air, such as hot water or steam heated coils or electrical elements;
>
> *Means for humidification;* and/or dehumidification;
>
> A *Control System* to regulate automatically the amount of cooling or warming.

VARIOUS SYSTEMS DESCRIBED

Central Plant (Fig. 17.1)—This type of system is suitable for air-conditioning large spaces, such as theatres, cinemas, restaurants, exhibition halls, or big factory spaces where no sub-division exists. The manner in which the various elements just referred to are incorporated in the plant will be obvious from the caption. It will be noted that in this example the cooling is performed by means of chilled water-cooling coils. The humidification is by means of a capillary cell air-washer. This unit would only run when humidification was required such as in winter.

Fig. 17.2.—This shows another version of a central plant in which the cooling coil is connected directly to the refrigerating plant and contains the actual refrigerating gas. On expansion of the gas leaving the condenser, cooling takes place and hence this system is known as a 'direct expansion system'. It is suitable for small to medium size plants. Humidification as shown in this figure is by means of direct spray into the air stream, but it might equally be in the form shown in Fig. 17.1.

In both Fig. 17.1 and Fig. 17.2 it will be noted that there is a separate exhaust fan shown exhausting from the ceiling of the room. This would apply particularly in cases where smoking takes place, such as in a restaurant, to remove smoke which might otherwise collect in a pocket at high level. Sometimes this exhaust can be designed to remove the quantity equivalent to the fresh-air intake, in which case the discharge shown to atmosphere from the return air fan would not be necessary.

Zoned System (Fig. 17.3)—Where there are a number of rooms or floors in a building to be served, it is necessary to consider means by which the varying heat gains in the different compartments can be dealt with. Some rooms may have some, and others not; some may be crowded and others empty; again, some may contain heat-producing equipment. Variations in requirements of this kind are the most common case with which air-conditioning has to deal, and for this the central system is unsuitable. One approach is the zoned system.

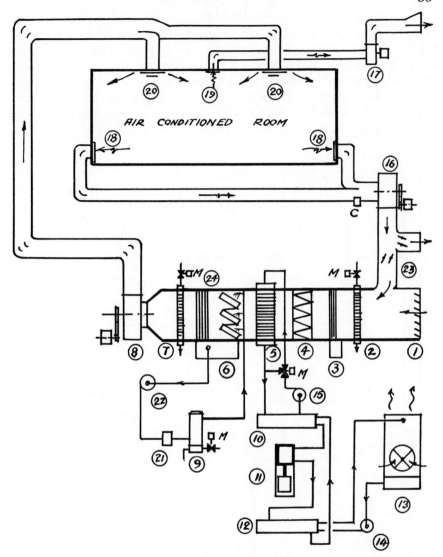

FIG. 17.1.—Diagram of Complete Air-Conditioning System Employing Air Washer and Dehumidifier.

1, Fresh air intake. 2, Air heater for fog elimination. 3, Self cleaning air filter. 4, Electrostatic or dry fabric type secondary filter. 5, Chilled water cooling and dehumidifying coil. 6, Air washer of high saturating efficiency. 7, Reheater. 8, Supply air fan. 9, Spray water heater. 10, Direct expansion water chiller. 11, Refrigerant compressor having steps of capacity reduction. 12, Shell and tube condenser. 13, Forced draught condenser water recooler. 14, Condenser water circulating pump. 15, Chilled water circulating pump. 16, Main extract fan. 17, Smoke extract fan. 18, Low level extract and recirculation air grilles. 19, Smoke extract grille. 20, Air inlet diffusers at ceiling. 21, Washer spray water strainer. 22, Washer spray water pump. 23, Recirculation duct. 24, Eliminators. M denotes motorized valves. C denotes temperature and humidity controls.

In this system the building is divided up into zones such that as nearly as possible similar conditions may be expected to exist. Each zone is provided with its own local recirculating fan and booster cooler or heater, and this unit receives fresh air supply conditioned to some average temperature and corrected for humidity by means of what is in effect a central plant. Such an arrangement is shown in Fig. 17.3. In this case the central plant is on the roof. The recirculating units are overhead adjacent to the corridor (Fig. 17.4), taking their return air therefrom. The distributing

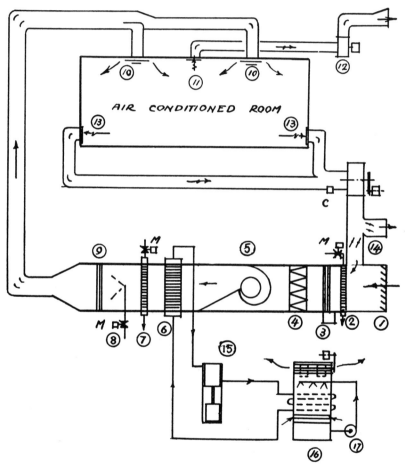

FIG. 17.2.—Diagram of Complete Air-Conditioning System
Employing Direct Expansion Cooling.

1, Fresh air intake. 2, Air heater for fog elimination. 3, Viscous filter. 4, Electrostatic or dry fabric secondary filter. 5, Inlet fan. 6, Direct expansion cooling coils. 7, Main air heater. 8, Humidifying water spray or steam jet. 9, Eliminator plates to remove surplus moisture. 10, Air inlet diffusers at ceiling. 11, Smoke extract grille. 12, Smoke extract fan. 13, Low level extract and recirculation grilles. 14, Recirculation duct. 15, Refrigerating compressor having capacity reduction. 16, Condenser cooler. 17, Evaporative cooling water pump. *M* denotes motorized valves. *C* denotes temperature and humidity controls.

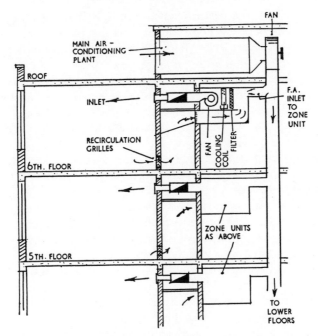

FIG. 17.3.—Zoned Air-Conditioning System for Multi-Storeyed Building.

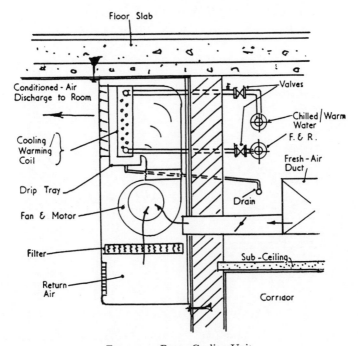

FIG. 17.4.—Room Cooling Unit.

2 A

ducts are run above the corridor false ceiling delivering into the various rooms, floor by floor, each floor constituting a separate zone. The return air from the rooms passes through grilles into the corridor which acts as the return-air collecting-duct.

In such a system, supposing one face of the building to be north and another south, it would be desirable to zone separately the two faces so that there would, in effect, be two recirculating units per floor. If the building were square, with four aspects, it would no doubt be desirable to divide the floors into four zones.

The cooling or heating booster coils would be served from circulating water mains, each coil being controlled locally according to the requirements of the zone served.

Fan Coil Units—If, in the zoned system, we consider each room as a separate zone, distributing ductwork from the recirculating unit can be eliminated, and we have what is known as the 'Fan Coil' system. Fan coil units may be arranged in a variety of ways; one is as shown in Fig. 17.4, where the unit is mounted overhead, and Fig. 17.5 shows a cabinet type for floor-mounting below a window. Units of this type have been developed with silent-running fans requiring little maintenance, and they are of wide application. It is usual to control these units thermostatically by switching the fan motor on and off, but an alternative is to control the water circulation leaving the fan running. Zoning of the water circulations is important, bearing in mind that under some conditions certain aspects of the building may require warming whilst others require cooling.

In both these examples it will be noted that the fresh air supply is from

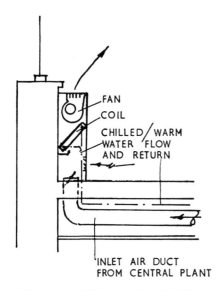

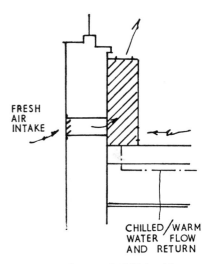

FIG. 17.5.—Cabinet type Fan Coil Unit.   FIG. 17.6.—Fan Coil Unit with Fresh Air Inlet.

Plate XVII. An example of clear open roof ventilation in the factory of Rockware Glass Ltd., Nottingley, Yorks. (see p. 334). (By courtesy of Colt Heating and Ventilation Ltd.)

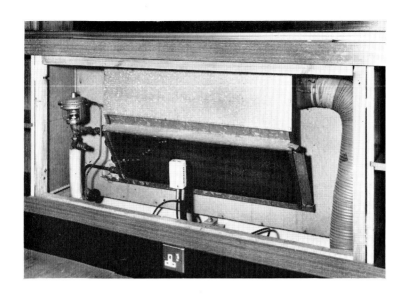

*Plate XVIII. Perimeter Induction Unit. Above: interior of unit.
Below: 1, conditioned air-delivery grille in sill; 2, return air grille;
3, casing to primary air duct and piping (see p. 358)*

a main fresh air duct which would be served from a central plant, as in the previous example.

An alternative and obviously cheaper method of application of fan coil units is where the fresh air supply is taken direct from outside and the central plant and ducting therefrom is eliminated. In this case, as shown in Fig. 17.6, each unit placed beneath the window has an aperture through the wall to outside, and the unit contains an air filter. Chilled or warmed water circulating piping is connected and the unit is under local thermostatic control. Humidity control is not specifically catered for.

Self-contained Air-Conditioners—This heading covers a separate field of air-conditioning. Unit air-conditioners are complete in themselves, containing compressor, fan, cooling and heating coils, filter and, sometimes, means for humidification. Fresh air can be introduced if required. Being

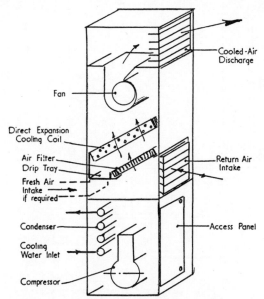

FIG. 17.7.—Self-Contained Unit Air-Conditioner.

of unit construction, alternatively described as 'Packaged', they are not tailor-made to a specific job and may thus be expected to be economical in first cost.

Unit air-conditioners of small size generally have the condenser of the refrigerator air-cooled, but in larger sizes the condenser may be water-cooled, in which case piping connections are required. Apart from this, the only services needed are an electric supply and a connection to drain to conduct away any moisture condensed out of the atmosphere during dehumidification. Compressors in most units are now hermetic and are therefore quiet in running.

Sizes vary from small units suitable for a single room, sometimes mounted in a window or in a cabinet, as in Fig. 17.7, and the range goes up to units of considerable size suitable for industrial application, in which case ducting may be connected for distribution.

Heat Recovery System—An interesting development of the unit air-conditioner system was first made available under the proprietary name 'Versatemp'. Fig. 17.8 illustrates diagrammatically the principle of this system. The various rooms throughout a building are each equipped with a unit containing a hermetic refrigerating compressor. The compressor,

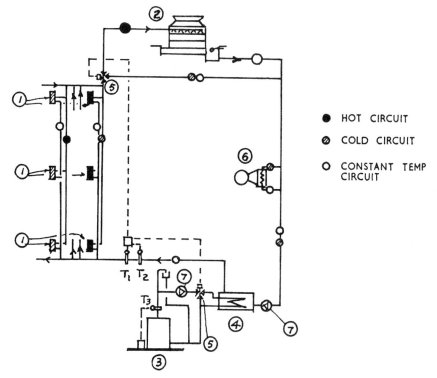

● HOT CIRCUIT

◐ COLD CIRCUIT

○ CONSTANT TEMP CIRCUIT

FIG. 17.8.—Diagram of Heat Recovery System

(1) Versatemp units	(5) Mixing valves	T_1, Thermostat
(2) Cooling tower	(6) Reclaim heat	T_2, Low limit stat
(3) Boiler	from exhaust etc.	T_3, Boiler control stat
(4) Heat exchanger	(7) Pumps	

with its evaporator and condenser, either cools the room or warms it according to the requirements of the control system. When operating as a cooler, the circulating water system removes heat from the condenser, the water in turn passing to the evaporative cooler on the roof as shown in the diagram. When serving as a heater, the functions of evaporator and cooler are reversed, and the circulating water removes the cooling effect. The circulating flow is kept at a constant temperature either by warming by means of the boiler and heat exchanger, or by the evaporative cooler. When cooling is not required, the evaporative cooler is bye-passed in the manner indicated. Heat may be regained from exhaust air or other possible sources. Economy of operating costs could result from this system, due to the heat pump effect.

Induction System—As the name implies, the principle of induction is employed in this system as a means of recirculation. Primary conditioned

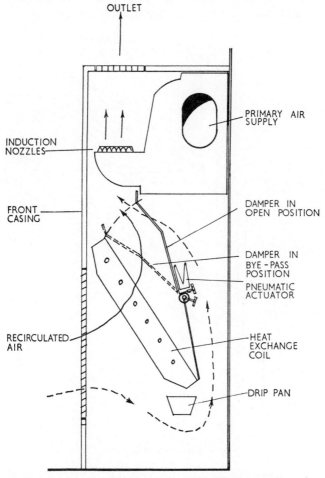

OUTLET

PRIMARY AIR SUPPLY

INDUCTION NOZZLES

FRONT CASING

DAMPER IN OPEN POSITION

DAMPER IN BYE-PASS POSITION

PNEUMATIC ACTUATOR

RECIRCULATED AIR

HEAT EXCHANGE COIL

DRIP PAN

FIG. 17.9 (*a*).—Detail of Induction Unit with Damper Control.

air from a central plant is supplied to each unit under pressure and the unit, generally placed below the window, contains a series of jets or nozzles as shown in Fig. 17.9 (*a*). The induced air from the room flows over the cooling or heating coils, and the mixture is delivered from the grille in the sill. The induction ratio is from three to one to six to one. The coils are fed with circulating water which, in the change-over system, is cooled in summer and warmed in winter. In Fig. 17.9 (*a*), control is achieved by an arrangement of dampers, such that the return air from the room is either drawn through the coils for heating or cooling, or the coils are by-passed to a greater or lesser extent. In the type shown in Fig. 17.9 (*b*), control is by variation of water flow: increase in water flow is required to lower temperature during the cooling cycle, and increase in flow is required to raise the temperature in the heating cycle. There is thus required some means of change-over of thermostat operation according to whether the winter or summer cycle is required. (See also Plate XVIII, facing p. 355).

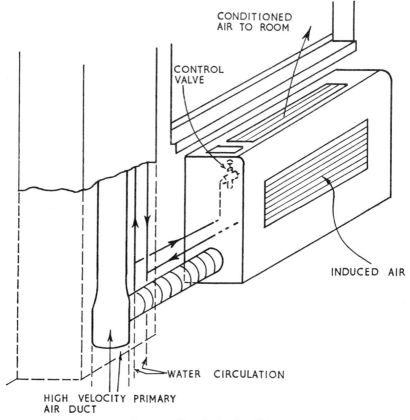

CONDITIONED AIR TO ROOM

CONTROL VALVE

INDUCED AIR

WATER CIRCULATION

HIGH VELOCITY PRIMARY AIR DUCT

FIG. 17.9 (*b*).—Induction Unit.

An alternative method avoids change-over by always circulating cool water through the coils of the induction units and varying the air temperature according to weather. Thus, in winter, the primary air is warm and its quantity and heat capacity must be adequate for warming the room. The water is cooled by a coil in the fresh air intake and enables control of the room temperature individually to be achieved. In summer, the primary air is maintained at about room temperature and the sensible cooling is dealt with by the induction unit coils now supplied with chilled water from the water-chiller.

A variation of the two-pipe induction system is the three-pipe, in which both warm and cool water are available at each unit, with a common return, and the control arrangement is so devised as to select from one or the other. Likewise, in the main system the return is diverted either to the cooling plant or to the heating plant, according to the mean temperature condition. The ideal system would be with four pipes, but it would obviously be correspondingly expensive.

Fig. 17.10 shows a type of induction unit continuous in length and low in height.

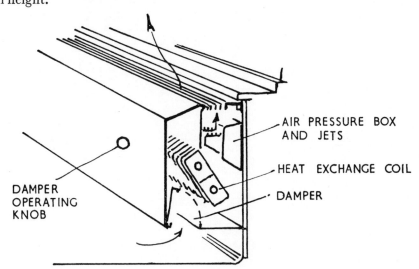

AIR PRESSURE BOX AND JETS

HEAT EXCHANGE COIL

DAMPER OPERATING KNOB

DAMPER

FIG.17.10.—Induction Unit; low model (Copperad Ltd., *Duct Line*).

Induction systems might be expected to be noisy, due to the high velocity air issuing from the jets, but the units have been developed with suitable acoustical treatment such that this disadvantage does not arise in practice. For reasons of economy, both in cost and space, ducts with this system are designed on high velocity principles. Air speeds of 15 to 25 m/s in the ducts are common, compared with the conventional speeds of the order of 5 m/s. Owing to the higher pressure at the fan, consequent on the use of high velocities, a silencer immediately after leaving the central plant is usually incorporated with further acoustic treatment in the ductwork distribution system as may be necessary.

Fig. 17.11 is a diagram of an induction system showing the primary conditioning plant, the primary ducting and the water circulation. The heat-exchanger shown is for warming the water circulated to the units, and this would be fed from a boiler or other heat source.

The induction system uses minimum primary air, generally of the order of $1\frac{1}{2}$ to 2 air changes/hour, and is generally considered most suitable for multi-storey office blocks or hotels where in either case there is a large number of separate rooms to be served on the perimeter of a building. Interior zones of such buildings or rooms where a higher rate of ventilation is required are usually dealt with independently by conventional systems.

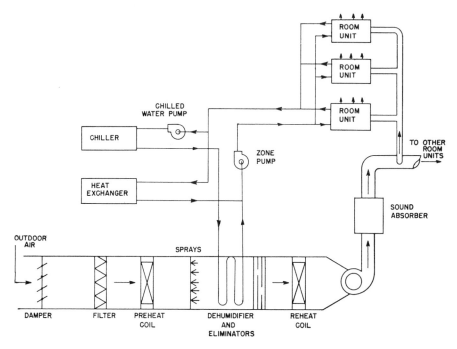

FIG. 17.11.—Diagram of typical Induction System.

Induction systems inherently cause any dust in the atmosphere of the room to be drawn in and over the finned coil surfaces, and, to prevent a build-up of deposit thereon, some form of coarse lint screen, easily removable for cleaning, is usually incorporated.

Dual-Duct System—In this system, water circulation throughout the building is eliminated completely. Use is made of two ducts, one conveying warm air and one conveying cool air, and each room contains a blender so arranged with air valves or dampers that all warm, all cool, or some mixture of both is delivered into the room. Fig. 17.12 illustrates such a unit, incorporated therein being a means for automatically regulating the total air-delivery, such that, regardless of variations in pressure in the system, each unit delivers its correct air-quantity. Referred to as 'constant volume control', it is an essential part of this system. (See Plate XX.)

Owing to the fact that air alone is employed, the air-quantity necessary to carry the cooling and heating load is greater than that used in an induction system. Air-delivery rates with the dual-duct system are frequently of the order of 5 or 6 changes per hour, compared with $1\frac{1}{2}$ to 2 with induction. This means that ducts are larger and fan powers greater. Air movement within the room from a dual-duct system is generally augmented to some extent by the natural induction of air delivered from the outlet, depending on the outlet velocity.

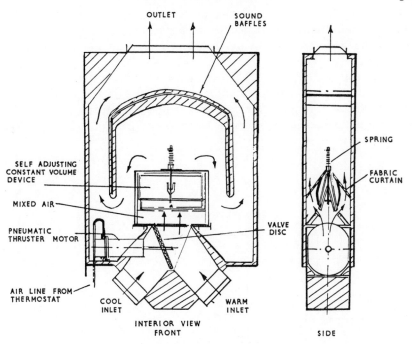

OUTLET   SOUND
           BAFFLES

SELF ADJUSTING
CONSTANT VOLUME
DEVICE

MIXED AIR

PNEUMATIC
THRUSTER MOTOR

AIR LINE FROM
THERMOSTAT

COOL
INLET

VALVE
DISC

WARM
INLET

INTERIOR VIEW
FRONT

SPRING

FABRIC
CURTAIN

SIDE

FIG. 17.12.—Above: Dual-Duct Blender Unit.
Below: Diagram showing duct connections.

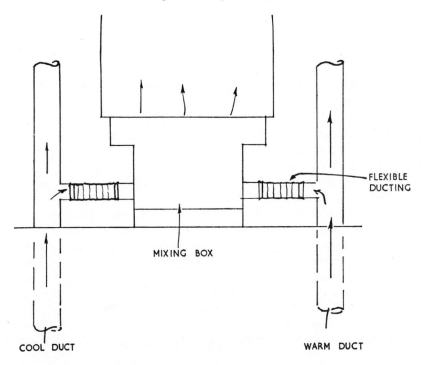

FLEXIBLE
DUCTING

MIXING BOX

COOL DUCT   WARM DUCT

Owing to the greater quantity of air in circulation throughout the building, dual-duct systems usually incorporate means for recirculation back to the main plant, and this involves return air ducts and shafts in some form.

It will thus be apparent that the dual-duct system is likely to occupy more duct space than the induction system, but on the other hand no accommodation is required for pipework. Ducts, both warm and cool, need to be insulated thermally, whereas with some forms of induction system this can be dispensed with.

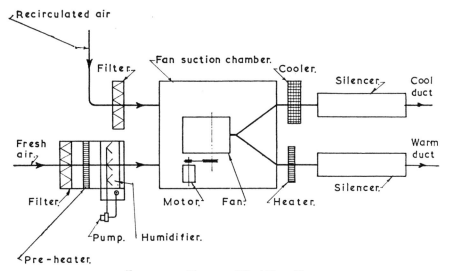

FIG. 17.13.—Diagram of Dual-Duct Plant.

An advantage of the dual-duct system is that any room may be warmed or cooled according to need without zoning or any problem of change-over thermostats.

Another advantage is that core areas of a building, or rooms requiring high rates of ventilation, may equally be served off the same system, no separate plants being necessary.

Fig. 17.13 shows the plant arrangement of a system in one form, though there are variations of this using two fans, one for the cool duct, one for the warm duct. (See Plates XIX and XXII, facing pp. 366, 464).

To avoid undue pressure differences in the duct system, due for instance to a greater number of the units taking warm air than cool air, static pressure control is usually incorporated so as to relieve the constant pressure devices in the units of too great a difference of pressure, such as might otherwise occur under conditions where the greater proportion of units are taking air from one duct rather than from both.

High velocities are used for air-distribution in the ducts, as in the case of the induction system. Acoustical problems are dealt with in like manner.

Variable Volume System—Considering internal zones of a building in which a good rate of ventilation is necessary, it is possible to give the necessary measure of control of temperature by the variation of volume of the incoming air. If there is a considerable amount of internal heat gain, more air will be required than with a low heat gain. To meet this situation the variable volume system has been developed which combines (as illustrated in Fig. 17.14) a self-energising method of control operated

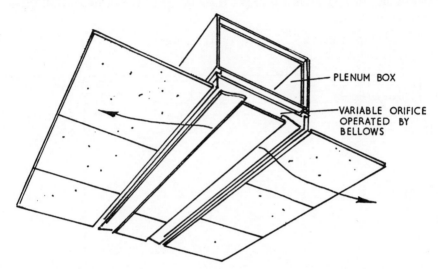

PLENUM BOX

VARIABLE ORIFICE OPERATED BY BELLOWS

FIG. 17.14.—Variable Volume Strip Inlet (Carlyle Air-Conditioning Ltd.)

according to the dictates of a room thermostat. Alternatively, control may be achieved by serving a number of strip inlets in a single space from a single silencer/damper box, the air supply volume to which is varied by a pneumatic thruster unit positioned by a room thermostat. This system provides a simple method of achieving room-by-room control with a single-duct system of distribution provided say 25 per cent heat gain is maintained such as by keeping the lighting on. (See Plate XXI, facing p. 367).

Panel Cooling—A fundamentally different approach to the whole matter of temperature control and ventilation, covered by the general term 'Air-Conditioning', is a system in which surfaces of the ceiling are cooled by chilled-water circulation for the removal of heat gains, leaving the air-distribution system the sole purpose of ventilation and humidity control. Such a system has been applied to the Shell building in London.

For a full description of the Shell Building system, a paper by Jamieson should be consulted.* The essential feature of the system is that the ventilation air shall be conditioned such that its dew-point is below the surface temperature of the cooling surfaces in the room, otherwise condensation might occur thereon. The cooling surfaces in the case in question comprise

* *Journal I.H.V.E.* Vol. 31, p. 1.

normal perforated aluminium trays with insulation above, clipped to a circulating water-grid fed with chilled water in summer and with warmed water for heating in winter. Air volumes are thereby reduced, as the sole purpose of the ventilation system is to provide air for occupancy, and the conditioning of same is to a set state. Fig. 17.15 gives an illustration taken from the above paper indicating the principles of the system. Bearing in mind that a ceiling treatment of this kind also incorporates acoustical absorption, an appraisal of overall costs needs to take this and all other factors into account.

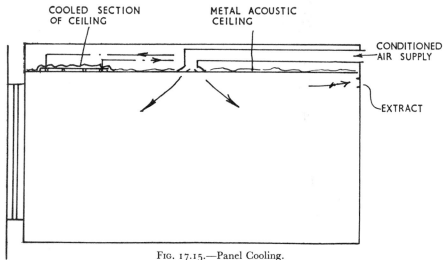

FIG. 17.15.—Panel Cooling.

PRINCIPLES OF AIR-CONDITIONING DESIGN

It is now proposed to consider the fundamental principles underlying air-conditioning design.

These principles are the same no matter what particular form the system may take, but the degree of accuracy necessary to be achieved will depend on the application and the sophistication of controls. For instance, a special research laboratory may require minimum tolerance in conditions, whereas less strict limits would be acceptable for comfort-conditioning in the case of a department store.

First to be considered is the general case, as applied to a central air-conditioning system for a single large space, and this is followed by some notes on how these general principles are applied to certain of the specific types of apparatus already discussed. Design data has been built up around each of the particular forms of equipment mentioned, and it would be beyond the scope of this book to explore each one in detail.

HEAT GAINS

Heat gains are the opposite of heat losses. In designing a heating system for a building we are chiefly concerned with the losses. When designing an air-conditioning system we are concerned with the gains, although, as the same system will probably serve for heating, the losses will also be involved.

Heat gains occur in warm weather due to heat transfer through the building fabric by conduction, by the effect of the sun, and by infiltration and ventilation. Internal heat gains occur irrespective of weather due to occupancy, lighting, and any other heat producing equipment within the space.

The gains so outlined are *sensible heat gains*, and cause a rise in temperature with any given air flow. The temperature rise which can be permitted may be limited to 6 to 8° C owing to the difficulty of mixing cool entering air with warmer room air without draught. The mass air flow required to maintain a desired temperature is simply arrived at by the sum:

$$\text{Mass flow air kg/s} = \frac{\text{heat gains kW}}{\text{design temperature rise °C} \times \text{specific heat capacity of air kJ/kg}}$$

Latent Heat Gains, on the other hand, do not cause temperature rise. These are due to moisture exhaled in the breath and by perspiration of occupants, by moist air entering the space due to infiltration, and by any steam- or vapour-producing equipment within the space. The latent gains are to be treated separately from the sensible gains; the mass flow of air determined for the latter will usually be found to cause only a small humidity increment, but in an extreme case limitation of the increment to an acceptable figure may override that sufficing for sensible gains.

Solar Gains—The most important heat gain is that from the sun. Heat from the sun on a clear day at midsummer falling on a horizontal surface is given as 910 W/m² in latitude 52° N. In addition, diffuse radiation from all parts of the sky is given as 50 W/m². Of these amounts,

(*a*) some is reflected back into space;

(*b*) some warms the surface of the material of roofs and walls whereby external air in contact removes heat by convection;

(*c*) some is transmitted through the material by conduction constituting a heat gain in the room;

(*d*) of that falling on windows, part is reflected back, part warms the glass, but the greater part passes through in the form of radiant heat, so elevating the temperature of surfaces of structure and contents.

Referring to these effects in greater detail:

(*a*) and (*b*). The proportion of solar heat absorbed depends on the nature and colour of the surface. Thus, if a perfect 'black body' is taken as 1, the absorbance of white is 0·3 to 0·5 and of a polished metallic surface 0·1 to 0·4. Most building materials such as brick and concrete will absorb 0·5 to 0·8.

(*c*) The rate of heat flow through the material will depend on its conductivity and use may be made of the usual U factors. The temperature difference outside to inside will, however, be influenced by the heat absorbed from the sun. If we consider first the design-air temperatures outside t_o and inside t_i and the sun-heat increment t_s due to radiation absorbed by the surface, the overall difference causing flow of heat through the material is then $(t_o - t_i) + t_s$. An instantaneous rise of temperature would occur with a material of negligible heat capacity but, for a material of some mass, a time lag is involved. If a material is of sufficient thickness and mass, it is obvious that, as the sun moves round in its path, radiation falling on a wall in the early part of the day may not penetrate to the inside until long after the surface is in shadow, when most likely the curious phenomenon will occur of heat stored in the core of the wall flowing out in both directions. In some cases the space may have ceased to be occupied before the peak heat gain arrives.

Table 17.1 gives sun-heat temperature increments for latitude 52° N (London) for the month of August, for various orientations and for two surface colours. Table 17.2 gives the time lag applicable to various materials and thicknesses. The decrement factor f depends on the thickness of the material and applies to the overall heat gain through the material, i.e.:

$$U\{(t_o - t_i) + t_s\}f$$

TABLE 17·1

SUN HEAT TEMPERATURE INCREMENTS (°C)

52° N—August

x = Dark Surface y = Light Surface

Position		British Summer Time (G.M.T. + 1).									
Orienta-tion	Surface	0800	0900	1000	1100	Noon	1300	1400	1500	1600	1700
Horiz.	x	4·4	10·1	15·4	18·8	21·1	21·9	21·1	18·8	15·4	10·1
Horiz.	y	0·8	4·2	6·9	8·9	10·2	10·6	10·2	8·9	6·9	4·2
N	x	2·4	3·4	4·3	4·8	5·2	5·3	5·2	4·8	4·3	3·4
N	y	1·3	1·9	2·4	2·7	2·9	3·0	2·9	2·7	2·4	1·9
NW	x	2·2	3·4	4·3	4·8	5·2	5·3	5·2	5·0	10·8	16·2
NW	y	1·2	1·9	2·4	2·7	2·9	3·0	2·9	2·8	6·0	9·0
W	x	2·2	3·4	4·3	4·8	5·2	6·2	14·6	22·7	27·6	29·3
W	y	1·2	1·9	2·4	2·7	2·9	3·4	8·1	12·6	15·3	16·3
SW	x	2·2	3·4	4·3	6·7	14·4	22·3	27·7	30·9	30·8	27·1
SW	y	1·2	1·9	2·4	3·7	8·0	12·4	15·4	17·2	17·1	15·1
S	x	4·5	11·9	19·1	24·3	27·9	29·3	27·9	24·3	19·1	11·9
S	y	2·5	6·6	10·6	13·5	15·5	16·3	15·5	13·5	10·6	6·6
SE	x	20·3	27·1	30·8	30·9	27·7	22·3	14·4	6·7	4·3	3·4
SE	y	11·3	15·1	17·1	17·2	15·4	12·4	8·0	3·7	2·4	1·9
E	x	25·5	29·3	27·6	22·7	14·6	6·2	5·2	4·8	4·3	3·4
E	y	14·2	16·3	15·3	12·6	8·1	3·4	2·9	2·7	2·4	1·9
NE	x	17·4	16·2	10·8	5·0	5·2	5·3	5·2	4·8	4·3	3·4
NE	y	9·7	9·0	6·0	2·8	2·9	3·0	2·9	2·7	2·4	1·9

Plate XIX. Air-conditioning plant showing cooling section, above, and heating section, below, of Dual-Duct system (see p. 362)

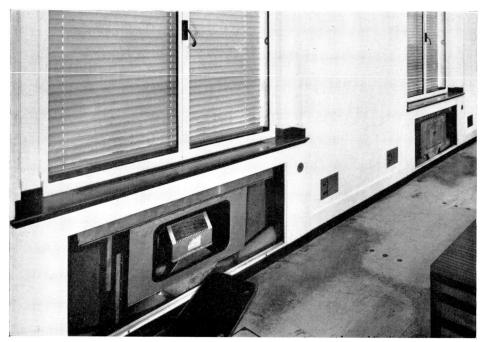

Plate XX. Examples of Dual-Duct blender units under windows (see p. 360)

*Plate XXI. A Variable Volume Control unit serving lighting trougher inlets.
Note the high velocity inlet and low velocity outlet from the box (see p. 363)*

TABLE 17·2

ADJUSTMENT FACTORS TO SOLAR GAINS THROUGH
SOLID MATERIALS

Construction	Adjustments	
	Time lag (hours)	Decrement Factor f
Light frame, sheeted and lined	$\frac{1}{2}$	1·0
Brickwork		
105 mm bare	$3\frac{1}{2}$	0·8
105 mm internally lined	4	0·7
220 mm bare	8	0·4
220 mm internally lined	$8\frac{1}{2}$	0·3
Concrete		
150 mm bare	5	0·6
150 mm internally lined	$5\frac{1}{2}$	0·5
150 mm externally lined	6	0·4
200 mm bare	6	0·5
200 mm internally lined	$6\frac{1}{2}$	0·4
200 mm externally lined	7	0·3

Thermal resistance of lining assumed to be less than
0·4 m °C/W

Flat roofs are subject to solar absorption throughout daylight hours, but, again, the greater the mass the longer is heat gain to the room deferred. Sun-heat temperature increments for flat roofs are given in Table 17.1. Insulation under a roof does not affect the time lag appreciably but it does influence the U value.

Heat gains through sloping roofs depend on direction of slope and pitch. The values of sun-heat increments for flat roofs may be adjusted according to the cosine of angle of incidence.

With regard to Tables 17.1 and 17.2, these are derived from *Guide* data and provide an approximate solution. For a more complete treatment applying to other latitudes and months of the year, the *Guide* should be consulted.

(*d*) (Solar gains through windows). Glass is by far the most important route by which solar heat enters a building and, as pointed out, it is without time lag. The effect in the room may not be instantaneous, however, depending on the nature and mass of internal structure, furniture or other contents. This subject has been explored by Loudon* in evolving a method of forecasting summertime temperatures in buildings without air-conditioning. The method involves the conception of environmental temperature (referred to on page 35) thus taking account of mean radiant temperature as well as air temperature. The method also involves use of the term 'admittance'. Admittance of a building component is a measure of its ability to smooth out diurnal temperature swings in the building.

* *I.H.V.E. Journal*, March, 1970.

Using the same conceptions, the *Guide* presents data for cooling load due to solar gains through windows on the basis of two forms of internal construction: *light* (i.e. demountable partitions and suspended ceilings) and *heavy* (i.e. brick or block partitions). Concrete floor slabs are assumed in both cases. Tables 17.3 and 17.4 are representative of part of these data for various orientations, seasons and times. Table 17.3 relates to unshaded windows or windows with external shades. Table 17.4 relates to windows with internal shadings. Table 17.5(*a*) gives shading factors applying to the former and Table 17.5(*b*) gives factors applying to the latter.

The data assume that design is on the basis of environmental temperature, but, if used for air temperature, a small inherent margin will be included.

Scattered or diffuse radiation from earth and sky is included in the values given in these Tables; but, where a ground floor abuts on to a

TABLE 17.3

Cooling Load due to Solar Gain

Through Vertical Glazing (12 hour plant operation)
for latitude 52° N. G.M.T. + 1. W/m².
(Without sun protection or where with heat-absorbing glass or
with external shades, using factors in Table 17.5(*a*))

		Light Construction						Heavy Construction					
Date	Orienta-tion	0800 hrs	1000	1200	1400	1600	1800	0800	1000	1200	1400	1600	1800 hrs
June 21	N	122	119	134	134	119	122	159	118	131	136	131	118
	NE	404	273	145	145	131	107	317	295	172	157	152	139
	E	510	478	285	155	140	116	373	431	348	190	185	172
	SE	360	460	405	221	125	101	263	377	391	308	181	168
	S	87	247	371	371	247	87	130	185	295	339	295	185
	SW	101	125	221	405	460	360	165	183	196	323	406	392
	W	116	140	155	285	478	510	186	204	217	222	380	463
	NW	107	131	145	145	273	404	151	169	182	187	203	326
Aug. and Apr.	N	47	73	87	87	73	47	67	68	82	87	82	68
	NE	282	178	97	97	82	56	190	217	107	112	107	93
	E	416	435	244	111	96	70	261	385	315	153	148	135
	SE	340	487	440	251	103	77	227	388	418	335	182	157
	S	102	306	440	440	306	102	137	233	353	400	353	233
	SW	77	103	252	440	487	340	146	166	191	343	426	396
	W	70	96	111	244	435	416	128	149	162	167	329	399
	NW	56	82	97	97	178	282	84	104	118	123	118	228
Sept. and Mar.	N	25	54	69	69	54	25	21	44	58	63	58	44
	NE	150	104	73	73	58	29	33	135	70	75	70	56
	E	259	378	213	86	71	42	71	300	266	114	109	94
	SE	241	472	453	273	84	56	112	346	411	337	189	135
	S	108	332	479	479	332	108	133	248	383	434	383	248
	SW	56	84	273	453	472	241	112	135	189	337	411	346
	W	42	71	86	213	378	259	72	94	109	114	266	300
	NW	29	58	73	73	104	150	33	56	70	75	70	135
Correction factor to peak load for 24 hour plant operation		0·96						0·80					

TABLE 17·4
COOLING LOAD DUE TO SOLAR GAIN
Through Vertical Glazing (12 hour plant operation)
for latitude 52° N. G.M.T. + 1. W/m².
(With internal sun protection using factors in Table 17.5(b))

Date	Orientation	Light Construction 0800 hrs	1000	1200	1400	1600	1800	Heavy Construction 0800	1000	1200	1400	1600	1800 hrs
June 21	N	73	71	81	81	71	73	73	72	81	81	72	73
	NE	261	170	82	82	72	56	252	168	85	85	76	60
	E	328	305	172	82	72	55	318	297	172	88	79	63
	SE	225	294	256	128	62	45	221	286	250	131	69	54
	S	40	151	236	236	151	40	47	150	230	230	150	47
	SW	45	62	128	256	294	225	54	69	131	250	286	221
	W	55	72	82	172	305	328	63	79	88	172	297	318
	NW	56	72	82	82	170	261	60	76	85	85	168	252
Aug. and Apr.	N	26	44	54	54	44	26	27	44	53	53	44	27
	NE	184	112	54	55	45	27	177	110	57	57	47	31
	E	268	281	149	56	46	28	259	272	148	62	52	36
	SE	210	312	280	149	46	28	208	303	273	151	55	38
	S	46	187	280	280	187	46	54	186	273	273	186	54
	SW	28	46	149	280	312	210	38	55	151	273	303	208
	W	28	46	56	149	281	268	36	52	62	148	272	259
	NW	27	45	55	55	112	184	31	47	57	57	110	177
Sept. and Mar.	N	13	33	43	43	33	13	14	33	42	42	33	14
	NE	97	66	44	44	33	14	94	64	44	44	34	16
	E	165	247	133	45	34	15	160	237	131	49	39	21
	SE	144	304	291	167	36	16	145	294	282	166	44	25
	S	48	203	305	305	203	48	57	202	297	297	202	57
	SW	16	36	167	291	304	144	25	44	166	282	294	145
	W	15	34	45	133	247	165	21	39	49	131	237	160
	NW	14	33	44	44	66	97	16	34	44	44	64	94
Correction factor to peak load for 24 hour plant operation		1·00						0·95					

TABLE 17·5
FACTORS FOR SHADING

(a) Applying to Table 17.3
Type of glass unshaded

	Glazing Single	Double
Clear - - - - - - -	1·00	0·88
Lightly heat absorbing - - - -	0·75	0·56
Densely heat absorbing - - - -	0·61	0·41
Heat reflecting gold (sealed unit when double) - - -	0·41	0·32
Lacquer coated, grey - - -	0·79	—
External shades		
Dark green open weave plastic - - -	0·33	0·26
Canvas roller blind - - - - -	0·22	0·16
White louvre sunbreaker, blades at 45° -	0·19	0·14
Dark green miniature louvre - - -	0·18	0·13

(b) Applying to Table 17.4

	Single	Double
Mid-pane Venetian blind white	—	0·60
Internal		
Dark green open weave plastic - -	1·32	1·25
White Venetian blind - - - -	1·00	1·02
White cotton curtain - - - -	0·77	0·82
Cream holland linen blind - - -	0·61	0·71

2B

pavement subject to solar gain, some addition to the window gain seems justified to allow for upward reflection. An addition of ten per cent is suggested.

It is perhaps worth noting that some modern all-glass buildings, even in Great Britain, are virtually 'un-airconditionable' (and in consequence virtually uninhabitable) unless measures are taken to exclude excessive solar gains.

External shades or sunbreaks are preferable to any form of internal shading, but they have in the past been considered as involving difficulties in maintenance; new materials and new methods of blind operation are beginning to solve this problem, however. Heat-reflecting glasses depend on finely divided metal incorporated in the glass and give a better performance than so-called 'heat-absorbing glass', due to the fact that they reflect radiation rather than absorb it. This feature of high reflection has, however, been known to cause serious complaint from occupants of neighbouring buildings.

In addition to solar gain by direct transmission of radiation, the air-to-air conductance must be allowed for, being the product of area, U value and air-temperature difference outside to inside. Double glazing in this case will show to advantage over single glazing. U factors may be taken as for heating, although some adjustment may be made for a higher external-surface resistance in summer.

Infiltration—This is the natural uncontrolled movement of air through an air-conditioned space due to opening of doors and minute crackage at fenestration. An allowance of one half air change per hour appears to cover this sort of leakage, but if double windows and impervious linings are used, this may be reduced to one quarter air change per hour. Special cases may occur in industrial buildings where doors must be open for transport of goods, etc. Air locks at all openings to unconditioned spaces are essential if conditions are not to go out of control.

Ventilation—The air-conditioning plant will of necessity be the means of introducing ventilation air, the rate depending on the number of occupants and whether smoking is permitted or not. Rates vary from a minimum of 0·005 m³/s (5 litre/s) per occupant, without smoking, to 0·018 m³/s (18 litre/s) assuming heavy smoking. Where contamination is concerned, as in a hospital operating theatre, 100 per cent fresh air is required: i.e. no recirculation. The same applies in any industrial application where fumes or dust are produced.

Another special case occurs in department stores owing to the very uncertain and variable number of occupants. Investigations carried out by one of the revisers* show that, from actual counts, concentrations vary from four persons per 10 m² to ten persons per 10 m² at peak periods. The fresh-air requirements is given at 5 litre/s per occupant.

The cooling load and heating load for ventilation air may be estimated on the basis of enthalpies, as is discussed later in this Chapter in the section on psychrometry.

* P. L. Martin: *I.H.V.E. Journal*, August, 1970.

Sun-Path Diagrams—These diagrams are so drawn as to enable altitude and azimuths to be read graphically for various periods of year and times of day. A sample diagram appears on page 493 for latitude 50° N. Similar diagrams for other latitudes are available in the form of overlays.*

A miniature plan of the building under consideration may be placed at the centre of the diagram, orientated correctly, and by inspection it is possible to note the likely peak incidence of solar radiation throughout the day on the various faces of the building. It will be clear that easterly faces will receive solar radiation first, the south-east and south next, meanwhile the east side will be going into shadow and so on.

For the purpose of assessing coincident cooling load, a schedule may be made of the various rooms or areas of the building and, taking each in turn, its own peak period may be established. From the diagram, coincident loads on various faces may be observed and, from this, the likely optimum throughout the day. This optimum may vary at different seasons of the year, as discussed in Chapter 22. Whilst this exercise is fairly straightforward in the case of a square or rectangular building, it becomes more complicated in a plan with re-entrant angles or internal courts where the building itself creates shadows on certain faces at certain periods. A model can be most helpful here. In the ultimate solution of a large scale project, calculation by computer is the obvious answer. Fig. 17.16 illustrates a

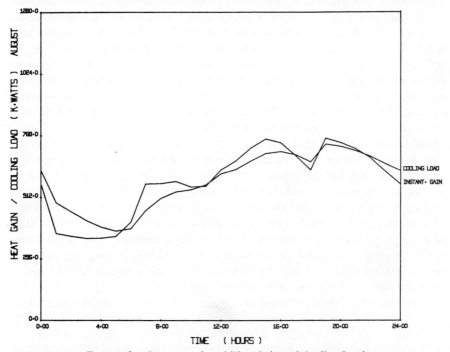

FIG. 17.16.—Computer plotted Heat Gain and Cooling Load.

* Petherbridge: *Sunpath diagrams and overlays*. Building Research Station, *Research Series, No. 39*.

computer plot output showing the cooling load for a complete building of about 750 kW loading, and its displacement from peak heat gain.*

A second use of the sun-path diagram is in the design of sunbreaks. The angles of incidence may be plotted on the plan and separate diagrams set up for the various aspects. Both horizontal and vertical sunbreaks may be employed, or a combination of both. Much architectural ingenuity is possible here. Adjustable sunbreaks are another possibility to overcome the disadvantage of completely obscuring the view. Such have indeed been used successfully in a building in Manchester.

The sun-path diagram may further be used to calculate solar gain from first principles, given the solar intensity, transmittance or absorbance of materials, and from the diagram determining the angles of incidence hour-by-hour and day-by-day. The task, however, is unnecessary if use is made of the data in the *Guide* or of the further boiled-down versions contained in the Tables in this chapter, which are admittedly limited to specific parameters. These are intended to indicate the process rather than the detail.

Design Temperatures, Summer—*External, U.K.*: A commonly accepted design condition is 30° C D.B., 20° C W.B. Infrequent occasions occur of higher temperatures, but are disregarded for design purposes. *Internal, U.K.*: An effective temperature of 19° C, represented by 21° D.B., 50 per cent sat. (50 per cent relative humidity, in common parlance).

Heat from Occupants—The heat emitted from the human body depends on activity (see p. 37), temperature, relative humidity and air-movement.

Fig. 17.17 shows, for a man seated at rest, how the sensible, latent and total heat varies with the dry bulb temperature for average humidities in still air. The sensible heat increases by about 20 per cent with air-movements up to 0·5 m/s. The latent heat is affected only slightly with air-movement under about 18° C, above which temperature the air-movement displaces the point at which the curve begins to rise, e.g. at 0·5 m/s the rise in evaporation rate begins at about 26° C.

Heat from Lights, Motors, etc.—This is dealt with fully on page 38. The energy input to any machine or heat-producing appliance within the space (including lighting) constitutes a heat gain and can be evaluated in kilowatts. Exceptions are where the energy produced, as in an electric motor, is dispersed outside the space—as possibly in a conveyor system or air compressor. In such cases only the motor inefficiency comes out in heat in the space. Another exception is where heat from lighting is removed by arranging extraction from. around the lighting fitting—a practice which is becoming increasingly common and upon which subject much research remains yet to be evaluated (see page 421).†

Having established the total heat gain in the room itself, the mass flow of air to be delivered to the room is found, as has already been shown, by

---

* The authors are indebted to Oscar Faber and Partners for the use of their computer program 'COLO' to produce this illustration.
† Bedocs and Hewitt: *I.H.V.E. Journal*, January, 1970.

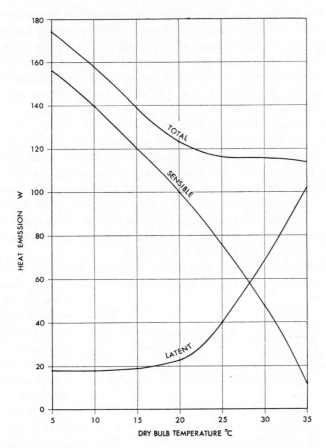

FIG. 17.17.—Latent, sensible, and total heat for a man seated at rest.

dividing this figure by the specific heat capacity of air × temperature rise.

PSYCHROMETRY

Psychrometry concerns the behaviour of mixtures of air and water vapour, and a knowledge of it is necessary in any air-conditioning calculations. The general principles were referred to on pp. 6 and 7. A complete study is outside the scope of this book and is dealt with in many text-books on thermodynamics.

The properties of a mixture of air and water vapour may be summarized as follows:

Dry-Bulb Temperature (D.B.) is the temperature of the air as indicated by an ordinary thermometer.

Wet-Bulb Temperature (W.B.) is a temperature related to the humidity of the air and to the dry-bulb temperature. If a thermometer has its bulb kept wet, as by a wick surrounding it, the evaporation of the water will take heat from the mercury which will consequently contract and indicate a lower temperature than in its dry-bulb condition. The rate of evaporation from the wetted bulb depends on the humidity of the air, i.e. very dry surrounding air will cause more rapid evaporation than moister air at the same dry-bulb temperature. The difference between dry- and wet-bulb temperatures can thus be used as a measure of the humidity. This difference is known as the *Wet-Bulb Depression.*

The rate of evaporation, and consequently the depression, depends also on whether the wetted bulb is exposed to still or moving air. In the latter case the film of moisture-laden air round the bulb is removed more rapidly than in the former, so that the depression is greater. Some systems of psychrometry adopt one condition as a standard, some the other. It is immaterial which is used provided the instruments are suitably calibrated, so that a given pair of dry- and wet-bulb readings can be accurately converted into a measure of the actual humidity.

Sling Wet Bulb is the temperature indicated by a sling or whirling psychrometer (Fig. 17.18), consisting of two thermometers mounted in a

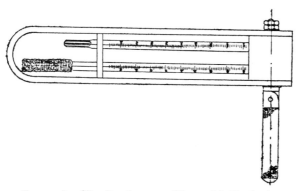

frame which can be whirled by hand. One thermometer has a wetted sleeve slipped over the bulb. Assmann has found that above an air-speed of 2·5 m/s and up to at least 45 m/s the depression is independent of the air-speed.

Another type of psychrometer is the Assmann type in

Fig. 17.18.—Sling Psychrometer (Negretti & Zambra).

which the air is drawn, by means of a small fan, over thermometers mounted in a tube. In this intrument the wet bulb is kept moist by means of a wick in a reservoir.

Screen Wet Bulb is the temperature indicated by a wet-bulb thermometer in stationary air, and is so called because the standard instrument is housed in a louvred box called a Stevenson Screen.

Another approach to measurement of moisture content is the conception known as *Temperature of Adiabatic Saturation.* This is the temperature which would be attained by moist air in intimate contact with a water surface, and in equilibrium conditions, assuming that no heat is gained or lost to an external source in the process of attaining equilibrium.

This process is a close approximation to that taking place in the wetted wick of a wet-bulb thermometer, and at ordinary atmospheric temperatures and humidities the Sling Wet Bulb and Adiabatic Saturation Temperature differ very little, which has led to the latter being adopted in many psychrometric systems in place of the true wet-bulb temperature.

This substitution has the great advantage that conditions of equal Adiabatic Saturation correspond to equal Total Heat, i.e. Adiabatic Saturation temperature becomes a criterion of Total Heat.

Dew-point Temperature is the temperature to which a mixture of air and water vapour must be reduced to produce condensation of the vapour. At the dew-point the air is said to be *saturated*, and in this condition the dry-bulb and wet-bulb temperatures both coincide and are equal to the dew-point temperature.

Moisture Content is the mass of water vapour contained in unit mass of the mixture. This is expressed in kg per kg of *dry air*, and not per unit of the mixture.

Vapour Pressure is the partial pressure exerted by the water vapour, and is expressed in millibars.

It follows from the above that the vapour pressure at any condition is dependent only upon the absolute humidity, and is independent of the pressure of the mixture.

Relative Humidity (R.H.) is the ratio of actual vapour pressure to the vapour pressure exerted when the air is saturated at the same temperature, expressed as a percentage. At the saturation temperature the relative humidity is 100 per cent, and dry bulb, wet bulb and dew-point are the same.

At ordinary humidities and at temperatures up to $40°$ C the vapour pressure is substantially proportional to the absolute humidity, from which it follows that the relative humidity, as defined above, is approximately equal to the ratio of the absolute humidity at the given condition to that of saturation at the same D.B. temperature. This ratio is referred to as *percentage saturation*.

Total Heat or Specific Enthalpy is the sum of the sensible and latent heats of the air and of the moisture contained in it at any given condition, reckoned from an arbitrary datum of $0°$ C. Specific enthalpies are referred to in kJ/kg of dry air.

Volume—The relationship between pressure, volume and temperature has been referred to on p. 13.

Specific volume is the volume of the mixture in m^3 per kg of dry air. Air containing water vapour is less dense than dry air, hence the volume occupied is greater per unit mass.

Table 17.6 gives data regarding volume, mass of saturated vapour, and total heat of dry and saturated air.

Barometric Pressure—The standard atmospheric pressure now adopted in SI units and in the *Guide* is, at sea level, $101\ 325$ N/m² $= 1013·25$ mbar $= 1·01325$ bar. The bar may thus be taken as the near equivalent of atmospheric pressure. The psychrometric properties of mixtures of air and water vapour given in the *Guide* and referred to later in this chapter may be taken as accurate for all practical purposes from 950 mbar to 1050 mbar. As a matter of interest the relationship between altitude and pressure is as follows:

Altitude (m)	500	1000	1500	2000	2500	3000
Barometric pressure (mbar)	950	895	845	795	745	700

TABLE 17·6

PROPERTIES OF DRY AND SATURATED AIR AT
VARIOUS TEMPERATURES

Temperature °C	Specific Volume m³ per kg Dry Air	Saturation Moisture content kg/kg Dry Air	Specific Enthalpy (Total Heat) kJ/kg Dry Air	
			Dry Air	Saturated Air
0	0·77	0·003 8	0·00	9·47
2	0·78	0·004 6	2·01	12·98
4	0·79	0·005 0	4·02	16·70
6	0·79	0·005 8	6·04	20·65
8	0·80	0·006 7	8·05	24·86
10	0·80	0·007 6	10·06	29·35
12	0·80	0·008 7	12·07	34·18
14	0·81	0·010 0	14·08	39·37
16	0·82	0·011 4	16·10	44·96
18	0·82	0·012 9	18·11	51·01
20	0·83	0·014 7	20·11	57·55
22	0·83	0·016 7	22·13	64·65
24	0·84	0·018 9	24·14	72·37
26	0·84	0·021 4	26·16	80·78
28	0·85	0·024 2	28·17	89·96
30	0·86	0·027 3	30·18	99·98
32	0·86	0·030 7	32·20	111·0
34	0·87	0·034 6	34·21	123·0
36	0·87	0·038 9	36·22	136·2
38	0·88	0·043 7	38·24	150·7
40	0·89	0·049 1	40·25	166·6

From the *I.H.V.E. Guide*: at atmospheric pressure of 1013·25 mbar

Psychrometric Chart—The relationships between the various properties of a mixture of air and water vapour may be expressed in the form of Tables, such as those published in the *I.H.V.E. Guide*, or the chart based thereon, as in Fig. 17.19.*

The method of using the chart is indicated by the diagram, Fig. 17.20. The chart is constructed as follows:

Dry Bulb—Evenly divided horizontal scale with divisions of 0·5° C. Read vertically to the required point.

Moisture Content—Evenly divided vertical scale reading in kg per kg dry air. Read horizontally.

---

* Copies of the chart in pad form are obtainable from the I.H.V.E., price 6op. Charts for non-standard pressures, computer drawn, are published by Troup Publications Ltd.

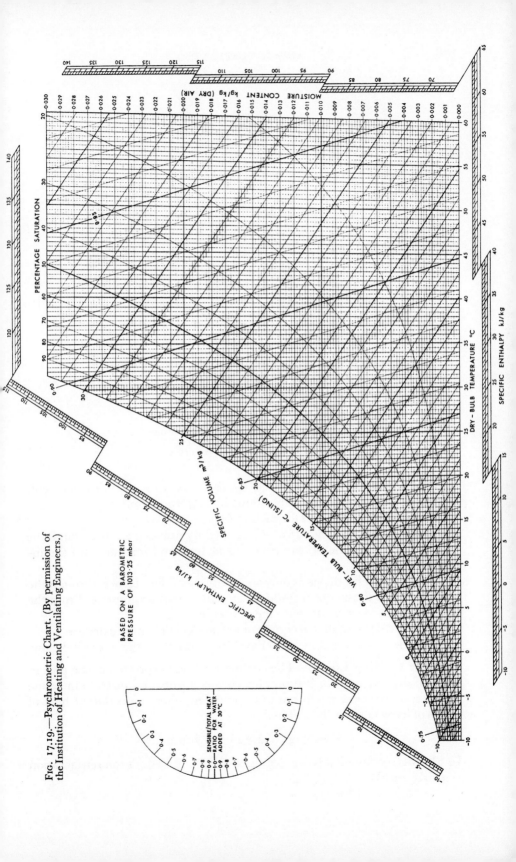

Fig. 17.19.—Psychrometric Chart. (By permission of the Institution of Heating and Ventilating Engineers.)

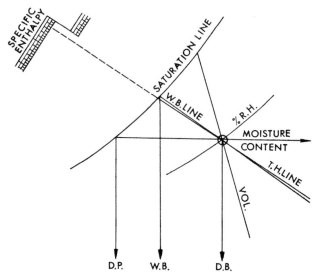

Fɪɢ. 17.20.—Example of the use of the Psychrometric Chart.

Point marked is 20° C D.B., 50% Sat. Other data
read from the chart is:

W.B.   = 13·9° C
D.P.   = 9·4° C
Moisture content = 0·0074 kg/kg dry air
T.H.   = 38·8 kJ/kg
Specific volume = 0·84 m³/kg dry air

Percentage Saturation—Indicated by the curves for each 10 per cent.
Intermediate percentages can be otained by interpolation.

Wet Bulb—Sloping lines for each 1° C.

Specific Enthalpy or Total Heat—Obtained by drawing a line through the
required point parallel to the nearest ᴛ.ʜ. line, and reading onto the inclined ᴛ.ʜ. scale.

Dew-point—This depends only on moisture content. Hence, through the
required point draw a horizontal line to cut the saturation line. From the
intersection read vertically onto the ᴅ.ʙ. scale.

Volume—Sloping lines of equal volume are given in m³/kg dry air.

The chart and Table 17.6 have been used in the *example* below.

Unit for Cooling Loads—The cooling load was expressed in the past in
terms of the 'ton of refrigeration', which was the heat required to melt one
U.S. ton of ice (i.e. 2000 lb.) at 0° C per twenty-four hours (latent heat of
fusion of ice = 144 Btu/lb.). Thus:

$$(2000 \times 144)/24 = 12\ 000 \text{ Btu/hour.}$$

In SI units, cooling loads are expressed simply in kW: (1 ton refrigeration
= nearly 3·5 kW).

AIR-CONDITIONING CALCULATIONS
1. DESIGN OF PLANT FOR SUMMER COOLING AND DEHUMIDIFICATION

Having arrived at the maximum hourly heat gain for the spaces to be conditioned, it is necessary to calculate the conditions to be maintained in the plant, and the capacity of the refrigeration unit, pump, etc.

Method—The following is a summary of the steps in the process which have to be calculated:

(a) Heat gains and solar gains through the structure.

(b) Infiltration gain.

(c) Internal heat gains.

(d) The mass flow of ventilation air and hence the cooling for same.

(e) The mass flow of air required for a given temperature rise, inlet to outlet; this is derived from the total sensible gains (a), (b) and (c).

(f) The recirculation mass flow, which equals (e) minus (d), and from whence the total heat of the mixture is found.

(g) The plant exit condition D.B., which is derived from room D.B. minus rise in room minus allowance for duct and plant gain. The moisture at plant exit derives from the moisture at room condition minus the increment due to occupants and infiltration.

(h) The apparatus dew-point, which follows from (g).

(j) The cooling load, which is equal to the product: mass air flow through plant and heat removed as between T.H. of entering mixture and T.H. at dew-point.

(k) The reheat load, refrigeration load and fan duties.

Example—

Basis of Design

Building	=	Public Hall
Seating capacity	=	500
Lighting	=	20 kW
Cube	=	3000 m³
Internal conditions summer	=	23° C D.B.
50% Sat	=	16·4° C W.B.
T.H.	=	45·79 kJ/kg
External conditions summer	=	30° C D.B.
		20° C W.B.
T.H.	=	56·71 kJ/kg
Fresh air per occupant	=	0·005 m³/s

Fig. 17.21 gives a plan and section of the building, its construction and orientation.

The orientation of the building presents two faces to solar gain.

From Table 17.4 it is clear that the SW glass will be subject to the greatest heat gain at 1600 hours in August and September.

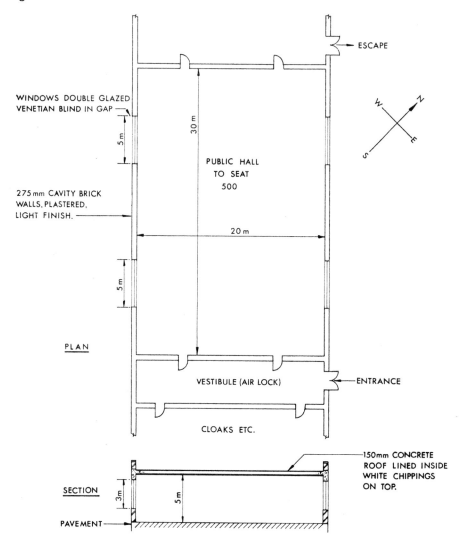

FIG. 17.21.—Plan, Section and Orientation of Building in the *Example*.

Cooling loads taken for August may be arrived at as follows:
Glazing. (Double glazing; Venetian blind in gap).

    Area NE = 30 m²
    Area SW = 30 m²

From Table 17.3 (internal construction taken as 'light')

Area NE	45 W/m²		
Area SW	312 W/m²		
Cooling load	30 × 45	=	1350 W
	30 × 312	=	9360 W

Allow reflected radiation from pavement SW
side 10%

$$= \quad \underline{936 \text{ W}}$$
$$11 \overline{646 \text{ W}}$$

Shade factor (Table 17.5(b))   0·6 W   7·00 kW

Air to Air. Take U = 3·0 W/m² °C

     Temp. diff.   7° C

     Area   60 m²   1·26 kW

$$\overline{8\cdot26 \text{ kW}}$$

Walls

Construction: 260 mm cavity, lined. Surface light, U = 1·5 W/m² °C.

From Table 17.2, time lag will be about $8\frac{1}{2}$ hours and decrement factor $f = 0\cdot3$.

Time considered for calculation is 1600 hrs.; so, allowing for time lag, we read sun heat increment $8\frac{1}{2}$ hours earlier (say 0800 hrs.).

Thus

N.E. Wall (from Table 17.1), increment = 17·4° C

Area = (30×5) – windows = 120 m²

$$\frac{120 \times 1\cdot5 \times 0\cdot3}{1000}\{(30-23)+17\cdot4\} \qquad = 1\cdot32 \text{ kW}$$

S.W. Wall (from Table 17.1), increment = 1·2° C

Area, as before, = 120 m²

$$\frac{120 \times 1\cdot5 \times 0\cdot3}{1000}\{(30-23)+1\cdot2\} \qquad = \underline{0\cdot44 \text{ kW}}$$

$$\overline{1\cdot76 \text{ kW}}$$

Roof

Construction: 150 mm concrete, lined internally.

Surface light, U = 2·0 W/m² °C

From Table 17.2, time lag will be about $5\frac{1}{2}$ hours and decrement factor $f = 0\cdot5$.

Time considered for calculation is 1600 hrs.; so, allowing for time lag, we read sun heat increment $5\frac{1}{2}$ hours earlier (say 1030 hrs.).

Thus, from Table 17.1, increment = 17·1° C

Area = $(30 \times 20) = 600$ m²

$$\frac{600 \times 2\cdot0 \times 0\cdot5}{1000}\{(30-23)+17\cdot1\} \qquad = \underline{14\cdot45 \text{ kW}}$$

Infiltration

Allow half airchange per hour

Cube = 3000 m³

Specific volume dry air at room condition 0·85 kg air

$$\frac{0\cdot5 \times 3000}{0\cdot85} = 1765 \text{ kg/hr}$$

T.H. dry air external   30·18

T.H. dry air internal   23·14

    difference   7·04 kJ/kg

$$\frac{1765 \times 7\cdot04}{3600} \qquad\qquad = 3\cdot45 \text{ kW}$$

Gain due to moisture diff. of T.H.s moisture

$$(56 \cdot 71 - 45 \cdot 79) \qquad = 10 \cdot 92$$

$$\frac{1765 \times (10 \cdot 92 - 7 \cdot 04)}{3600} \qquad\qquad = \underline{1 \cdot 90 \text{ kW}}$$

Internal Heat Gains

500 occupants @ 88 W sensible (from Fig. 17.16)	= 44 kW	
20 kW lighting	= 20 kW	64 kW

Gain due to moisture

500 occupants @ 30 W   =   15 kW

Total heat gains sensible

$8 \cdot 26 + 1 \cdot 76 + 14 \cdot 45 + 3 \cdot 45 + 64 \cdot 00$ =   91·92 kW

Total gains due to moisture

$1 \cdot 90 + 15 \cdot 00$ =   16·90 kW

(*Note:* Heat gains due to occupants assume they are seated at rest. If for instance they were dancing, the heat gain would increase, but the number would be less.)

Ventilation Air

Fresh air 500 occupants @ 0·005 m³/s   =   2·5 m³/s

$$\frac{2 \cdot 5}{0 \cdot 85} \qquad\qquad = \qquad 3 \cdot 0 \text{ kg/s}$$

Sensible heat gain, by difference, T.H.s as dry, 7·04 kJ/kg

$3 \cdot 0 \times 7 \cdot 04$ =   21·12 kW

Due to moisture, 3·88 kJ/kg

$3 \cdot 0 \times 3 \cdot 88$ =   11·64 kW

Total Load

It is of interest to consider the relation of the components of the total load for, whilst no building can be said to be 'typical', the significance of the internal gains and ventilation gains is worth noting as is the ratio between the sensible gains and latent gains.

Sensible Gains	%	%	%
Solar			
Glazing	4·9		
Walls	0·7		
Roof	7·2	12·8	
Air-to-Air			
Glazing	0·9		
Walls	0·6		
Roof	2·9	4·4	
Internal			
Lighting	14·0		
Occupancy	31·0	45·0	
Ventilation			
Infiltration	2·5		
Fresh Air Supply	15·0	17·5	79·7*

* This total does not include any reheat, fan, pump or other plant gains which may be quite significant.

Latent Gains
Internal
    Occupancy   10·6
Ventilation
    Infiltration   1·3
    Fresh Air Supply   8·4   9·7   20·3
                                        100·0

Air Quantity

Permissible temp. rise between room inlet and
outlet assume $= 8°$ C
Room inlet temp. then $= 23 - 8 = 15°$ C
Specific heat capacity air take as 1·0 kJ/kg

Then air required $= \dfrac{91·92}{8 \times 1}$ $= 11·5$ kg/s

Specific volume $= 0·82$ m³/kg
Fan duty $= 11·5 \times 0·82$ $= 9·5$ m³/s (9500 litre/s)

Air change rate $= \dfrac{9·5 \times 3600}{3000} = 11·4$ air changes/hr

Of the mass of air provided, 11·5 kg/s, 3·0 is fresh (26%), 8·5 recirculated
(74%)

T.H. of mixture
    Room $= 45·79 \times 8·5 = 392$
    Fresh $= 56·71 \times 3·0 = \underline{170}$
                       $\underline{562} \div 11·5 = 49·0$ kJ/kg

Plant Exit Condition

This will differ from the room inlet by heat gain through the duct walls
(thermally insulated to reduce it as much as possible) and also by heat gain from
the energy input by the fan. These may be assessed in detail but for present
purposes will be taken as $12\frac{1}{2}$ per cent of sensible gain, thus accounting for $1°$ C
rise.

Plant exit condition then $= 15 - 1$ $= \underline{14°}$ C D.B.
Moisture.   Room condition $= 0·0089$ kg/kg

Increment in T.H. $= \dfrac{16·9}{11·5} = 1·47$ kJ/kg

                      $\div$ latent heat 2450 kJ/kg $= \underline{0·0006}$ kg/kg
Thus, air leaving plant $= \underline{0·0083}$ kg/kg
    at $14°$ C $= 83\%$ sat
                              $= 12·4°$ C W.B.
    T.H. $= 35$ kJ/kg

Apparatus Dew-point

Moisture content of 0·0083 corresponds to a dew-point of $11·2°$ C
    T.H. $= 32·0$ kJ/kg

Cooling Load

Mass flow air through plant		$= 11 \cdot 5$ kg/s
T.H. mixture	49 kJ/kg	
T.H. dew-point	32	
difference	$= \overline{17}$	
$11 \cdot 5 \times 17$		$= \underline{195 \cdot 5}$ kW

Washer data

Outlet D.P.	$11 \cdot 2°$ C
Differential water outlet to D.P.	$2 \cdot 0°$
	$\overline{9 \cdot 2°}$
Allow temp. rise in washer	$3 \cdot 0°$
Chilled water inlet temp.	$\overline{6 \cdot 2°}$ C
Water mass flow through sprays	
$\dfrac{195 \cdot 5}{3 \cdot 0 \times 4 \cdot 2}$	$= 15 \cdot 5$ kg/s
	$= 15 \cdot 5$ litre/s

Chilled water from refrigerating plant (allowing 10% heat gain in pump and piping) say $5 \cdot 7°$ C.

Reheat Load

Apparatus dew-point T.H.	$= 32$ kJ/kg
Plant outlet condition T.H.	$= \underline{35}$
difference	$= \underline{3}$
Air mass flow	$= 11 \cdot 5$ kg/s
Heat input $11 \cdot 5 \times 3$	$= \underline{34 \cdot 5}$ kW

Refrigeration Load

Cooling load as above	$195 \cdot 5$ kW
Heat gain in pump and piping, say 10%	$\underline{19 \cdot 5}$
Compressor duty	$= 215 \cdot 0$ kW*

Fan Duties

Inlet fan as above	$= \underline{9 \cdot 5 \text{ m}^3/\text{s}}$ (9500 litre/s)
Exhaust fan (equal to fresh air) $3 \times 0 \cdot 85$	$= 2 \cdot 55 \text{ m}^3/\text{s}$ (2550 litre/s)
Recirculation fan (if provided) $8 \cdot 5 \times 0 \cdot 85$	$= 7 \cdot 2 \text{ m}^3/\text{s}$ (7200 litre/s)
Total extract	$= 9 \cdot 75 \text{ m}^3/\text{s}$ (9750 litre/s)

(The increase over inlet is due to the rise in temperature. In practice, some excess of inlet over extract would be allowed to reduce infiltration).

Fig. 17.22 shows, graphically, the psychrometric changes taking place in the above example on a small area of the chart. Fig. 17.23 illustrates the conditions and other data on a plant diagram.

* In old units = 61·5 tons R.

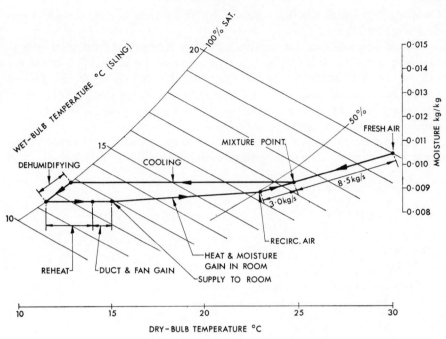

Fig. 17.22.—Diagram of Psychrometric Changes (see *Example*).

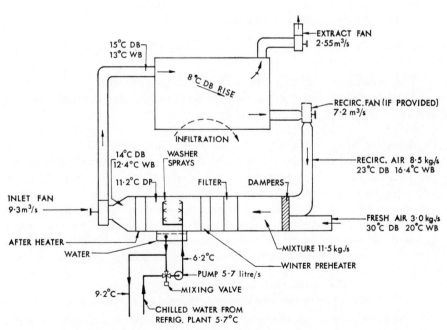

Fig. 17.23.—Air-Conditioning Plant showing conditions when cooling (see *Example*).

2 C   K.H.A.C.

Cooling by Cold Coils—Cooling by this method has already been described (see Fig. 17.2).

To avoid cooling all the air down to a low dew-point for dehumidification and subsequently reheating, advantage may be taken of a characteristic of coil surfaces whereby moisture can be deposited on the chilled surface, without the bulk of the air being reduced to the same temperature. This is achieved by keeping the surfaces at the lowest possible temperature consistent with absence of freezing.

In the example above, if chilled water coil surface be used instead of an air washer, the calculation of refrigeration load would be as follows:

Mixed air inlet to coil
    3 kg at 30° C   = 90
    8·5 kg at 23° C   = 196·5

                               $186·5 \div 11·5 = 25°$ C

    T.H.   = 49 kJ/kg
    D.P.   = 13° C
Outlet from coil (as above)
    D.B.   = 14° C
    W.B.   = 12·4° C
    T.H.   = 35 kJ/kg

    Diff T.H. = 14 kJ/kg
Mass flow   = 11·5 kg/s
Cooling load = 11·5 × 14   = 162 kW
    Heat gain in pump and piping 10% = 16 kW

Compressor duty   = 178 kW

This compares with 215 kW with the washer system.

Coils cooled by direct expansion would be calculated similarly, except that the surface temperature would be lower.

Cooling by Sprayed Coils—In this system unchilled and recirculated water is sprayed over the chilled coil surface, thus combining both the previous methods. The water serves to flush the coils of dirt accumulations but adds to resistance to air flow. The balance between the respective duties performed by the coil surface and the water spray are inherent in a particular design, depending on coil surface area, fin spacing and water quantity. It has the merit of compactness compared with a washer, and provides means for winter humidification which cooling coils alone do not.

AIR-CONDITIONING CALCULATIONS (*Continued*)

2. DESIGN OF PLANT FOR WINTER WARMING AND HUMIDIFYING

The operation in winter may be with partial recirculation, as in summer, or with 100 per cent fresh air. The plant might be arranged so as to be suitable for the latter method of operation, in average spring and autumn weather, when neither cooling nor heating is required. In cold weather recirculation might be used for economy of running.

The following calculation assumes 100 per cent fresh air at 4° C, below which it is assumed that recirculation would be used. This is often done by an outside control operating motorized dampers to vary the proportion of fresh to recirculated air.

Example—

Basis of Design

Fresh air	4° C sat.		T.H.	17 kJ/kg
Room (fabric losses offset by direct heating: room empty)				
	20° C 60% sat.		T.H.	42 kJ/kg
D.P. of room condition	= 12° C		T.H.	34 kJ/kg
Mass flow air as for cooling			=	11·5 kg/s

Plant Duty

Pre-heater or spray-water heater

$(34 - 17) \times 11·5$ = 195 kW

Note: The washer performs adiabatically; thus T.H. at entry is the same as at leaving, following a constant wet bulb line of 12° C.

After-heater

$(42 - 34) \times 11·5$ = 92 kW

287 kW

Moisture added

Room condition	0·0088 kg/kg	
Outside ,,	0·0050 ,,	
	0·0038 ,,	
11·5 × 0·0038	= 0·0444 kg/s	
	= 160 litre/hr	

(The above assumes the air washer 100% efficient, i.e. saturating the air at the dew-point. Some allowance for inefficiency would in practice be made depending on the form of washer).

Steam Humidification—Where cooling is by cooling coils, humidity may be increased by direct injection of vapour into the air stream.

Where some slight trace of odour is objectionable, the vapour should be generated indirectly by electric immersion heater, hot water or steam coil; steam direct from a steam boiler is liable to carry smell. Steam in contact with hot copper or steel acquires smell, but tinned copper seems to leave vapour odourless. The method of injection may be by one of the proprietary perforated pipes with the steam jacket designed to pass only dry steam. The point of injection should be after the cooling and heating coils.

In the example above, the element generating vapour would have an input (taking latent heat at 2500) of

$$0.0444 \times 2500 = 111 \text{ kW}$$

Air heater. Temp. rise 16° C
Specific heat capacity air 1·0
Mass flow 11·5 kg/s

$$11.5 \times 16 \times 1 = 184 \text{ kW}$$

Total heat requirement   295 kW

This compares with 287 kW in the previous example, the difference being due to roundings up.

AIR-CONDITIONING SYSTEMS—BASIS OF CALCULATIONS APPLYING TO OTHER SYSTEMS

Fan Coil, ducted fresh air—The central plant supplying fresh air is designed to deal with:

(*a*) moisture adjustment up or down from outside to room condition;

(*b*) temperature adjustment to supply constant temperature in summer and raised in winter.

Coils in units then cater for cooling or warming requirements only according to sensible heat gains or losses of the room.

Unit Air-Conditioners, recirculating—Air handling, capacity over design temperature range is calculated to meet a suitable differential such as 8° C. The cooling coil then deals with the sensible heat load of room plus dehumidification if any.

Warming is designed to meet winter heat losses plus humidification if any.

Unit Air-Conditioner with fresh air intake—The cooling coil deals with:

(*a*) sensible heat gains of room

(*b*) cooling of fresh air intake

(*c*) dehumidification of fresh air

(*d*) dehumidification additionally owing to moisture gain if any.

Warming is designed to meet:

(*a*) room heat losses

(*b*) warming of fresh air to room temperature

(*c*) evaporation of moisture for humidification where such is provided.

Induction System—Cooling/warming coils are designed to deal with sensible heat gains or losses of rooms.

Primary air, conditioned in the central plant, is designed to provide supply at a scheduled temperature and moisture content according to external conditions.

Dual-Duct System—Cool air duct—air quantity and temperature differential are designed to suit total anticipated peak load of:

(a) recirculated air bringing with it heat and moisture gains of building;

(b) fresh air.

Warm air duct—temperature is designed to meet heat losses for maintaining building at desired temperature. Volume is usually allowed at 75 per cent of inlet.

AUTOMATIC CONTROLS

Automatic controls are an essential part of any air-conditioning system. The wide variety of purposes for which such plants are required, the number of different systems which it is possible to select, and the multitude of types of control equipment available require a complete book in themselves to cover adequately.

It is possible here only to indicate the main principles involved, and some of the chief component items of apparatus commonly used. From this it will be possible to understand some of the problems concerned and how they may be tackled. Other permutations and combinations of system and equipment may be built up to suit any variety of circumstances.

Five main types of plant will be considered and reference to the diagrams will be necessary along with the following descriptions.

Fig. 17.24—This is the type of plant envisaged in the cooling and heating calculations taken earlier.

Summer Control

Required dew-point is controlled by low limit 'stat $L.L.T._1$, operating, mixing valve in washer pump suction whereby water is drawn from the cooled water supply (from the evaporator) or from the washer tank, in the requisite proportions to give the resultant mixture at the correct temperature.

Should the hall be empty (or partially so) the humidity gain will be less, and a higher dew-point will suffice, hence a humidistat H. is placed in the recirculation duct (equivalent to placing it in the room) so as to shut off the mixing valve against the cooled water inlet at a higher temperature.

Final temperature leaving fan is controlled by low limit 'stat $L.L.T._2$, admitting heat to the heater through $T.V._2$, whenever the air temperature drops below the designed condition and cutting off heat when it rises above.

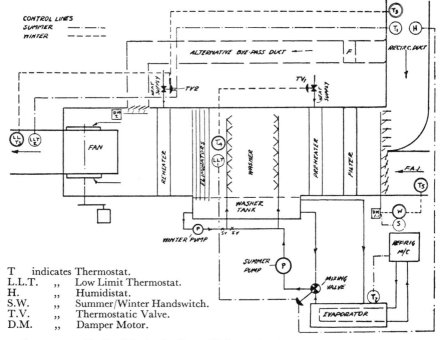

T indicates Thermostat.
L.L.T. ,, Low Limit Thermostat.
H. ,, Humidistat.
S.W. ,, Summer/Winter Handswitch.
T.V. ,, Thermostatic Valve.
D.M. ,, Damper Motor.

FIG. 17.24.—Air-Conditioning by Spray Washer and Dehumidifier. Summer and Winter Control Diagram.

Again, should the heat gain in the hall be insufficient to maintain the return air at the required temperature, thermostat T. will come into play and admit heat via $T.V._2$, thus over-riding $L.L.T._2$.

To maintain a constant cooled water outlet temperature from the evaporator, $T._2$ stops and starts the refrigerating compressor, or in some cases may control its speed.

The proportions of recirculated to fresh air as designed will be controlled by the louvres shown, adjusted by the damper motor $D.M._1$. This can most simply be controlled by the hand change-over switch S.W., for operation when summer cooling begins, being subsequently thrown over to the winter position. In some control lay-outs this switch is operated automatically so as to make most use of outside air when conditions permit, so reducing refrigeration running period.

If, instead of the reheater, a bye-pass duct is employed to bring the air to required outlet temperature, the controls operating $T.V._2$ would instead operate damper motor $D.M._2$, to adjust the quantity of bye-pass air admitted.

Winter Control

Humidification is controlled from the thermostat $T._4$, giving a constant dew-point, by admitting or closing the heat supply to the preheater via $T.V._1$.

The final temperature delivered is controlled by the low limit 'stat L.L.T.$_3$, and return air thermostat T.$_3$ acting in conjunction as explained above.

Separate thermostats are shown for winter, and are generally selected by a switch on the control panel for summer and winter operation, whichever is required being brought into use.

When the fresh air temperature falls below some predetermined point, such as 4° C, economy of heat requires the fresh air to be shut down to the minimum. This is achieved by the thermostat T.$_5$ operating the damper motor D.M.$_1$ when S.W. is set to the winter position.

Fig. 17.25 shows a zoned system served from a central plant, as described on page 350. In this case the central plant delivers a constant outlet condition and the zones are controlled individually.

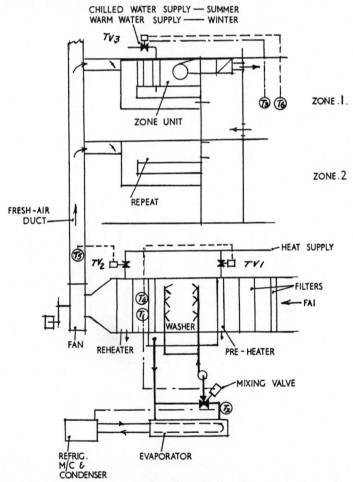

CHILLED WATER SUPPLY — SUMMER
WARM WATER SUPPLY —— WINTER

TV3

ZONE.1.

ZONE UNIT

ZONE.2

REPEAT

FRESH-AIR DUCT

HEAT SUPPLY

TV2   TV1

FILTERS

FAI

WASHER

FAN   REHEATER   PRE-HEATER

MIXING VALVE

REFRIG. M/C & CONDENSER   EVAPORATOR

FIG. 17.25.—Air-Conditioning by Central Plant with Zone Control
Diagram of Controls.

Summer Control.

Constant dew-point outlet is controlled by $T_{.1}$, operating the mixing valve in cooled water line as before.

The refrigerator is controlled for constant water temperature by $T_{.2}$.

Re-cooling occurs at each zone according to requirements. $T_{.3}$ in the zoned room, or in the return duct, controls $T.V_{.3}$, so varying the quantity of chilled water supplied to the coil.

Winter Control.

Moisture Content is controlled by constant dew-point thermostat $T_{.4}$, operating $T.V_{.1}$.

Final temperature from the plant in a zoned system is generally set at 10 to 12° C, and is controlled by 'stat $T_{.5}$ operating valve $T.V_{.2}$.

The reheating in each zone is controlled by the winter 'stat $T_{.6}$ operating $T.V_{.3}$ as before, but in reverse. $T_{.6}$ naturally has a different setting from $T_{.3}$.

Changeover switches from winter to summer operation would form part of the system.

Fig. 17.26—Direct expansion system with separate latent and sensible coolers.

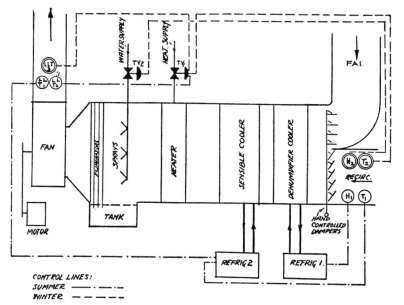

CONTROL LINES:
SUMMER —— .—— .——
WINTER —— —— ——

| T. | indicates Thermostat. | H. | indicates Humidistat. |
| L.L.T. | ,, Low Limit Thermostat. | T.V. | ,, Thermostatic Valve. |

FIG. 17.26.—Air-Conditioning using Direct Expansion. Summer and Winter Controls Diagram.

Summer Control.

Dehumidification is controlled by humidistat $H_{.1}$ in return duct stopping and starting compressor, or in some cases controlling refrigerant admission or cooled brine admission from a central plant.

Should this reduce the air temperature too greatly, low limit 'stat L.L.T.$_2$ admits heat to the heater.

Sensible cooling is controlled by T.$_1$ and L.L.T.$_1$, acting in conjunction as before described; and operating starting and stopping of refrigerator or valve in cooling supply.

An interlock would in practice be provided to prevent the sensible cooler functioning whilst heat is being supplied to the heater.

An alternative arrangement to that shewn in Fig. 17.25 is where the sensible cooler is placed first and the dehumidifier cooler second; or another alternative is to arrange both sets of coils in parallel.

Winter Control.

Heat supply through T.V.$_1$ is controlled by T.$_2$ and L.L.T.$_3$, acting conjointly as previously described.

Humidity is supplied by sprays direct from a cold water line, or alternatively by steam injection and is controlled on or off by T.V.$_2$ operated by humidistat H.$_2$.

Changeover switches for summer to winter control would be provided.

Induction System Controls—Fig. 17.27 illustrates the controls of the non-changeover type. Water is kept constant at say 10° C both in summer and winter by means of the mixing valve in the 'secondary' water circuit as shown. The water is cooled by a water-chiller in summer, but in winter by the entering fresh air. The whole of the chilled water is circulated through the cooling coils without control, thus ensuring that in summer the dew-point (about 7° C) of the primary air to the units is below the temperature of the coils in the room units, in order to prevent the latter from gathering condensation.

The primary air temperature is maintained at about 16° C in summer but in winter is varied according to the weather, by the outside stat shown, up to about 60° C.

To maintain warming during winter nights without the fan running, the calorifier is brought into service allowing the units to act as natural convectors.

Each induction unit is this system has a direct-acting control valve in the water circuit with the actuating element in the induction air-stream from the room. Thus, in summer, assuming no heat gains in the room, the valve will be closed, the air only maintaining room temperature. When sun or internal heat gains cause the room temperature to rise, the valve will open and the coil will lower the circulating air temperature. In winter the primary air will warm the room, but, should the temperature mount too high, the control valve will open to admit cool water, so correcting the temperature. To avoid variation in pressure, should a large number of control valves close, a constant pressure differential control valve is included as a bye-pass, as shown.

Moisture content is controlled in summer by the chilled coils and in

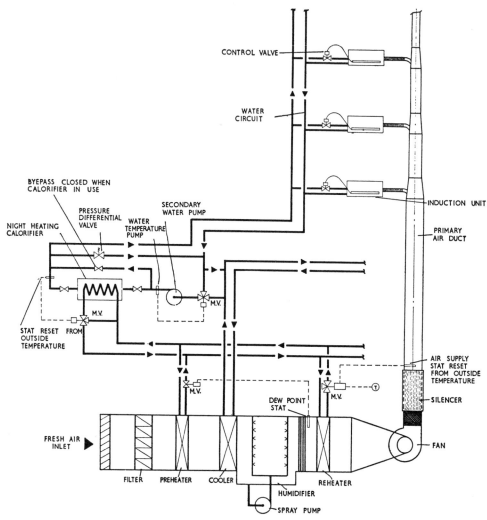

FIG. 17.27.—Induction System: Principle of Controls.

winter by the humidifier, to a constant dew-point. Alternatively, a sprayed coil system is used, controlled in the same manner.

Dual-Duct System Controls. Fig. 17.28 illustrates the principle of the control system applying to a dual-duct system where two fans are included, one for the warm duct and one for the cool.

The temperature of the cool duct is varied according to the external temperature from say 7° C in summer to say 16° C in winter. In summer it is cooled by the chilled water coil: in winter the heater comes into use as necessary.

The temperature of the warm duct is also varied according to the outside temperature, from say 21° C in summer to say 38° C in winter.

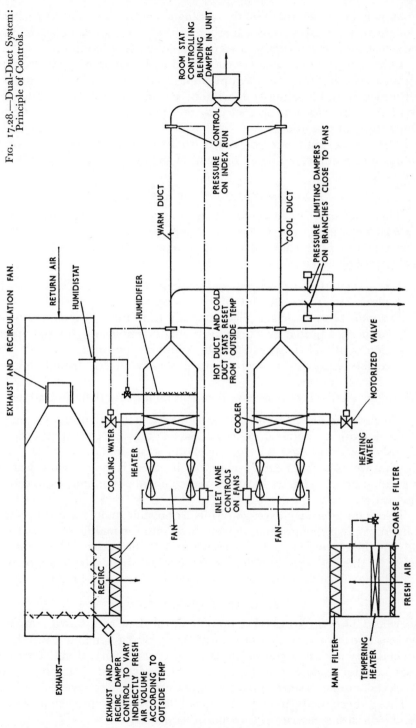

FIG. 17.28.—Dual-Duct System: Principle of Controls.

It will be noted that, in the arrangement shown, the return air is brought into the fan suction-chamber on the side nearest the warm duct, the return air being normally nearer to the warm duct condition.

Other controls indicated are concerned with maintaining air pressures within certain limits by damper operation. Dehumidification in summer is performed by the cooling coil, without specific control, and, in winter, humidity is increased as necessary by the sprays or steam jets controlled from return air condition.

The blender units are controlled from the rooms individually, or in groups thermostatically, by adjustment of the internal mixing damper or dampers.

AUTOMATIC CONTROL APPARATUS

There are three main classes of controls: electrically operated, pneumatically operated, and hydraulic or oil operated. The choice will be determined by preference based on experience, size of job, cost, personnel in charge of plant, etc.

Electrical controls are best where inexperienced operators supervise the plant, but are equally suitable on any plant.

Pneumatic controls need skilled attention from time to time, and when properly adjusted give perhaps the most simple and gradual operation of all. They are often to be found on industrial plants for this reason, especially as such plants generally already include a compressed air system.

Hydraulic and oil controls have their special uses but their limitations.

The positions in which controlling elements, such as thermostatic bulbs, are fixed, require careful selection. Air passing through a plant may 'layer' seriously. Near the edges of a duct or plant it may similarly be different in temperature from that near the core. Again, there is the matter of heat or cold radiation from heater batteries and cooling coils or even from cooled spray water.

To obtain as nearly average results as possible, temperature sensitive or hygroscopic elements should extend well into the middle of the stream, and not merely project a few inches through the casing. The elements should be shielded against radiation by polished shields, left open or perforated away from the radiation, or they may be in a bleed-off duct.

ELECTRICAL CONTROLS

Electrical Thermostats—For on-off control these comprise a temperature sensitive element, either of volatile liquid tilting a mercury switch, or bi-metallic type closing point contacts. Fig. 17.29, 2-wire, simply opens or closes a circuit to stop or start a motor. The 3-wire type opens one circuit and closes another to start a motor or operate a valve in reverse.

These thermostats may be arranged to give a step-by-step action, opening or closing a series of contacts in sequence.

Modulating type (Fig. 17.30)—
This works in conjunction with a
modulating valve or damper motor
as described later. The thermostat
operates the moving arm of a potentiometer in a 'bridge' circuit.

These systems should not be
confused with electronic systems of
control referred to later.

Electrical Humidistat (Fig. 17.31)
—The hygroscopic material is either
a bundle of hairs maintained in a
state of tension, or a length of specially sensitive wood or pine-cone fibre.
The expansion on increase of percentage saturation is arranged to open or
close contacts or operate a potentiometer, as in the case of a thermostat.

Another form of humidity control
is by wet and dry bulb thermostats
working on a differential principle.
These control for a constant differ-

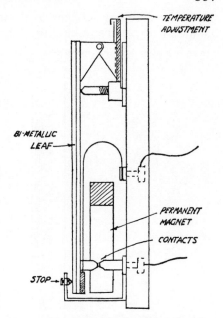

FIG. 17.29.—Principle of Operation of
Electrical Bi-metallic On-Off Thermostat.

ence which corresponds over small ranges with the relative humidity. The
wet bulb is kept wet by a wick dipping in a reservoir of distilled water.

More sophisticated humidity controls convert change of moisture

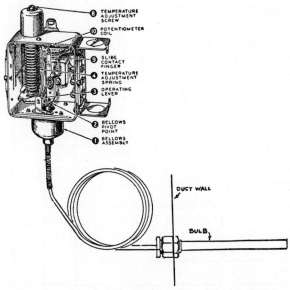

FIG. 17.30 —Electrical Duct Thermostat, Modulating Type.

content directly into change of electrical resistance and are used in combination with electronic circuits.

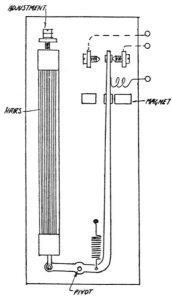

FIG. 17.31.—Principle of Operation of
Electrical Humidistat. Two-way Type.

Contactors—When a large machine such as a refrigeration compressor
motor has to be started, it is necessary to interpose between the thermostat
and the heavy current switchgear a device which will operate with the
small current handled by the 'stat. This is accomplished by a contactor,
which is a common piece of electrical gear. The small current serves to
close the solenoid circuit, which in turn pulls in the main switch.

Step Controller (Fig. 17.32)—It is sometimes necessary to operate from

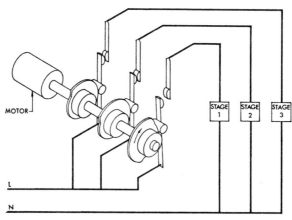

FIG. 17.32.—Step Controller

one thermostat a series of switches in sequence, as when two or three refrigeration compressors have to be started in turn with increases of load, or when the air heating is accomplished by electrical heaters requiring to be switched on in steps. For this purpose a step controller may be used as shown in the Figure. The switches are operated by cams fixed on a shaft driven through gearing from a small motor controlled by the thermostat or humidistat of modulating or step-by-step type.

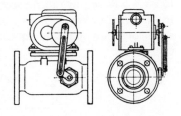

FIG. 17.33.—On-Off Motorized Valve for Steam.

EQUIPMENT CONTROLLED: ELECTRICAL

Electrically-operated Valves—These control steam or water admitted to heaters and coolers, or refrigerant admitted to coils.

Fig. 17.33—This shows an *on–off valve* with motor operation. A three-wire 'stat is required. The motor contains limit switches which, at the end of half a revolution, cut off and re-set for reversal.

Another type is the *stalling motor*, which opens on making of circuit and closes on opening of circuit. It is self-closing on current failure, a desirable feature in some cases. It requires a two-wire 'stat. Alternatively it may be arranged to operate in the opposite direction, i.e. opening when the switch closes and vice versa.

Fig. 17.34—This shows a *modulating valve* operated from a modulating thermostat or humidistat. Its purpose is to adjust the valve to any intermediate position as required by the conditions. With each movement of the

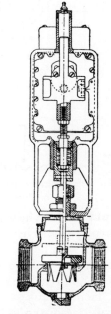

FIG. 17.34.—Modulating Electrically-Operated Valve.

thermostat arm the valve 'inches' either more open or closed, where it remains until another change takes place. The system operates at low-voltage through a transformer, the circuit being as Fig. 17.35.

Fig. 17.36—This shows a modulating three-way *mixing valve*, the function of which has already been discussed. This is of necessity a valve which requires gradual adjustment over its range, and the modulating system above described will normally be used with it.

Electrically-operated Dampers (Fig. 17.37)—For operating louvres or dampers, a *damper motor* is required. The power required to move dampers may be considerable and the motor must be chosen to suit. Sometimes the louvres will require to be sectionalized, each section being worked by one motor with linkage. Again, such damper motors may be on-off, i.e. open-shut, or they may be modulated to give settings in any intermediate position.

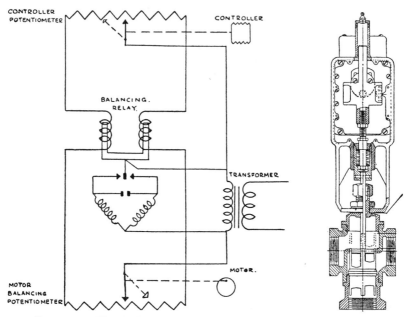

FIG. 17.35.—Electrical Circuit for Potentiometer Modulating Control.

FIG. 17.36.—Modulating 3-Way Mixing Valve.

Control Panel—The wiring to electrical controls on a big scheme becomes very involved, and it is usual to centralize all switches, pilot lights, etc., on a panel with labelling, so that the operation can be seen at a glance. This panel may also accommodate remote temperature and humidity indicators and recorders if required. A compact example is shown in Plate XXIV, facing p. 465.

Electronic Controls—A development of the electrical modulating system is the electronic system, which dispenses with moving parts in the thermostats, such as sliding potentiometers. One such system is shown diagrammatically in Fig. 17.38. The temperature detecting element is, in effect, a resistance thermometer. Variations in resistance due to change of temperature cause a change in grid current in an electronic valve circuit, causing a much greater change in current passed by the valve. This may be magnified again by further valves and applied to operation of motorized valves, damper motors, etc. More recent developments replace valves with transistors.

FIG. 17.37.—Electric Stalling Motor for Damper Operation.

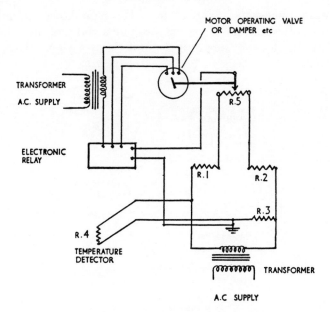

FIG. 17.38.—Diagram of Electronic Control System.
R1, R2, R3, are fixed resistances in bridge circuit. R4 is variable resistance
depending on temperature. R5 is adjusted by motor to preserve balance.

A further development is the use of the *Thermistor* element for temperature control in an electronic circuit. A Thermistor is a resistance material
having a negative characteristic; whereas most materials increase in resistance with increase in temperature, a Thermistor decreases in resistance.
It furthermore gives a greater change in resistance for a given temperature
change than other forms of element, and less magnification is necessary.

PNEUMATIC CONTROLS

The compressed air for these controls, if derived from a central supply,
is usually taken through a reducing valve at about 1 bar gauge. If the
supply is provided independently a small air compressor and storage
cylinder are required, automatically maintaining a constant pressure of
perhaps 4 bar gauge and supplying again through a reducing valve so as
to give a good storage of air for sudden demands.

The piping has to be carefully run to provide adequate drainage of the
water which condenses out of the air. Traps are generally fitted at low
points, and other steps taken to ensure dryness. Water may completely
upset operation as also will oil—hence the use of oil-free compressors.

Where several plants exist, a central compressor may be used piped to
each unit at high pressure, each having its own storage cylinder and
reducing valve.

2D

Thermostat (pneumatic) (Fig. 17.39)—The type shown operates on the different rate of expansion of two metals, one a rod and the other a tube surrounding it. This movement opens and closes a pilot valve to control the main unit.

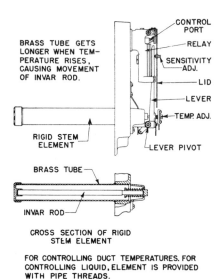

FIG. 17.39.—Pneumatic Rigid Stem Insertion Thermostat.

Humidistat (pneumatic)—One type operates on the wet and dry bulb principle, comprising two normal thermostats connected so that when their differential changes the pilot valve is caused to open or shut. Alternatively a hygroscopic material (Fig. 17.40) may be used.

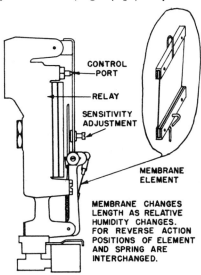

FIG. 17.40.—Pneumatic Membrane Type Humidistat.

On-off Controls (pneumatic)—The method of operation is the same as the above, but the pilot valve is so arranged as to remain open or shut against a spring, or with a lost-motion device until beyond a certain point of movement it closes or opens suddenly.

Equipment Controlled (pneumatic) (Fig. 17.41)—*Pneumatic valves* operate by means of a diaphragm or copper bellows connected to the valve spindle. The air supply connects to the top of this and has a small orifice plate, or needle valve, allowing only a small quantity of air to pass. The diaphragm top also connects to the thermostat. When the pilot valve in the latter shuts, the air pressure builds up on top of the dia-phragm and depresses the valve spindle to close or (in the reverse-acting type) to open the valve. When the pilot valve opens, the pressure on the diaphragm is released and the spring around the valve spindle forces the valve up again.

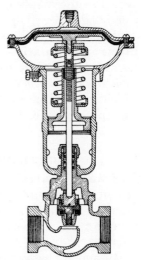

FIG. 17.41.—Typical Pneu-matic Diaphragm Valve.

The air discharged from the pilot valve dis-charges to atmosphere. Any movement of the pilot valve causes the main valve to find a similar intermediate position, so giving a 'floating' con-trol over the complete range.

A pneumatic *mixing valve* operates in the same way, as will be appreciated.

Fig. 17.42—Pneumatic *damper motors* work on the same principle as for valves, except that the diaphragm is larger to give the necessary force to operate the damper.

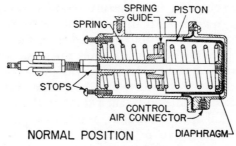

NORMAL POSITION

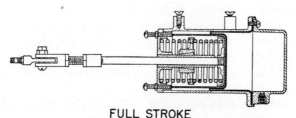

FULL STROKE

FIG. 17.42.—Piston Pneumatic Damper Operator

Pneumatic Fittings—Each valve should have a control cock and pressure gauge. These are usually centralized on a board along with the thermometers and other instruments as with the electrical system, so that the operator can see the condition at a glance. The pressure gauges are connected to the diaphragm top in each case, so giving an indication as to whether the valve is open, shut, or in a mid-position. A main pressure gauge on the air supply will show whether or not the compressor is functioning.

Electro-pneumatic—A combination of electrical and pneumatic controls may be useful where room control is required. Pneumatic room 'stats are sometimes cumbersome, and may be replaced by electrical instruments which in turn control the air supply to pneumatic valves, etc., as before. This special application will be apparent, and need not be discussed further.

MISCELLANEOUS CONTROLS AND INSTRUMENTS

Hydraulic Controls—Operate like pneumatic ones, except that water at a steady pressure replaces air; the water from the pilot valves is led to drains. The disadvantage of water is that the small ports and orifices tend to close up with deposit and require cleaning at more frequent intervals.

In another system the water is replaced with oil, pumped by a small pump, and returned from the pilot valve to a tank to be used again.

Static Pressure Control—This is a device to adjust main air flow to suit variable flow of a number of branches. If, for instance, volume control is applied to a zoned system to meet variable occupancy in the different zones, the main fan volume must be controlled; otherwise any branches left full open will receive a great excess of air if others are shut down. The static pressure controller is a pressure sensitive device connected to the main trunking system, and arranged, either by electric or pneumatic means, to regulate a set of louvre dampers in the fan discharge. Other methods have been devised for achieving the same object, notably *constant velocity control* which is preferable in some circumstances.

Design of Valves—For control of steam, water, refrigerant, etc., valves however operated require to be such that the fluid controlled is metered gradually from full open to full shut, as nearly as possible in a straight line fashion. The ordinary gate or globe valve does not obey such a requirement, but on closing allows full volume to pass until within about 20 per cent of closing, when the volume begins suddenly to fall off.

The desideratum is achieved by adopting valves of a Vee-port design, and such are commonly included by makers of this equipment.

As already explained, other varieties of automatic controls abound for which makers' publications and other technical literature should be consulted.

Instruments for Air-Conditioning—The results being achieved by an air-conditioning plant necessitate some supervisory equipment.

For a small direct expansion cooling plant for air-conditioning, say a small restaurant, perhaps nothing more than a thermometer on inlet and outlet is required.

For a large plant, or a combination of several plants in a building, more extensive instruments are advisable.

In general it may be said that wherever a thermostat exists in the system it is desirable to accompany it by a thermometer—this will be an aid to adjustment. Similarly, where a humidistat exists or a thermostat controlling humidity in some way, a wet bulb thermometer or hygrometer is desirable. As stated earlier, these are best of dial pattern for reading on the control panel.

For a multi-plant job, the essential outlet conditions may be transmitted electrically to dials or recorders on a central control panel. The positioning of the bulbs requires care, as in the case of thermostats.

A view of a comprehensive control and instrument panel is shown in Plate XXIV, as already mentioned.

CHAPTER 18

Air Distribution

HAVING DECIDED ON THE type of air system to be employed, and made the calculations of air quantity and temperature, it is now necessary to consider in more detail the different systems of air distribution which may be utilized, and to discuss the characteristics and performance of the apparatus employed. Most of the discussion which follows applies equally to ventilating or air-conditioning systems.

AIR DISTRIBUTION

Successful air distribution requires that an even supply of air over the whole area be given without direct impingement on the occupants and without stagnant pockets, at the same time creating sufficient air movement to cause a feeling of freshness.

This indicates what is probably the key to the problem of successful distribution: that unduly low velocities of inlet are to be avoided, just as much as excessively high ones; and that distribution above head level not directly discharging towards the occupants will give the necessary air movement to ensure proper distribution over the whole area, and without draughts.

There are four general methods of air distribution.

(1) Upward.
(2) Downward.
(3) Mixed upward and downward.
(4) Crosswise.

The choice of system will depend on:

(a) Whether straight ventilation or complete air-conditioning is employed.
(b) The size, height and type of building or room.
(c) The position of occupants and/or heat sources.
(d) The location of the ventilating plant, and economy of duct design.

Upward System—The air is introduced at low level and exhausted at high level as in Fig. 18.1 which shows a section through an auditorium, with mushroom inlets under the seats and riser gratings (see p. 423) in the gallery risers. The air is exhausted around the central laylight in the roof. This system was designed so as to be reversible, i.e. it may be worked as a downward system.

When working upwards the air appears to be somewhat 'dead', due to the very low velocity of inlet (about 0·5 m/s) necessary with floor outlets

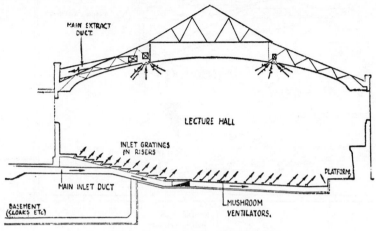

FIG. 18.1.—Example of Upward Ventilation.

to prevent draughts. When working downwards more turbulence is set up in the air stream, with a greater feeling of freshness.

The upward system is not, however, confined to one with floor inlets. The inlets may equally well be in the side walls, with extract in the ceiling

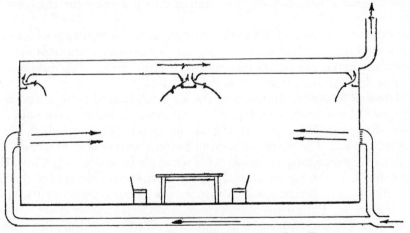

FIG. 18.2.—Upward System applied to Committee Room.

as before. Such a system for a committee room of a town hall is shown in Fig. 18.2, and another example for a theatre in Fig. 18.3.

The limitation of an upward system is that in a large hall it may be difficult to get the air to carry right across without its picking up heat *en route* and rising before it reaches the centre.

The upward system is generally used with straight ventilation. When the air is cooled, as in a complete air-conditioning system, it will tend to

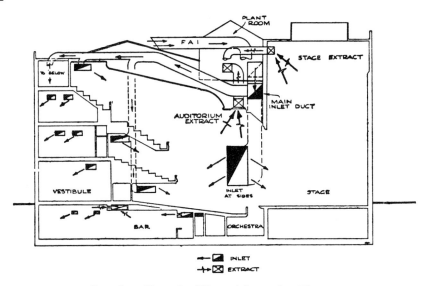

FIG. 18.3.—Example of Upward System in a Theatre.

fall too early, before diffusion, and thus cause cold draughts. The upward system also lends itself to simple extract by propeller fans in the roof in the case of a hall, factory, etc., and is thus generally the cheapest to install.

Another application of the upward system is the swimming pool hall, as shown in Fig. 18.4. Here the supply air is introduced through special plastic discharge spouts situated below a large area of glazing and is exhausted by specially treated roof-exhaust units.

Downward System—In this type the air is introduced at high level and exhausted at low level, as in Fig. 18.5. It is most commonly used with full air-conditioning where, due to the air being admitted cooled, it has a tendency to fall. The object of distribution in this case is so to diffuse the inlet that the incoming air mixes with room air before falling. Thus, the inlets shown in the figure as discharging downwards, in practice deliver horizontally at sufficient speed to ensure that the air completely traverses the auditorium. Turbulence is thus caused with the desirable effect already mentioned. On a smaller scale, as applied to an office building, this system appears as in Fig. 18.6.

Provided the height of room is normal, the extract opening may be at high level as in Fig. 18.7. Short circuiting is avoided by the velocity of the inlet air carrying over to the far side of the room. Another possible arrangement is a variation of this, namely, 'downward-upward', as in Fig. 18.8.

An application of downward inlet with both downward and upward extract, suitable for rooms of greater height, is shown in Fig. 18.9. This is

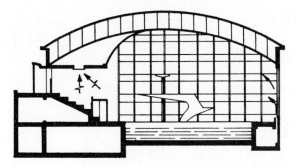

Cross Section through Pool

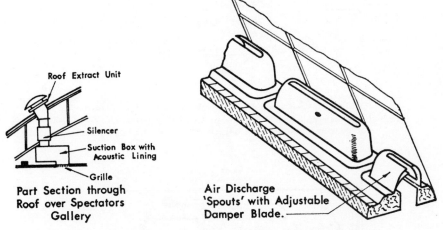

Roof Extract Unit

Silencer

Suction Box with
Acoustic Lining

Grille

Part Section through
Roof over Spectators
Gallery

Air Discharge
'Spouts' with Adjustable
Damper Blade.

FIG. 18.4.—Upward System as applied to a Swimming Pool.

usually adopted where smoking occurs and it is necessary to provide some top extract to remove the smoke. In this case the top extract is discharged to atmosphere by a separate fan and the low-level extract constitutes the recirculated air. Care must be taken to avoid placing low-level extract grilles near where people sit. If there is no alternative position, they must be at a very low velocity (about 0·75 m/s) through the free area, and well spaced out, so that big volumes do not occur at any one point.

Mixed Upward and Downward—Such a system is shown in Fig. 18.10. The principle will be clear from what has been described above. It is in effect an upward system giving good turbulence above head-level, with about 25 per cent extract at low level. The remainder of the extract is exhausted normally at the ceiling.

Crosswise Ventilation—This arrangement may sometimes be necessary where dictates of planning preclude more orthodox solutions. Air is introduced near the ceiling on one side of a long low room with smooth flush

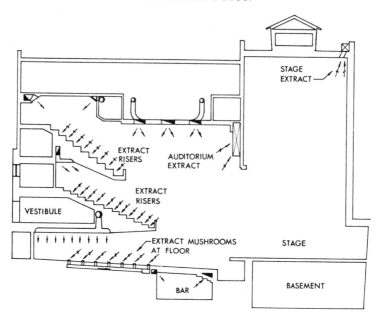

FIG. 18.5.—Example of a Downward System in a Theatre.
The high level exhaust labelled 'Auditorium Extract' only comes into use when the
fire curtain is lowered, the normal extracts then being shut off.

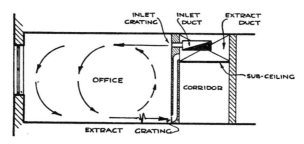

FIG. 18.6.—Wall Gratings for Downward Distribution in Office Building.

ceiling, and is exhausted at the opposite end at the same level (see Fig.
18.11). The inlet is at a high speed and strong secondary currents are set
up in the reverse direction at the lower levels, as shown. It is these secondary
currents which are important in any distribution system, causing turbulence as already mentioned.

AIR DISTRIBUTION FOR AIR-CONDITIONING

Air-conditioning usually involves the handling of large air quantities,
and one of the chief problems of successful design is how to introduce and
extract these quantities without giving rise to complaints of draught or of

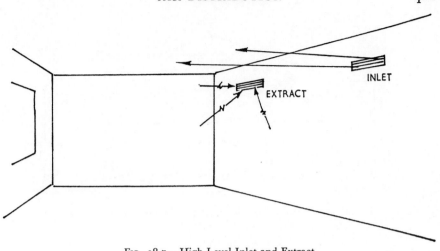

Fig. 18.7.—High Level Inlet and Extract.

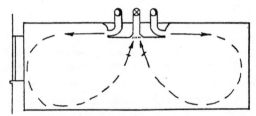

Fig. 18.8.—Downward-upward System.

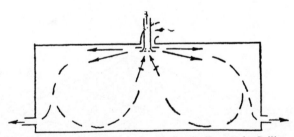

Fig. 18.9.—Downward System with Smoke Extract in Ceiling.

causing noise. A great variety of methods have been adopted and there are available innumerable devices to suit different conditions or architectural tastes. Some of these are described below.

Perforated Ceilings—Probably the most completely diffused form of inlet is the system which employs a perforated ceiling as shown in Fig. 18.12. In this case air is discharged into the space above the ceiling divided up so as to ensure uniformity of distribution, and the air enters the room through the perforations. Where extremely large air quantities are involved, the whole ceiling may be used, but in the normal case certain areas

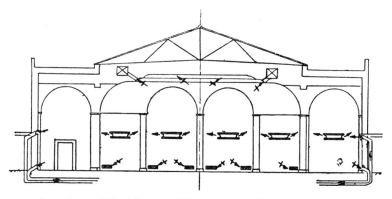

Fig. 18.10.—Mixed System of Distribution applied to a Restaurant.

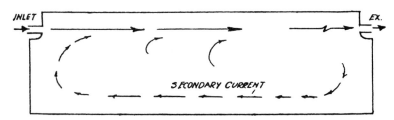

Fig. 18.11.—Crosswise Distribution.

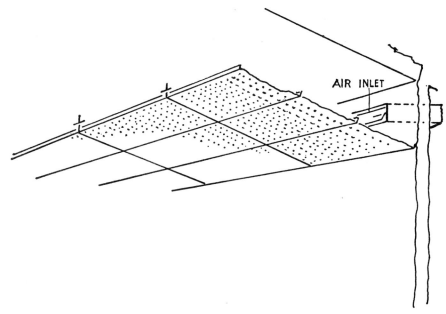

Fig. 18.12.—Perforated Ceiling Inlet.

of panels serve as inlets, although this may produce the effect of 'pattern staining' in course of time. The ceiling inlet system destroys all turbulence, which may be a good thing in some cases but not in others. This system has been applied successfully in shops and department stores.

Line Diffusers—The type of inlet diffuser shown in Fig. 18.13 is a neat and efficient way of introducing air at the ceiling. Being extended if necessary across the full width of a room, it gives good diffusion, and at the same time good entrainment.

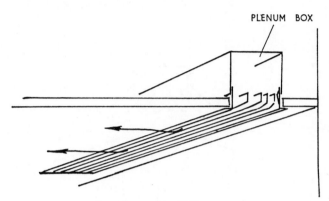

PLENUM BOX

FIG. 18.13.—Line Diffuser.

The same principle is sometimes used around the four sides of ceiling lighting panels, thereby combining two functions in one.

Ceiling Diffusers—The type of ceiling diffuser, shown in Fig. 18.14, has certain interesting characteristics. Fig. 18.15 illustrates the kind of flow pattern to be expected from such a unit and it will be noted that entrainment air is drawn up in the centre immediately under the diffuser, being caught up in the nearly horizontal delivery from the unit itself. Due to the fact that air is discharged radially, the velocity of inlet falls off rapidly as the distance from the centre increases, and hence this kind of unit may be used with temperature differentials up to 20° C.

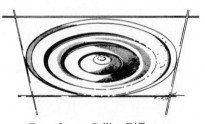

FIG. 18.14.—Ceiling Diffuser.

There is a great variety of makes of this type of ceiling diffuser, all under different trade names, and some with special features, such as:

> An adjustable arrangement whereby the inner cones can be raised or lowered in relation to the periphery, so causing a variation in the flow pattern to be achieved. Instead of the bulk of the air travelling horizontally, it is possible by this kind of adjustment to make it discharge vertically downwards or at any inter-

mediate flow pattern desired. This might be of advantage in cases where it is desirable to cause the air to descend quickly, such as in a hot kitchen, rather than it should be dispersed and lost at high level.

In another type, the unit is square instead of circular, this sometimes being necessary for architectural reasons.

In other examples, the diffuser is flush with the ceiling instead of projecting.

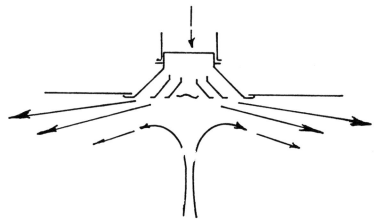

FIG. 18.15.—Flow Pattern from Ceiling Diffuser.

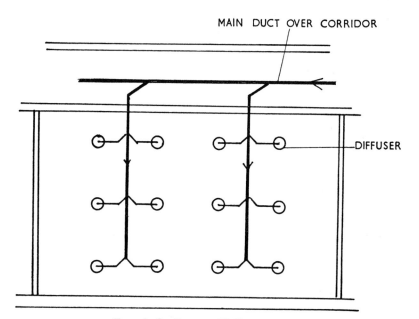

FIG. 18.16.—Layout of Ceiling Diffusers.

A typical layout of ceiling diffusers is shown in Fig. 18.16. It will be noted that use is made of the false ceiling space for the concealment of the connecting ducts, the final connections to the diffusers being of flexible ducting.

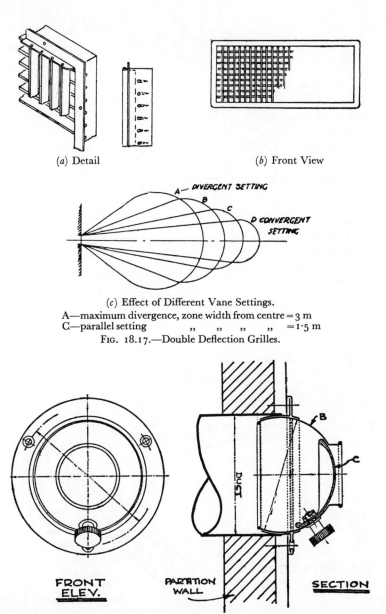

(a) Detail   (b) Front View

(c) Effect of Different Vane Settings.
A—maximum divergence, zone width from centre = 3 m
C—parallel setting ,, ,, ,, ,, = 1·5 m
Fig. 18.17.—Double Deflection Grilles.

FRONT ELEV. PARTITION WALL SECTION

Fig. 18.18.—Punkah Type of High-Velocity Directional Inlet Opening. The shield C controls the volume delivered, and the direction of the jet is varied by moving the ball B in its spherical seating.

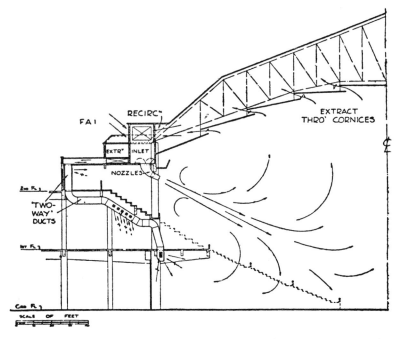

FIG. 18.19(a)—Earls Court (Hall B).

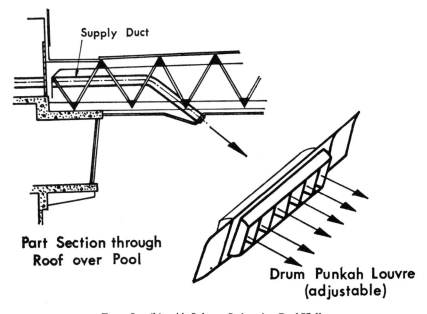

Part Section through Roof over Pool

Drum Punkah Louvre (adjustable)

FIG. 18.19(b)—Air Inlet to Swimming Pool Hall.

Side Wall Inlets—Where there is no false ceiling or other means of introducing the air through ceiling diffusers, it is necessary to adopt side wall inlets, this usually taking the form of a series of grilles distributed at intervals along the inner partition wall with ducts in the corridor false-ceiling behind. Each inlet is equipped with a grille, so designed as to enable the requisite quantity of air to be introduced without draught. The commonest, and probably most effective, form of grille now in use is that known as the 'double-deflection' type, as shown in Fig. 18.17. In this form of grille there are two sets of adjustable louvres, one controlling the air delivery in the vertical plane and one in the horizontal plane. The vanes are usually independently adjustable by means of a special tool, and when once set are not altered. A variety of flow patterns can be achieved according to the width of room, aspect ratio and velocity, such as the patterns indicated in Fig. 18.17(c). Here, it will be noted, by setting the vertical vanes in a diverging manner, the air is fanned out with a much shorter length of delivery.

Nozzles—Introduction of air by nozzles is usually confined to small compartments such as ship's cabins and to aircraft, where it is desired to obtain the maximum effect under the direct and easy control of the occupant, both in the matter of quantity and direction. Nozzles of this type are known as 'punkah louvres' as in Fig. 18.18. Nozzles may however be a useful manner of introducing large quantities of air in some vast arena or high bay factory, where it would be impossible to conceive of too much air movement being created. In effect, the more the better. An example of this kind is illustrated in Fig. 18.19(a), this being the Earls Court Exhibition. Nozzles were specially designed of the type shown in Fig. 18.20 where it was necessary to achieve a throw of about 40 m from the nozzles to the breathing zone. In this case each nozzle handles 4 m³/s at a velocity of 11·5 m per second, is 635 mm in diameter, and requires a static pressure of 115 N/m². Fig. 18.19(b) shows another but similar approach where 'drum' type punkah louvres have been used to introduce large air quantities into a swimming pool hall.

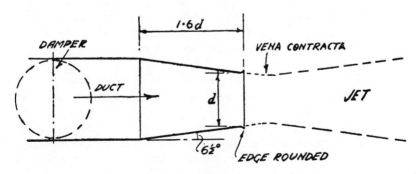

FIG. 18.20.—Detail of Nozzle.

Volume Control—Means for controlling as accurately as possible the volume of air issuing from each grille, diffuser or other inlet, are essential. Probably the commonest form of control is the opposed blade damper such as shown in Fig. 18.21. For ceiling diffusers and the like, a variety of multi-louvre dampers are available or plain butterfly dampers in the ducts may be used. Any damper is a potential source of noise and its location and duty require careful consideration if such is to be avoided.

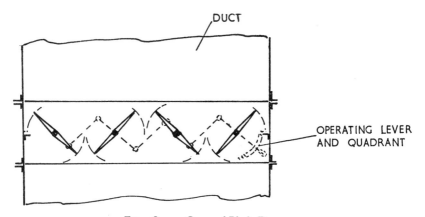

FIG. 18.21.—Opposed Blade Damper.

Grille Velocities—The selection of the best number and size of inlet grilles, diffusers and the like, involves a choice to meet a number of variables at the same time:

(1) Air velocities produced at head or foot level must not exceed 0·15 to 0·20 m/s if draughts are to be avoided.

(2) If the air is entering cooler than the room, as in an air-conditioning installation, there will be a tendency to fall, best avoided by keeping a reasonable entering velocity with the object of inducing air entrainment from the room; the air mixture then being warmer will have less tendency to fall.

(3) The velocity of inlet must not be so high that the air will impinge on the wall opposite, thereby causing undue turbulence.

(4) The distance apart of inlets, particularly in the case of ceiling diffusers, must be such that the streams from two adjacent units do not collide with such force that a strong downward current results.

(5) The velocity selected for the grille or diffuser must be such that the sound level produced therefrom is below the design standard for the room.

(6) Appearance, layout and pattern are all architectural matters, which must also as a rule be taken into account and this sometimes creates a difficulty where the desired spacing or size is not acceptable on these grounds.

The selection makes use of manufacturers' data which is now freely available for all types of equipment and, by a study of which, a variety of alternative solutions to a particular problem can perhaps be put down with the object of a final selection being made best suited to meet all the other conditions. The reader is referred to such data for all this information, but a few sample grille sizes are given in Table 18.1. It will be clear that

TABLE 18.1

SOME SAMPLE INLET GRILLE SIZES

Type	Air quantity litre/s				
	50	100	200	300	500
	dimensions in mm.				
Perforated ceiling panels (face size) - - - - -	300 × 150	300 × 300	500 × 500	600 × 600	1200 × 600
Line diffuser (duct size) - -	750 × 50	1000 × 90	2000 × 90	3000 × 90	2400 × 165
Ceiling diffuser (Anemostat)					
(neck dia.) - - - -	100	150	200	250	380
(overall dia.) - - - -	330	330	450	600	860
Side Wall grille double deflection					
type (face size) - - -	200 × 150	300 × 200	450 × 250	450 × 400	660 × 450

Note: Above are all selected at approximately the same sound level of 40 dB.

with the large number of variables, air-distribution design is perhaps more an art than a science, but on it, to a large measure, depends the success or otherwise of any air-conditioning installation.

High-Velocity Supply Fittings—The use of high velocity air-distribution in ducts has been referred to in connection with the induction and dual-duct systems. High velocity air-distribution may also be the most practical and economical method both in cost and space for use with any otherwise normal ventilation or air-conditioning system where extensive ductwork is involved. This is particularly the case in multi-storey buildings where duct sizes are much reduced by use of higher velocities. Duct velocities up to 30 m/s may be used, although it is more usual to limit them to the range of 15 to 20 m/s.

At the terminal end, it is necessary to break down the high velocity to low velocity for introduction into the room, and a silencing or attenuating box in some form is required. Fig. 18.22 illustrates examples of this kind:

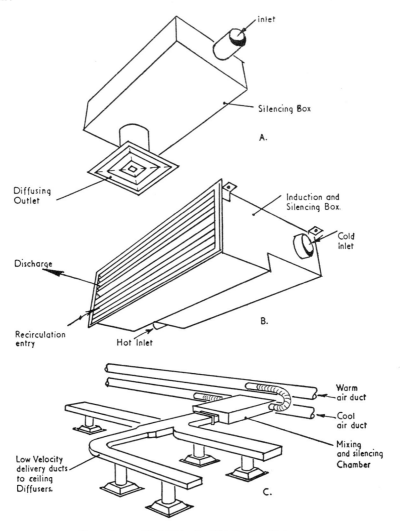

FIG. 18.22.—High Velocity Distribution Units.

A. shows a silencing box in which there are a number of baffles with acoustical treatment terminating in a ceiling diffuser.

B. shows a unit suitable for a dual-duct installation, though it would be equally suitable for a single inlet. In this case the unit would be mounted above the false ceiling of a corridor, discharging through the grille in the side wall.

C. shows a single mixing and silencing box served off high velocity dual-ducts. The outlet from the box, sometimes described as an 'octopus' is at low velocity connecting to a number of ceiling diffusers in the manner shown. All ducts and the mixing box would be above the false ceiling which would require to be removable for access to the mixing device.

Combined Lighting and Air Distribution—Mention has previously been made (p. 372) of the use of special fittings which enable some part of the heat generated by lighting to be dealt with at source before it enters the room. Early types of such fittings used 'boots' mounted to conventional lighting enclosures, as Fig. 18.23(*a*), but more sophisticated arrangements have since been developed where the lighting and air-handling functions have been considered in an integrated design, as Fig. 18.23(*b*) and (*c*). The performance of the lighting apparatus can be improved by such arrangements due to temperature control of the tubes, and the heat entering the room may be reduced very considerably.

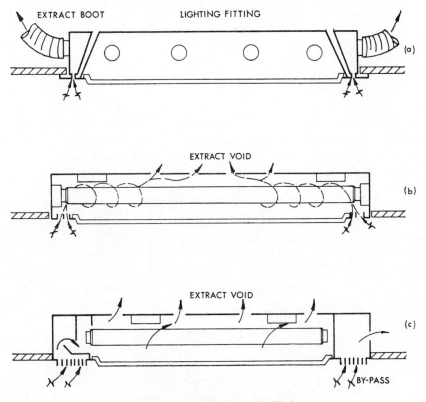

Fig. 18.23.—Air-Handling Light Fittings.

Extract or Return Air Grilles—The particular form of grille for extract is unimportant. It may, for instance, be of egg-crate plastic within an aluminium frame as Fig. 18.24, or louvred, or any design giving the required free area. Dust collects on extract gratings on the outside, and close mesh or closely placed slats are undesirable as they quickly block up and impede the air flow.

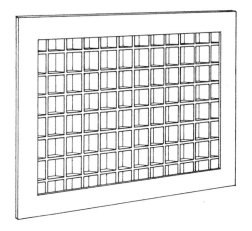

FIG. 18.24.—Standard Egg-crate Plastic Register.

Where extraction takes place naturally into a corridor, as in an office building, it is necessary to use a light-trap grille to avoid direct vision. Fig. 18. 25 shows a double-louvred fitting suitable for this purpose.

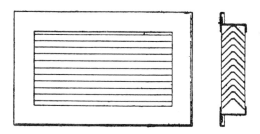

FIG. 18.25.—Double Louvred Natural Outlet.

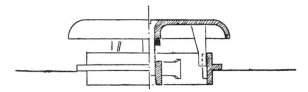

FIG. 18.26.—Mushroom Floor Ventilator.

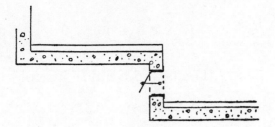

FIG. 18.27.—Gallery Riser Vent.

Details of the mushroom ventilator and gallery-riser vent referred to earlier are shown in Figs. 18.26 and 18.27. These types are more commonly used as extracts, though they may be used as inlets at low velocity.

CHAPTER 19

Fans, Ducts and Sound Control

THE FAN IS THE one item of equipment which every mechanical ventilation and air-conditioning system has in common. A fan is simply a device for impelling air through the ducts or channels and other resistances forming part of the distribution system. It takes the form of a series of blades attached to a shaft rotated by a motor or other source of power. The blades are either in the plane of a disc (propeller fan) or in the form of a drum (centrifugal fan).

There is as yet no other practicable commercial method of moving air for ventilation purposes, but fans in general suffer from various disadvantages such as low efficiency and noise. It is the latter which is probably the most troublesome to designers, and to which much careful attention must be given if silent running is to be achieved in the system as a whole: factors involved are air speeds, fan speeds, duct design, materials of construction and acoustical treatment of ducts, and provision for absorption of vibration.

FAN TYPES AND PERFORMANCE

Static Pressure—The purpose of any fan is to move air. When air is moved in a duct or through a filtering, heating, cooling or washing plant, a resistance to flow is set up.

The air is slightly compressed by the fan on its outlet side, so setting up a *static pressure* in the duct or plant. This pressure is tending to 'burst the duct', and may be read by means of a U-tube partly filled with water, connected at right angles to the air stream at any point in the duct. It is called a 'side gauge' (Fig. 19.1 (*a*)).

On the suction side of the fan the static pressure is negative, tending to collapse the duct.

As the air proceeds along the duct from the fan its compression is gradually released until at the end of the duct open to atmosphere the air is at atmospheric pressure. This falling away of the static pressure proportionately with the length of travel is called the *resistance* of the duct. Similarly all obstructions, such as heaters, filters, etc., cause a loss of pressure when air is passing through them.

It should be noted that as the static pressure becomes reduced, the air in effect expands such that pressure × volume = a constant (or nearly so, as explained earlier). This expansion therefore signifies an increase in velocity of the air if the size of the duct is unchanged.

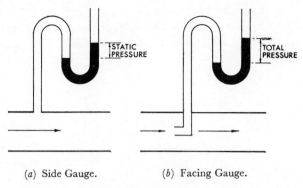

(a) Side Gauge. (b) Facing Gauge.

Fig. 19.1.—Gauges.

Velocity Pressure—The fan, in addition to generating static pressure, supplies the force to accelerate the air and give it velocity. This force is termed the *velocity pressure*, and is proportional to the square of the velocity. It is measured by a **U**-tube connected to a pipe facing the direction of air flow in a duct, etc. It is called a 'facing gauge' (see Fig. 19.1 (*b*)). But, obviously, the pressure so measured will in addition include the static pressure which occurs throughout the duct, as mentioned earlier, and the reading so obtained will thus be the *total* pressure. Thus, the velocity pressure alone may be found by deducting the static pressure from the total pressure reading, or by connecting one side of the **U**-tube to the facing gauge and the other to the side gauge provided the two gauges are at the same point.

If a fan discharges into an expanding duct (Fig. 19.2), the velocity will obviously decrease as the distance from the fan increases, and at the same time the velocity pressure will be converted into static pressure (not at 100 per cent efficiency, but about 75 per cent if the expansion is sufficiently

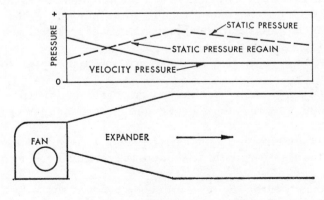

Fig. 19.2.

gradual). If the fan discharges into a large box (Fig. 19.3), from which at some point a duct connects, the fan velocity pressure will be entirely lost in eddies, and at the duct entrance must be recreated by a corresponding reduction in static pressure.

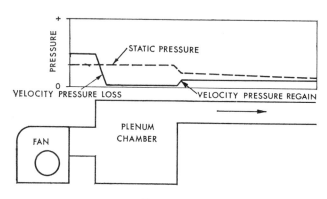

FIG. 19.3.

Total Pressure—The sum of the static and velocity pressure is called the *total pressure*.

Fan Pressure—In all air flow considerations as affecting resistances of ducts, plant, etc., it is the static pressure alone which concerns us, as this is the pressure which changes with such restrictions. It is the static pressure set up by a fan which is, therefore, a criterion of its performance. The velocity pressure, if taken as supplementing the fan duty, may be more misleading than useful, owing to the uncertainty of friction losses which occur at points of varying velocities. The velocity pressure is more generally not recovered, though sufficient must remain at the duct termination to eject the air at the required speed.

Where, however, by careful design of the fan discharge expander, the velocity pressure is converted to static pressure (probably to the extent of about 75 per cent), this additional pressure may be reckoned as augmenting the static pressure of the fan.

The pressure generated by a fan will be appreciated from Fig. 19.4.

The total fan pressure is defined as the algebraic difference between the mean total pressure at the fan outlet and the mean total pressure at the fan inlet.

The total pressure on the suction side, as will be seen from this Figure, is TP_S, i.e. the negative pressure AO minus the velocity pressure equivalent to AB.

The total pressure TP_D on the discharge side is similarly the static pressure OC plus the velocity pressure CD.

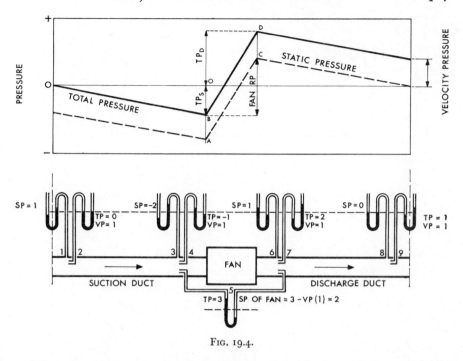

FIG. 19.4.

If, as stated above, we are concerned only with the resistance pressure set up by the fan, we find that this is the difference in pressure of points B and C.

We can arrive at the static pressure by measuring the total pressure of the fan as by the **U**-tube 5, and deducting therefrom the velocity pressure as given by difference between gauges 6 and 7. Such a method is valid only if the velocity in suction and discharge ducts is the same.

The other **U**-tubes indicate the pressures at the various points along suction and discharge ducts, and their meaning will be apparent.

If a fan has suction ducting only, the static pressure produced for overcoming friction of ducting will be represented by OB (a negative pressure), since the discharge will be at atmospheric pressure.

Similarly, if the fan has only discharge ducting, the static pressure will be represented by OC, the suction being at atmospheric pressure.

Fan Characteristics—A comparison of the operation of fans of various types is best understood by studying their characteristic curves. For this purpose consider a fan connected to a duct with an adjustable orifice at the end, as in Fig. 19.5 (a). Pressures are measured by water gauges connected to a standard Pitot tube, the end of which is shown in (b). The perforated portion gives the static pressure, and the facing tube the total pressure.

If the fan is running with the orifice shut no air will be delivered. Static pressure will be a maximum, and velocity pressure nil. As the orifice is

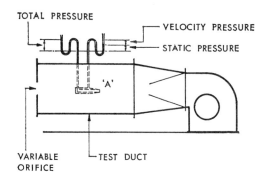

TOTAL PRESSURE

VELOCITY PRESSURE

STATIC PRESSURE

'A'

VARIABLE ORIFICE

TEST DUCT

(*a*) General Arrangement.   (*b*) Detail of Pitot Tube 'A'.

FIG. 19.5.—Fan Test.

opened s.p. will fall and v.p. increase until when full open s.p. will be small and v.p. a maximum. Over this range the power required to drive the fan will have increased from minimum to maximum, and perhaps will fall away as the total pressure falls off. The power may be arrived at as follows:

Air power (W)

$$= \text{Volume (m}^3/\text{s)} \times \text{total pressure set up by fan (N/m}^2)$$

The mechanical efficiency will be Air power/Fan power, and will depend on design, type of fan, speed, and proportion of full discharge: in the days of Imperial measure it was easier to distinguish between these criteria, but we shall no doubt become accustomed to the new terminology as manufacturers produce catalogues in coherent terms (as they, too, come to understand the S.I. system.*)

If the static pressure is used, the efficiency derived will be *static efficiency*; if the total pressure is used the efficiency will be *total efficiency*.

The standard air† for testing fans is taken at a temperature of 20° C, a density of 1·2 kg/m³ and a pressure of 101 325 N/m² (1·013 bar). Any fan at constant speed will deliver a constant volume at any temperature; as the temperature varies the density will increase or decrease proportionately with the absolute temperature, hence the power input will vary in the same ratio. With increase of temperature the power will be reduced and vice versa. Similarly, decrease of atmospheric pressure (as in the case of a fan working at high altitude) will cause a reduction in power and conversely.

If a fan running at a certain speed is increased to some higher speed, the system remaining the same, the volume will increase directly as the

---

* See *Engineering Language of the Future*. Published by the I.H.V.E. (Free on request.)
† For details of standard fan testing see B.S. 848.

speed; the total head will increase in the ratio of the speeds squared; the power input will increase in the ratio of the speeds cubed.

Characteristic Curves—Fans are of two main types, each with sub-divisions as follows:

1. Centrifugal type:
   - (*a*) Multivane, forward bladed.
   - (*b*)   „   radial   „
   - (*c*)   „   backward „
   - (*d*) Paddle wheel.
2. Propeller type:
   - (*a*) Ordinary propeller or disc fan.
   - (*b*) Axial flow.

The three types of runner, (*a*), (*b*) and (*c*), are shown in Fig. 19.6.

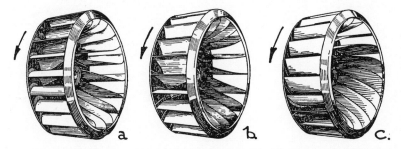

Fig. 19.6.—Fan Impellers: (*a*) Forward Bladed; (*b*) Radial Bladed; (*c*) Backward Bladed.

We will consider their characteristic curves and discuss their applications. The curves for static pressure, power input, and efficiency (static) are drawn from tests at constant speed, as already described. The base of the curve is percentage of full opening of the orifice. The vertical scale is percentage of pressure, efficiency or power input.

Fig. 19.7 (*a*)–(*e*) gives typical curves for the five main types. Paddle-wheel fans are not given, as they are now little used in ventilating work on account of their noise, being confined chiefly to dust removal and industrial uses.

Forward Curved—It will be noted that the forward-bladed centrifugal fan most commonly used in ventilation reaches a maximum efficiency at about 50 per cent opening, where at the same time the static pressure is fairly high. Fans are generally selected to work near this point. It will also be observed that the power curve rises continuously. Thus, if in a duct system the pressure loss is less than calculated, the air delivered will be more and the power more, which will lead to overloading of the motor.

Backward Curved—This type of fan runs at a higher speed to achieve the same output as a forward curved. The efficiency reaches a maximum at

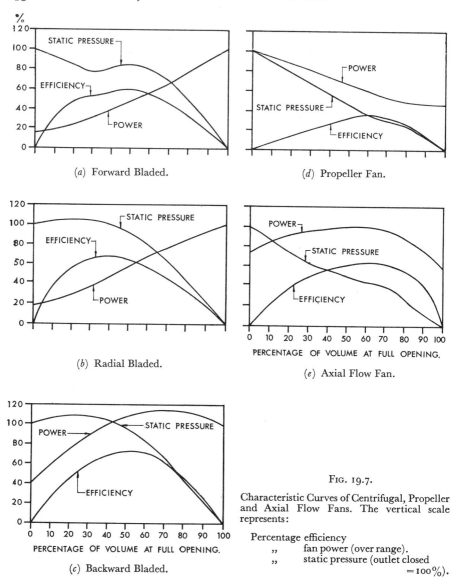

(a) Forward Bladed.

(d) Propeller Fan.

(b) Radial Bladed.

(e) Axial Flow Fan.

(c) Backward Bladed.

FIG. 19.7.

Characteristic Curves of Centrifugal, Propeller and Axial Flow Fans. The vertical scale represents:

Percentage efficiency
  „   fan power (over range).
  „   static pressure (outlet closed
                                    = 100%).

about the same point, and the power, after reaching a peak, begins to fall. This is called a *self-limiting characteristic*, and means that if the motor is installed large enough to cover this peak it cannot be overloaded. This is often useful in cases where the pressure is variable or indeterminate. The pressure curve is smooth without the dip of the forward curved; for this reason this kind of fan is to be preferred where two fans are working in parallel. The forward curved type under such conditions is apt to hunt from one peak to the next, so that one fan may take more than its share of the load and the other much less. The backward-bladed fan is often made with aerofoil blades, so raising efficiency. It is much used in high velocity systems where high pressures are required. (See Plate XXII, facing p. 464.)

Radial—This type is in effect similar to a backward curved in some respects, but has not the power-limiting characteristic. It is not commonly used.

Ordinary Propeller—From the curves it will be observed how the pressure falls away continuously, and the static efficiency reaches but a low figure. Thus, this type is unsuitable where any considerable run of ducting is used. To generate any appreciable pressure its speed becomes unduly high, and hence the fan is noisy. Its main purpose is for free air discharge where its velocity curve would rise towards a maximum at full opening. It should be noted that the fan power is a maximum at closed discharge, and, as the motors supplied with these fans are not usually rated to work at such a condition, the closing or baffling of the discharge or suction may cause overloading.

Axial Flow—This type shows a great improvement over the ordinary propeller fan, both as regards efficiency and pressure. The fan-power curve is self-limiting. Hence, these fans may safely be used in conjunction with a system of ductwork, being often more convenient than a centrifugal, particularly for exhausting. They run at higher speed than a centrifugal fan to produce a given pressure, and are liable to be more noisy: this may be overcome, to some extent, by the use of a silencer.

Fan Arrangements and Drives—Centrifugal fans may be *open* or *cased*. When open they can only be used for exhausting, and the discharge is tangential from the perimeter of the impeller, as might be suitable in a large roof turret.

The usual arrangement is the cased type, and the suction is then either on one side, as in Fig. 19.8 (single inlet), or both sides as Fig. 19.9 (double inlet). The single inlet is the more usual. The double inlet double-width fan is useful where large volumes are concerned, as it gives double the capacity of the single inlet with the same height of casing.

Fans are now invariably driven by electric motor. Fig. 19.10 shows a typical arrangement with the fan impeller mounted on a shaft extension of the motor. This is a compact arrangement, but generally used for small or medium-sized fans only.

Fig. 19.11 shows a motor direct-coupled to a fan with a flexible coupling. The fan shaft runs in its own bearings independently of the motor. This is obviously to be preferred for heavy duty and for large fans. The motor can be removed and replaced without affecting the fan.

Fig. 19.12 illustrates a motor driving a fan by vee-belts. This arrangement has the great advantage that the *motor* speed may be a standard, such as 16 rps* or 24 rps, whilst the *fan* speed is that best suited to the duty. A further advantage is that if on testing it is found that the pressure

* rps = revolutions per second.

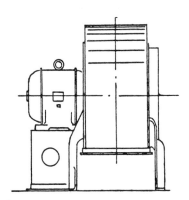

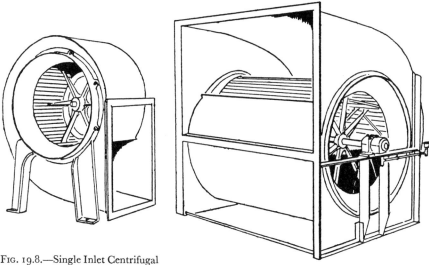

FIG. 19.8.—Single Inlet Centrifugal
Fan.

FIG. 19.9.—Double Inlet Centrifugal Fan.

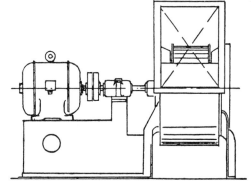

FIG. 19.10.—Centrifugal Fan with
Close-coupled Motor.

FIG. 19.11.—Centrifugal Fan with Flexible
Coupling to Motor.

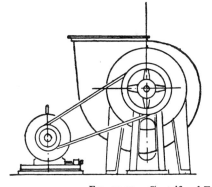

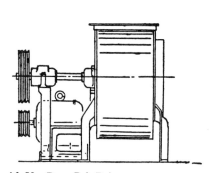

FIG. 19.12.—Centrifugal Fan with Vee Rope Belt Drive.

loss of the system is less or more than allowed for, the fan duty can be corrected by merely changing the pulleys.

It will be noted that in the illustrations different positions of the discharge openings in relation to the suction eye of the fan in each case are given. It is possible to obtain a fan with its discharge at any angle, vertical, horizontal top, horizontal bottom, downwards, and intermediately at an angle of 45°.

Axial Flow Fan Arrangements—This type of fan, it is stated, can be designed to give static pressures up to 500 N/m² within the limits of quiet running. The maximum peripheral speed should not normally exceed 37 m/s for this condition.

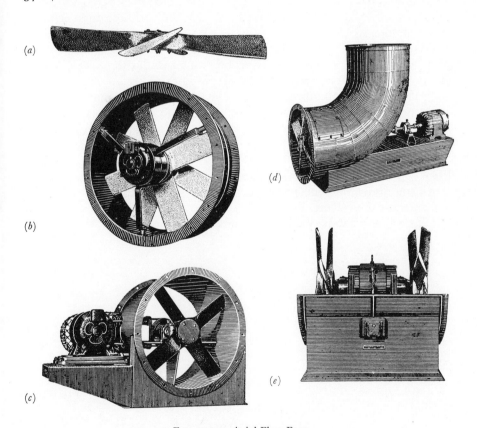

FIG. 19.13.—Axial Flow Fans.

These fans can be built in one, two or three stages, to obtain increased pressure, the volume remaining the same. Alternatively they may have two blades made counter-rotating.

Such fans, illustrated in Fig. 19.13 (a)–(e), may have blades of specially treated wood, similar to an aeroplane propeller or of die-cast aluminium; these are shaped to aerofoil section (see (a)), and may be designed for any angle and any number of blades from two to twelve.

(b) shows a detail of the arrangement with eight blades, driven from direct-coupled motor in the duct. (c) shows drive by external motor with vee belt, suitable where steam or fumes have to be removed. (d) shows a fan in a bend driven by direct-coupled external motor; this type is described as an *acid-bend* fan, bend and fan being treated with anticorrosive protection. This is suitable for many ventilation applications in chemical plants, etc., but it may also be used for extraction, say, of kitchen exhaust fumes or laundry steam. (e) shows a two-stage fan driven by a common motor, mounted in a duct, the top half of which has been removed.

Other makes of axial flow fans include streamlined nose and tail pieces to the motors.

Fan Duties—The range of fan types, speeds, pressures, and volumes is too great for any indication to be given here of sizes, duties, power requirements, etc., or of the problem of motor types suitable for fan drives.

Enquiries to fan-makers should give the fullest information possible about any system, as there are many hidden points to be watched in the selection of fans which render mere catalogue reference insufficient.

MEASUREMENT OF AIR FLOW

Pitot Tube—Reference has been made to the Pitot tube as a means of measuring static and velocity pressure. It may be used in its latter capacity to determine the velocity in a duct.

The relationship between speed and pressure is:

$$p_v = 0.5 \, \rho v^2$$

where p_v = Velocity pressure N/m^2

ρ = density of fluid kg/m^3

v = Velocity m/s

for standard air $\rho = 1.2$

thus $p_v = 0.6 \, v^2$

Fig. 19.14 gives the relationship graphically for speeds encountered in ventilation work. For other temperatures and pressures

$$\rho = 1.2 \left(\frac{\text{Working pressure } N/m^2}{101\,325} \right) \left(\frac{293}{273 + t} \right).$$

where t = temperature of air °C and working pressure = atmospheric pressure ± static pressure.

Micromanometer—The pressures at low velocities are slight, as will be noted, and for the purpose of reading them an ordinary **U**-tube or inclined gauge is too insensitive. The micromanometer, of which one type is shown in Fig. 19.15, is therefore necessary. In this an extended **U**-tube is used, being tilted by a micro-adjustment. The level is viewed through the magnifying eyepiece against a crosswire. The coarse reading is taken on the side scale, and the fine reading on the rotating dial, one revolution of which corresponds to one division of the scale. The liquid may be alcohol

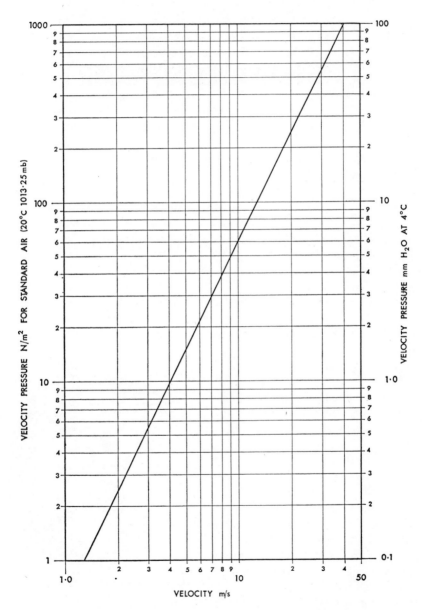

Fig. 19.14.—Velocity Pressure related to Air-Speed for Standard Air.

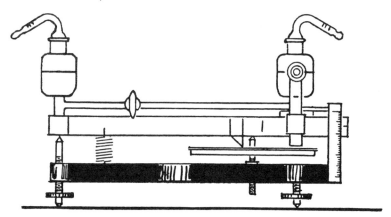

FIG. 19.15.—Tilting Micromanometer.

and the scale is calibrated in N/m^2. Other still more sensitive instruments are made.

When measuring air speeds in a duct, however, the velocity varies across the duct, being greatest at the centre and least at the periphery. Thus, in a round duct it is necessary to take readings at a number of points in concentric rings of approximately equal area. The average speed multiplied by the area of the duct will give the volume of air passing. In the case of a rectangular duct the method is similar, except that the duct is divided into equal rectangles.

Anemometer—The anemometer (Fig. 19.16) measures air speed by vanes which revolve as the air impinges on them. The dials, which are calibrated in metres, serve only to count the revolutions over a given time, such as one minute, taken by a stop-watch.

The lever at the top enables the gearing to be engaged and disengaged from the vane at the start and finish of the time. The knob shown on top is depressed to cancel the reading and return the dials to zero.

The instrument requires to be calibrated periodically, and a correction factor applied. The standard instrument is too insensitive for use below about 0.5 m/s, but other slow-speed types may be used down to about 0.15 m/s. Above about 15 m/s a high speed type is used.

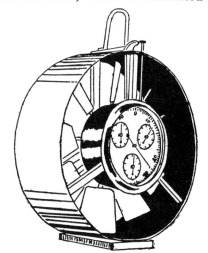

FIG. 19.16.—Vane Anemometer.

For general work it is a useful instrument, chiefly for measuring the air speed from or to gratings, etc. In such case the instrument is placed about one inch from the grille and the speed is multiplied by the whole face area (regardless of grating-free area) to obtain the volume in m³/s. Readings are taken at various points on the grille and averaged. It is not so useful for measurements in ducts, as the anemometer has to be introduced through a hole in the duct wall, and it is then difficult to manipulate.

It must be admitted that the anemometer requires very careful handling, and in the hands of an unskilled operator entirely erroneous results can be obtained, particularly when measuring over a large grille.

The Velometer—An instrument of later development, the 'Velometer', is shown in Fig. 19.17. It relies on the speed of air to deflect a shutter against a spring. The shutter causes the needle to move over the sector-shaped dial, reading direct in metres per second. It requires no timing.

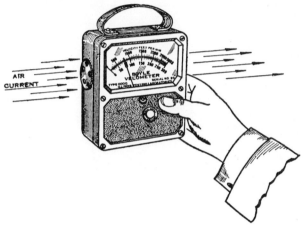

FIG. 19.17.—Velometer.

When used direct, as in the illustration, for measuring speeds from or to gratings at low velocities, it reads from zero to 1·5 m/s. Higher ranges of speeds, 0–15 and 5–30 m/s, are covered by means of adaptors screwed into the aperture at one end. These may have rubber tube connection for reading at a distance, which, in turn, may be used with various special mouthpieces fitted to the end of the rubber tube to read velocities in ducts, etc. The same instrument may be used in place of a Pitot tube to give static pressure readings.

Altogether this is a most versatile instrument for giving a quick check on adjustments or for exploring velocities over an area.

AIR DUCTS

Ducts for ventilation and air-conditioning are commonly constructed either of galvanized sheet steel or of 'Builders' Work'. The principles to be followed in the design of such ducts are:

(*a*) Avoidance of sudden restrictions or enlargements, or any arrangement producing abrupt changes of velocity.

(b) Bends to be kept to a minimum, but where required should have a centre-line radius not less than one and a half times the diameter or width of duct at the bend. Alternatively, deflectors should be fitted into the duct, of the type shown in Fig. 19.18, which permit of right-angled turns being used with much reduced pressure loss, and reduction of noise.

(c) Where branches occur, they should be taken off at a gradual angle before turning.

(d) Sharp edges should be avoided, as these will be the cause of noise which may travel a considerable distance through the duct system.

(e) Rectangular ducts should be as nearly as possible square, the more they depart therefrom the more uneconomic they become.

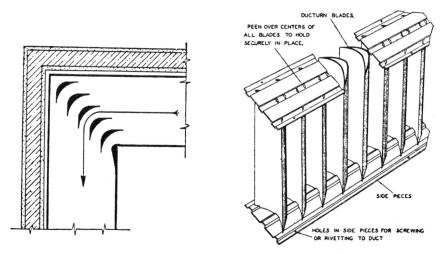

FIG. 19.18.—Standard Duct Turn.

Galvanised Steel Ducts—The gauges of metal usually used are set out in two well presented documents* issued by the H.V.C.A. together with much information regarding constructional techniques for both low-pressure and high-pressure systems.

Ducts of galvanised sheet steel are often priced by weight. With the Imperial system, it was necessary to use complicated tabulated data to determine the weight of sheets manufactured to Birmingham gauge, or whatever other archaic system of measurement was employed, to describe the metal thickness. With the SI system, sheet will be made to millimetre

* Metric Specifications DW/121 and DW/132. Heating and Ventilating Contractors Association.

dimensions and the mass of metal will thus merely be thickness times area times 7800 kg/m³ (for steel) or whatever other specific mass is appropriate.

Rectangular ducts are made up out of flat sheets by bending, folding and riveting and are erected in sections with slip joints or angle ring joints bolted. Larger sizes tend to drum and adequate stiffening is necessary. Sometimes a 'diamond break' is used to assist in stiffening, as Fig. 19.19. Where rectangular ducts are used in high velocity systems, stiffening is particularly important using bracing angles, tie angles, and internal tie rods for larger sizes.

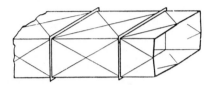

FIG. 19.19.—Diamond-break Stiffening of Duct Walls.

Circular ducts may equally be formed from flat sheet with folded or riveted seams and are inherently free from risk of drumming. They have been used in the past chiefly in industrial applications.

The advent of circular ducts made from galvanized strip steel with a special locking seam, as in Fig. 19.20, has brought a return to their use more generally. Sizes are available from 75 mm up to 800 mm diameter. Ducts of this type are particularly suitable for high velocity systems due to their rigidity, 'deadness' and air-tightness. Ranges of standard tees, bends, reducers and other fittings are made and these, together with lengths of straight ducting cut off on site as required, facilitate ease of erection compared with the tailor-made methods previously adopted. Jointing is by means of a special mastic, the whole joint being bound with a prepared sealing tape.

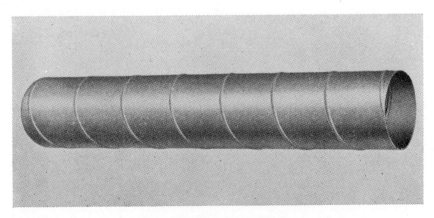

FIG. 19.20.—Spiral-wound Duct.

Flat Oval or spirally wound rectangular ducts are currently coming into increasing use in air-conditioning. These are formed from spirally wound circular ducts to the profile illustrated in Fig. 19.21 and are available in sizes from 150 mm × 550 mm to 500 mm × 980 mm. A limited range of standard tees, bends and reducers is made. Such ducts have the advantage of being mass produced and can be used in conjunction with circular sections where space is limited.

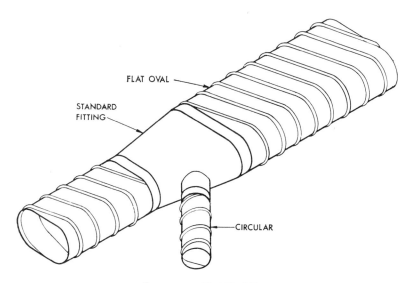

FLAT OVAL

STANDARD FITTING

CIRCULAR

FIG. 19.21.—Flat Oval Duct.

Builders' Work Ducts—It is often possible, in the design of a building, to construct main ducts and large rising shafts as part of the fabric. Where this can be done it has the following advantages:

(*a*) Rigidity, and hence reduction of noise transmission.

(*b*) Permanence.

(*c*) Cheapness, in that the space very often exists in any case, and only requires to be made smooth and air-tight.

(*d*) Reduction of heat gains and losses, due to the heavy construction as compared with metal.

(*e*) Accessibility for cleaning.

A disadvantage is the 'flywheel' effect of heavy constructions, and hence for applications where careful control of the air temperature is required, such ducts must be insulated *on the inside* and great care is needed to ensure air-tightness.

Fig. 19.22 shows an example of a duct constructed over a basement corridor and serving as a main distribution trunk from the plant to rising shafts about the building.

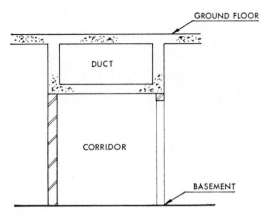

GROUND FLOOR

DUCT

CORRIDOR

BASEMENT

FIG. 19.22.—Main Duct over Basement Corridor.

Ducts of Other Materials—Other materials used for ducts are:

(a) *Asbestos Cement*, which is preformed in sections and erected like a metal duct. It is more commonly used for fume extraction and tends to be expensive due to the cost of the moulds required for fittings and special sections.

(b) *Glass fibre resin bonded*, can be moulded to any desired shape, and can be cut and adapted at site, and jointed to form a smooth continuous duct.

(c) *P.V.C.* is chiefly used for chemical fume extraction.

(d) *Copper, Aluminium, Stainless Steel*. Ducts constructed of these materials are used in special cases where permanence or a high degree of finish is required, or where special corrosion problems arise.

(e) *Welded Sheet Steel* is mainly used in industrial work, or where the air-tightness of the joints is of great importance; such ducts may be 'galvanised after made'.

Duct Sizing for Normal Velocities—The most convenient method of sizing ducts is on the *Equal Pressure-loss* basis. For this purpose use may be made of Figs. 19.23 and 19.24 and Tables 19.1, 19.2 and 19.3.

The following procedure is adopted:

(1) On the plan of the duct system it is necessary to mark the volume of air to be delivered or exhausted at each outlet. These must be totalled back to the fan, including the sums brought in at each branch. These volumes are conveniently expressed in m³/s or litre/s.

(2) Establish the maximum velocity in the main duct leaving or entering the fan according to the type of building, etc., from Table 19.2. The velocities given in this table are arbitrary. The higher the velocity the greater the noise, hence low velocities are desirable for buildings where

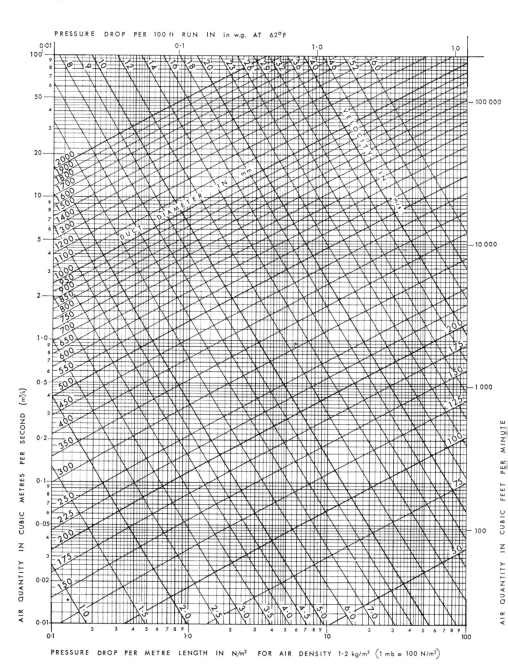

PRESSURE DROP PER 100 ft RUN IN in w.g. AT 62°F

PRESSURE DROP PER METRE LENGTH IN N/m² FOR AIR DENSITY 1·2 kg/m³ $\left(1 \text{ mb} = 100 \text{ N/m}^2\right)$

FIG. 19.23.—Duct Sizing Chart (by permission of the *I.H.V.E.*).

silence is required. Some experience is necessary in interpreting this aspect of the matter.

(3) From Fig. 19.23 select the velocity line for the main duct, and find the point of intersection with the air volume carried in the main. The vertical line passing through this point will give the pressure loss per metre run of duct by reference to the bottom scale, and at the same time the diameter of main duct, assumed to be circular.

(4) The sizes of the subsequent sections of the main duct on to the end are then read off at the point of intersection of the appropriate horizontal volume lines with the same vertical resistance line as for (3). It will be observed that as the volume reduces, the velocity also becomes less.

(5) The total pressure loss of the main duct is then calculated by tabulating thus:

(a) Section (Ref. Letter or No.)	(b) Volume m^3/s	(c) Duct Dia- meter mm	(d) Length m	(e) Resist- ance per m from Chart N/m^2	(f) Pressure loss for Length (d) N/m^2	(g) Single Resist- ances N/m^2	(h) Total for Section N/m^2	(i) Progres- sive Total N/m^2

The values of single resistances (for bends, etc.), column (g), are obtained on the velocity pressure (v.p.) method. The pressure corresponding to any velocity is taken from Fig. 19.14. The loss in terms of v.p. for each particular bend, reducer, etc., is taken from Fig. 19.24 (p. 444).

(6) The branches are then dealt with thus:

Pressure available for branch = (total pressure loss of main duct) – (pressure loss from fan to branch, column (i)).

The length of the branch to its end is then measured and the loss per m length established.

With this new loss, duct sizes for each volume are read off (Fig. 19.23) as before.

If this gives too high a velocity, a lower pressure loss must be selected, and the surplus pressure can then only be absorbed by dampering.

(7) Sizes for the whole system in terms of circular ducts will then be marked on the plan.

If rectangular ducts are used instead of round, the equivalent sizes to give equal pressure loss per m run may be taken from Table 19.1. The rectangular sizes may be selected to suit the positions which the ducts must occupy. To simplify this process, however, use may be made of the I.H.V.E. Guide chart of air flows in rectangular ducts of standard sizes.

	Duct Element	Circular			Rectangular	
		$\frac{R}{D}$	90°	135°	$\frac{W}{D}=·5$	$\frac{W}{D}=1$ to 4
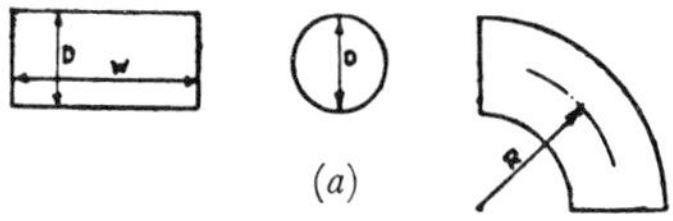 (*a*)	Bends - -	4	·12	·06	·15	·10
		2	·15	·08	·17	·13
		1	·25	·15	·28	·20
		·75	·35	·22	·47	·33
		·5	·6	·4	1·3	·95
					Square	
					90°	135°
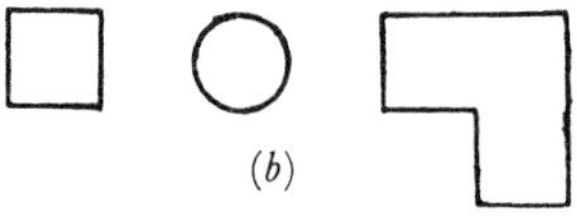 (*b*)	Elbows -	—	·85	·43	1·25	·65
		$\frac{R}{D}$		90° 1 Vane		90° 2 Vanes
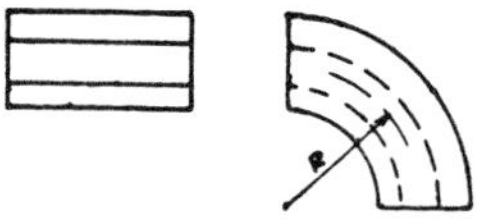 (*c*)	Bends with Deflector Vanes	1·5		·11		·15
		1·0		·13		·10
		·75		·16		·12
		·5		·70		·45
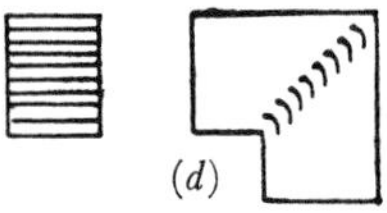 (*d*)	Elbow with 'Duct Turns'	Single blade vanes -				·35
		Streamlined double vanes - - -				·10
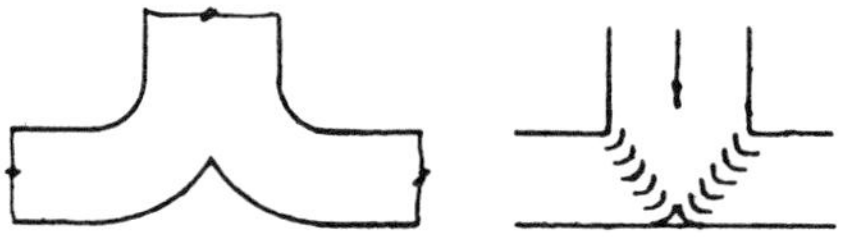 (*e*)	Tees or Twin Elbows	Take as for corresponding Bend or Elbow				
		α		V.P. in branch ×		
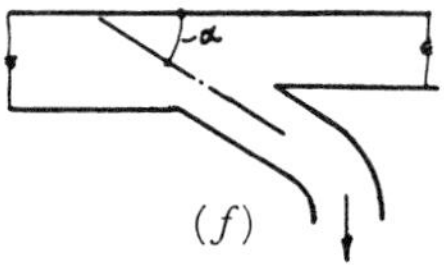 (*f*)	Branch Inclined	15°		·09	Plus loss in bend	
		30°		·17		
		45°		·22		
		60°		·44		
		$\frac{R}{D}$		V.P. in branch × 90°		135°
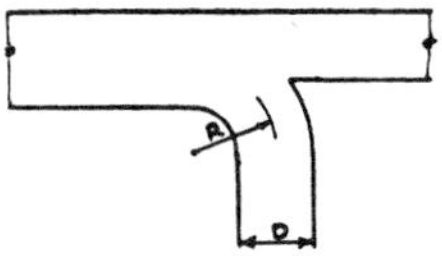 (*g*)	Branch Radius Type. Circular or Square	2		·2		·13
		1		·32		·22
		·75		·47		·34
		·5		·75		·55

FIG. 19.24.
Velocity Pressure Loss Factors for Single Duct Resistances.

	Duct Element	α	V.P. in.	Factor
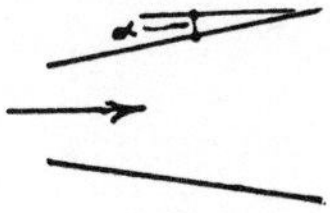 (h)	Expander (Square or Circ.)	60° 30° 10°	Entrance ,, ,,	·75 ·6 ·2
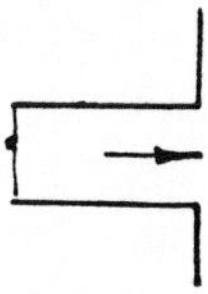 (i)	Sudden Expansion	—	Entrance	1·0
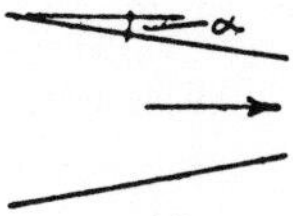 (j)	Contraction	60° 30° 10°	Exit ,, ,,	·25 ·10 ·02
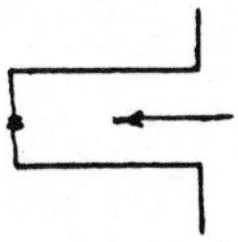 (k)	Sudden Contraction	—	Exit	·5
	Gratings: 50% free area	—	Free area	1·5
 (l)	Louvres: 90% free area 70% ,, ,,	45° 45°	Free area ,, ,,	·5 ·75
	In General -	Abrupt changes in direction or velocity		1·0
		Easy changes in direction or velocity		·5

FIG. 19.24.—*Continued.*

TABLE 19.1

RECTANGULAR DUCTS AND EQUIVALENT DIAMETERS (mm) FOR EQUAL VOLUME AND PRESSURE DROP

Dimension of side b	Dimension of side of duct a									
	100	125	150	175	200	225	250	300	350	400
100	110	123	134	145	154	163	171	185	199	211
150	134	151	165	178	190	202	212	231	248	264
200	154	173	190	206	220	233	246	269	289	308
250		192	212	230	246	261	275	301	325	346
300			231	251	269	286	301	330	357	381
350				269	289	308	325	357	385	412
400					308	328	346	381	412	441
450						346	366	403	436	467
500							385	424	459	492

$$d = 1 \cdot 265 \left[\frac{(ab)^3}{a+b} \right]^{0 \cdot 2}$$

	450	500	600	700	800	900	1000	1100	1200
400	467	492	537	578	616	650	683	713	741
500	492	551	603	650	693	733	770	804	837
600	537	603	661	713	761	806	848	887	924
700		650	713	771	824	873	919	962	1002
800			761	824	881	934	984	1031	1075
900				873	934	991	1044	1095	1142
1000					984	1044	1101	1155	1205
1100						1095	1155	1211	1265
1200							1205	1265	1322

	1300	1400	1500	1600	1700	1800	1900	2000
1000	1254	1299	1343	1385	1426	1465	1503	1539
1200	1375	1427	1476	1523	1568	1612	1654	1695
1400	1486	1452	1596	1648	1697	1745	1792	1837
1600		1648	1706	1762	1816	1868	1919	1968
1800			1808	1868	1926	1982	2036	2089
2000				1968	2029	2089	2147	2203
2200					2126	2189	2250	2310
2400						2284	2348	2411

Table 19.1 is based on the formula stated where

$$d = \text{diameter of duct.}$$
$$a \text{ and } b = \text{dimensions of rectangular duct.}$$

It is advantageous when sizing a long run of ducting with air supply grilles on its full length, to take static regain into account. This is done by so reducing the velocity in the duct in stages that the regain of static pressure compensates for the frictional loss in the duct to that point. This gives approximately equal pressures at each grille, and assists regulation. The regain is usually taken as 50 per cent of the change in velocity head.

(8) Where ducts are made of other materials, the resistance will be less or more than that of galvanized steel ducts as given in Table 19.3 (p. 448).

TABLE 19.2

MAXIMUM DUCT VELOCITIES FOR CONVENTIONAL SYSTEMS FOR VARIOUS TYPES OF BUILDINGS

	Main Ducts and Shafts	Branch Ducts
	m/s	m/s
Hospitals, Concert Halls, Theatres, Libraries, Film Studios, etc. - -	5	4
Cinemas, Restaurants, Assembly Halls, etc. - - - - -	7.5	5
General Offices, Dance Halls, Shops, Exhibition Halls, etc. - -	9	6
Factories, Workshops, etc. - - - - - - - -	10 to 12	7.5

Fan Pressure—The static pressure to be set up by the fan is the sum of all resistances to air flow throughout the system as follows:

Loss
N/m^2

(1) Fresh air intake louvres or grille (estimate on v.p. method, Fig. 19.14) - ....

(2) Fresh air intake duct (size and pressure loss established as for other ducts). This must include loss in single resistances as before - - - - - ....

(3) Pressure loss of:
    (a) air filter if any - - - - - - - - - - ....
    (b) heater or heaters - - - - - - - - - - ....
    (c) washer if any - - - - - - - - - - ....
    (d) cooling coils if any - - - - - - - - - ....
        (The above will be determined from makers' data.)

(4) Convergence of fan suction - - - - - - - - ....
    (Use v.p. method.)

(5) Divergence of fan delivery - - - - - - - - ....
    (Use v.p. method.)

(6) Changes of section, if any, through plant - - - - - - - ....
    (Use v.p. method.)

(7) Pressure loss of duct system estimated as already tabulated - - ....

(8) Pressure loss of final outlet grille - - - - - - - - ....

        Static pressure of Fan - - - - - - - ....

TABLE 19.3

FACTORS FOR DUCTS OF OTHER MATERIALS

(Basis: Taking resistance from Fig. 19.23 as 1·0 for galvanized steel, the pressure loss should be multiplied by the appropriate factor given below.)

Smooth Copper, Aluminium, or PVC, etc.	-	-	-	-	-	0·8
Smooth Cement	-	-	-	-	-	1·2
Rough Concrete or Good Brick	-	-	-	-	-	1·5
Rough Brick or Pre-cast Concrete	-	-	-	-	-	2·0

Extract System—Exactly the same summation of pressure loss is arrived at for an extract system as for an inlet. Items (3) and (6) in the above calculation will disappear, and in place of item (1) must be included the pressure loss of the discharge duct and cowl or other outlet.

Ducts for High Velocity Systems—Velocities over about 10 m/s may be described as high velocity. Such systems work up to 20 m/s and are applicable to induction and dual-duct systems of air-conditioning where adequate provision for silencing is included.

Duct sizing may follow the method of equal pressure drop previously discussed for conventional velocity systems.

Circular ducts should be used in preference to rectangular, to avoid drumming and to facilitate airtightness. Adequate sealing of joints is necessary. Branches are usually taken off at 45° and sharp edges are to be avoided. Spiral formed ducts referred to on p. 439 are commonly used for high velocity systems.

Thermal Insulation of Ducts—Ducts conveying cooled air in an air-conditioning installation are usually insulated on the outside so as to reduce temperature rise to a minimum, also to prevent surface condensation from the surrounding air. Such insulation must be 'vapour sealed', i.e. sealed against ingress of atmospheric air, otherwise the insulation will become quickly waterlogged and useless. Ducts conveying warm air in most cases equally require insulation to reduce loss of heat. Where ducts are run in the conditioned space, insulation may be unnecessary.

Transmission of heat through bare metal duct walls in W/m^2 per degree difference varies according to air velocity as follows:

Air velocity m/s	Transmission factor U (still air outside) $W/m^2 \, °C$
2	5·6
4	6·2
6	6·5
8	6·8
10	7·0
12·5	7·1
15 and over	7·2

The effect of insulation is to reduce the loss such that air velocity in the duct affects the U value only to a minor extent, and the following coefficients may be adopted as covering the loss at any practical air speed:

Metal duct with glass fibre insulation mm	$\dfrac{U}{\text{W/m}^2\,°\text{C}}$
25 thickness	1·4
38 ,,	1·0
50 ,,	0·7

The economic thickness can be estimated for any particular case.

Materials for duct insulation must be such as not to support fire, and cork (at one time in common use) is no longer acceptable. Insulating materials for use in ventilating systems are now subject to Fire Test*: some suitable ones are: glass fibre neoprene-faced, polyurethane and mineral wool.

The gain or loss of temperature in air conveyed in a length of duct may be calculated from the following:

let mean temperature difference between inside and out-
    side of duct $= \theta$
transmission coefficient of bare or insulated duct
    W/m² °C $= U$
length of duct in m $= L$
perimeter of duct in m $= P$
then heat loss in watts $= PLU\theta$
let mass of air carried by duct in kg/s $= M$
and specific heat capacity of air in kJ/kg °C $= 1·0$
thus temperature difference between ends of duct in
    question in °C $= \dfrac{PLU\theta}{M}$

This is sufficiently true for practical purposes, though not strictly so when the difference of temperature along the duct is of such a magnitude as to bring the internal temperature of the duct closer and closer to that outside.

Where insulating material is used as a duct lining it may serve a double purpose, thermally and acoustically.

Dampers in Ducts—Control of the air volume is effected by means of dampers, which are either:

(a) Permanently set when the installation is tested, to give the designed volumes in each branch. These are commonly of butterfly or sliding type with some form of locking device.

(b) 'Controllable' dampers for use in air-conditioned installations, and adjusted, either manually or automatically, to suit varying conditions. Such dampers are generally of louvred type as in Fig. 19.25.

Multi-louvre dampers having adjacent louvres contra-rotating, are much superior, from the regulating point of view, to the commonly used type where all vanes rotate in the same direction (see Fig. 18.20, p. 417).

* G.L.C. publication: *Report of Working Party on Fire Prevention, No. 4250*. County Hall.

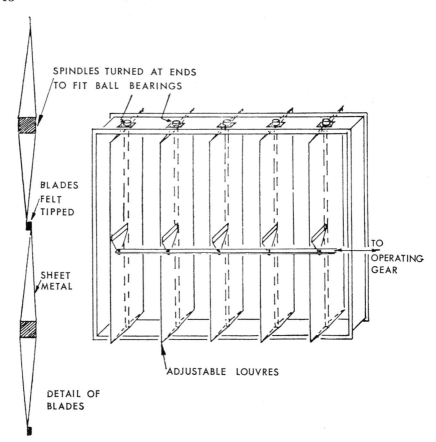

FIG. 19.25.—Louvered Dampers.

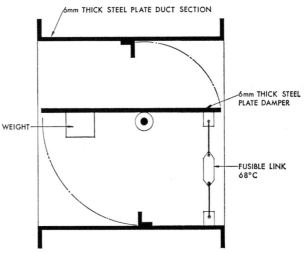

FIG.19.26.—Horizontal Fire Damper.

Where tight shut-off is essential, it is necessary for the dampers to be felt-tipped and to close on to a felted frame.

Fire Dampers—Where ducts pass through walls and floor slabs, fire authorities commonly require the insertion of fire dampers. A fire damper is shown in Fig. 19.26 and consists of a heavy steel casing and damper normally kept open by the fusible link. In the event of fire the link melts and the damper closes. Sensitive electronic detectors may also be used to close dampers. Access is needed to dampers for resetting, etc.

SOUND CONTROL

The increasing amount of mechanical plant in buildings brings with it the problem of noise. We here confine our attention to equipment such as fans, compressors, pumps and boilers, and the channels by which sound may be conveyed therefrom to other parts of the building.

Sound—It is necessary first to consider the mechanism by which sound is transmitted.

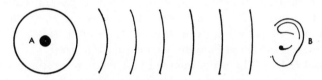

FIG. 19.27.—Sound transmission.

Vibrations at a point A set in motion the molecules of adjacent air such that a series of waves of compression and rarefaction radiate from the source spherically in all directions, a part reaching the ear at point B where the eardrum is set in like motion and the brain interprets the sensation as a sound, pleasant or otherwise. A *noise* may be defined as an unpleasant sound, or one that is overloud, or monotonous or erratic.

The speed of sound depends on the properties of the medium, but may be taken at 331·46 m/s at stp.

Frequency—Sound from point A will most likely be of more than one frequency. The deepest note which the human ear can detect is about 15 Hz, and the highest about 20,000 Hz. Frequencies are divided into octaves, each octave being double that of the lower one. The frequency of Middle C is 262 Hz; thus C of the octave above is 524 Hz and so on. Most sound sources set up harmonics, i.e. higher frequencies, being multiples of the fundamental.

Amplitude—The amount of energy imparted to the air at point A is termed the *Sound Power*, and is referred to in terms of *watts*. Sound power may be visualised as amplitude at the source.

The amplitude of sound X is greater than Y: X is a loud sound and transmits more energy than soft sound Y. The energy involved is represented by watt × time. (See Fig. 19.28).

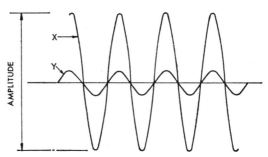

FIG. 19.28.—Amplitudes.

Sound power cannot be measured as such; it can be inferred mathematically from the sound received at a known distance in a sound-proof room.

Sound Received—The power of sound received is termed *Sound Intensity* and is measured in watt/m². Although the sound produced at point *A* cannot be measured directly, the sound received by the ear—or by a sound meter—at point *B* can be expressed in terms of pressure variation.

The amount of variation in pressure of the sound waves reaching the receiver—termed the *Sound Pressure*—is measured in N/m².

The Decibel (dB)—The units referred to above are absolute values and are not convenient for general use. Units of reference have therefore been invented and are termed *bel* and *decibel* (10 dB = 1 bel). They are logarithmic, so that if two sounds differ in magnitude by 1 bel, the magnitude of one is ten times that of the other. Similarly, if the difference is 1 decibel, the magnitude of one will be about 26 per cent greater.

The decibel may be defined as $10 \times \log_{10}$ of the ratio between two absolute values such as levels of sound power, sound intensity and pressure.

Sound Power level (with reference to 10^{-12} watt)

$$= 10 \log_{10} \frac{W}{10^{-12}} \text{ dB}$$

where W = actual sound power.

Sound Intensity level (with reference to 10^{-12} watt/m², taken to be threshold of hearing)

$$= 10 \log_{10} \frac{I}{10^{-12}} \text{ dB}$$

where I = actual sound intensity.

Sound Pressure level (with reference to 2×10^{-5} N/m² which corresponds to the reference intensity 10^{-12} watt)

$$= 20 \log_{10} \frac{P}{2 \times 10^{-5}}$$

where P = actual sound pressure level.

Decibels cannot be added arithmetically, but if two sounds of known decibel rating are taken separately, the decibel rating of the two taken together will be:

$$\log_{10} \text{ (antilog of sound 1 in dB + antilog of sound 2 in dB)}$$

A graph published in the *I.H.V.E. Guide* simplifies this task.

The Phon—This unit is based on the subjective reaction of the ear to loudness and it may be related to sound pressure level in dB by means of

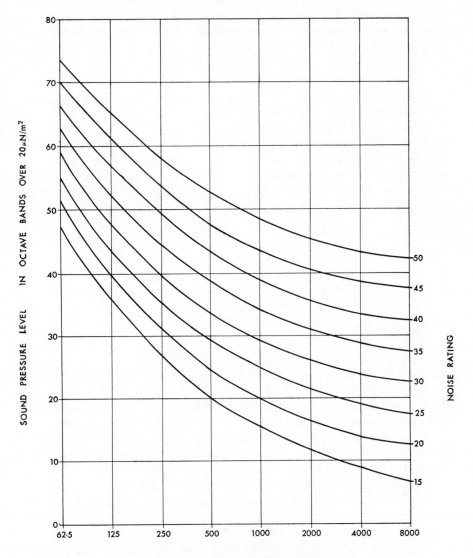

MID-FREQUENCIES OF OCTAVE BANDS Hz

FIG. 19.29.—Noise Ratings.

data which have been derived by experiments on individuals. Some values on this scale are:

Threshold of hearing -　　0　　phons
Whispering　　-　　-　　20　　,,
Average room　　-　　-　　40　　,,
Busy street -　　-　　-　　50 to 70　　,,
Pneumatic drill -　　-　　100　　,,

Noise Criteria—It is common experience that the human ear can tolerate noise of low pitch much more than of high pitch, even though of equal intensity. Thus it is necessary to consider pitch in relation to sound measurement and for this purpose the spectrum of frequencies has been split up into 8 octave bands.

Noise Criteria and Noise Rating curves (NC and NR) have been established on a subjective basis and each is given a number corresponding to the sound pressure level in decibels in the sixth octave band (1200–2400 Hz). NC curves have been commonly used in the past but, with the advent of the SI system, the NR curves more familiar in Europe are being adopted. For most practical purposes the two sets of curves may be regarded as interchangeable. Fig. 19.29 shows NR plots of equal tolerance for the eight frequency bands. Acceptable NR values may be classified as follows:

NR 25　　Very quiet
NR 30　　Normal living space
NR 35　　Spaces with some activity
NR 40　　Busy spaces
NR 45　　Light industry
NR 50　　Heavy industry

SOUND MEASUREMENT

Sound Meter—In one form this consists of a microphone coupled to a sensitive milliameter and battery such that variations in sound pressure cause corresponding deflections of the needle of the instrument. By calibrating the dial in decibels a direct reading in dB can be obtained.

As pointed out, however, the ear does not evaluate sounds of different pitch in a linear manner and hence scales have been weighted in an attempt to correct for this by adding filters to the circuit.

An unweighted scale is termed a Linear or flat scale.

Scale C is as Linear, but very high and very low frequencies are suppressed.

Scale B is as C but with more low frequencies suppressed.

Scale A is further weighted to exclude all low frequency sound intensities under about 55 dB.

Sound Analyser—In effect this is a sound meter, but so equipped with filters that measurements of sound pressure at any range of frequency is possible thus enabling a sound 'spectrum' to be built up.

In using the three scales A, B and C for subjective assessment, scale A would be appropriate for low noise levels, bearing in mind the greater tolerance of the ear to low pitch sounds. Scale B is for medium levels, and C for loud noises.

The greater the difference between A and C scale readings, the greater the importance of the low frequency component.

The Linear scale would be used to obtain a basis for calculation of absolute values.

Noise of Fans—The noise produced by fans is often quoted in makers' lists in decibels. It will be apparent from what has been stated that such a rating can only be a measure of the sound intensity level at a given point and the level will vary according to the distance at which the measurement

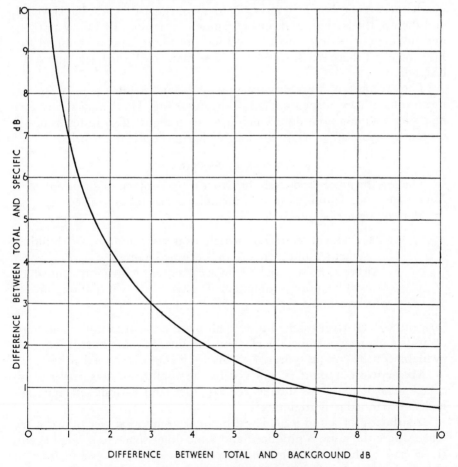

DIFFERENCE BETWEEN TOTAL AND BACKGROUND dB

Fig. 19.30.—Subtraction of decibel Values.

is taken and other conditions. The sound spectrum is also highly important. It is to be assumed that the test takes place in an anechoic room, i.e. without reflection or extraneous noise. Given the sound power level of the fan at each frequency range, and ignoring any difference between suction and discharge, it is possible to determine the attenuation resulting from ducts, changes of direction, grilles and the like. If the result shows that too high a sound level from the fan remains, it is necessary to consider means for greater absorption as referred to later.

Noise in Rooms—The measurement of noise in rooms should be in conformity with some standard. Readings are commonly taken:

> not more than 1·2 m from floor;
> not less than 1 m distance from any grille or other ventilation device;
> and not more than 1 m/s air speed.

There is the question of background noise which is never absent except in a sound-proof room. It is necessary to measure the dB of the background, and then of the background plus the noise in question. By taking antilogs and the log of the difference, the dB level of the noise can be found (see Fig. 19.30). If there is a difference of more than 10 dB the background can be ignored.

The ear must, of course, always remain the final arbiter, as instrumental measurement, however careful, may be misleading. There may for instance be a single monotonous note which has only a slight effect in terms of dB. Or there may be an intermittent noise which goes quite unrecorded.

NOISE DISPERSAL

The less the noise produced, the easier is the problem of disposing of it. Nevertheless, all plant is liable to cause noise in greater or less degree.

Plant noises are of two kinds:

(a) Mechanical vibration which is transmitted to the building structure and thence to the occupied rooms.

(b) Airborne noise which likewise can enter the structure but which will be more troublesome if conveyed by ventilation ducts to the rooms.

(a) can be dealt with by mounting such plant on anti-vibration supports comprising springs, rubber blocks and the like. Manufacturers' data are available for the various types of systems. (b) is considered further here.

Absorption of sound is achieved by the application to surfaces of so-called acoustic materials such as acoustic tiles, and various soft materials such as expanded polyurethane.

Insulation of sound is achieved by interposing a barrier between the space where the noise is produced and some other space to be kept silent. In the case of the plant room mentioned in (b) above, the sound may be absorbed by covering the wall, ceiling and other surfaces with suitable material; but this is generally friable and liable to damage, hence the use of insulation might be preferable between say two faces of a hollow partition, or on top of a false ceiling. Usually this problem is not intractable and is covered under the general heading of 'Building Insulation' dealt with elsewhere.

Ducts—Considering now the transmission of noise by ducts, we may assume that the noise produced by a fan is conveyed equally via the suction and delivery. The air in passing through the system of ducts will lose some of the noise by attenuation, that is by the pressure waves being damped down due to friction on the duct walls, and at changes of direction.

Secondary noise may be generated in the ductwork by sharp edges, and particularly by dampers.

At the terminal end the grille or diffuser may also generate noise, and this at a point where no treatment is possible; hence correct selection of type and velocity of such items is inherent in good acoustical design.

When the air enters the room from the grille, it expands into free space which again accounts for some attenuation.

The attenuation in ducts depends on length and size: the longer the duct the greater the attenuation; the larger the duct the less the attenuation. The amount of attenuation may be calculated from the data given in Table 19.4 and the accompanying notes, but for fuller treatment consult the *I.H.V.E. Guide*.

TABLE 19.4
ATTENUATION OF PLAIN DUCTS

		Attenuation dB/m run
Straight ducts	150 × 150 mm	0·3
,, ,,	600 × 600 mm	0·15
,, ,,	1800 × 1800 mm	0·03
		dB/Bend
Round bends	75–375 mm depth	2
,, ,,	375–900 mm	1·5
,, ,,	over 900 mm	1
Branches	Acoustic energy is divided in ratio of duct areas.	

Where it is clear that residual sound will be too great, ducts may be lined with absorption material either throughout or for certain lengths, as Fig. 19.31.

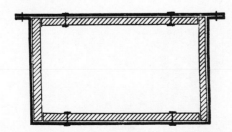

FIG. 19.31.—Method of Fixing Insulation in Metal Duct.

An approximate prediction for attenuation by use of an absorption lining to ducts is given by the formula

$$dB/m \ run = \left(\frac{D}{A}\right)\alpha^{1\cdot4}$$

where D = perimeter of lined duct in m.

A = free area of duct in m².

α = absorption coefficient of lining material (Table 19.5).

This expression holds for lined lengths up to about 2 m only and has been found to overestimate performance at high frequencies.

TABLE 19.5

ABSORPTION COEFFICIENTS OF SOME LINING MATERIALS

	Absorption coefficient at			
	125 Hz	500 Hz	2000 Hz	4000 Hz
Fibreglass resin bonded mat				
25 mm thick	0·1	0·55	0·75	0·8
50 mm thick	0·2	0·70	0·75	0·8
Stillite slabs - 25 mm thick	0·05	0·45	0·80	0·75
Expanded Polyurethane				
25 mm thick	0·25	0·85	0·90	0·90
Fibreboard perforated tiles - -	0·10	0·4	0·45	0·5
Burgess tiles backed fibreglass -	0·10	0·60	0·80	0·90

Alternatively, or in addition, specially designed silencers may be used inserted in the run of ducting. These are rated according to dB loss, but the specification of performance should be based on 'insertion loss' not on a test in free space. Silencers may be built up by splitters or egg crates, as in Fig. 19.32, and their design may be considered as if each section were a small duct lined with absorption material. As attenuation is inversely

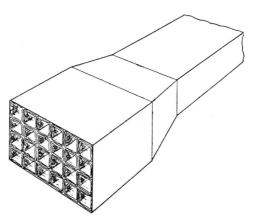

FIG. 19.32.—Sound-absorber for use in ducts, using Acoustically Absorbent Honeycombs.

proportional to the cross-sectional area of the duct, it follows that the smaller the sub-divisions of the absorber the shorter the length required. On the other hand, any absorption device increases resistance to air flow and hence a reasonable balance must be kept.

Another application for silencers occurs where a number of rooms are connected to a common duct system either inlet or extract, and where it is essential that speech or sounds in one room are not heard in other rooms. In Fig. 19.33 (*a*), a sound in room A has but a short path to rooms B and C. In Fig. 19.33 (*b*), the branch serving each room is treated acoustically, thereby avoiding any cross-talk.

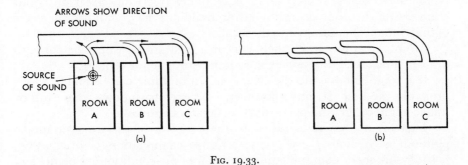

FIG. 19.33.

High-Velocity Systems—The high pressures much involved in high velocity systems necessitate special attention to acoustical design. The usual procedure is to provide a long silencer in the fan-discharge for absorption of fan noise, especially at the lower frequencies. Noise generated in ducts is dealt with by lining of selected lengths. With the induction system the jets have an important attenuation effect, but some acoustical treatment of the cabinet is often provided. In the case of the dual-duct system the blending cabinet is always acoustically treated. Where this is used to serve low velocity ducts, a silencer after the blender is desirable.

CHAPTER 20

Air Filters, Heaters and Coolers

It might be thought that the purest supply of air would be obtainable at the roof of a building, but experience shows that in many instances this is not the case. At roof-level air is certainly free from street dust, but chimneys of the same or neighbouring buildings may, with certain states of the wind, deliver fumes into the intake.

An intake at low level, such as near a busy street, would be liable to draw in much road dust and exhaust fumes from motor vehicles.

If a point half-way up the elevation of a building can be found, this is probably the best. It must, however, be clear of windows where fire or smoke might occur, and particularly of lavatory windows.

There is no general solution to the problem of the fresh-air intake position, as obviously every case requires examination of orientation, possible sources of contamination, position of air-conditioning plant, and so on.

AIR FILTRATION

Air Contaminants—Atmospheric air is contaminated by a variety of particles, such as soot, ash, pollens, mould spores, fibrous materials, dust, grit and disintegrated rubber from roads, metallic dusts and bacteria. The heavier particles may be such that under calm conditions they will settle out of their own volition. These are termed 'temporary'. The smokes, gases and lighter particulate matter remain in suspension and are termed 'permanent'.

The unit of measurement for dust particles is the micron; one millimeter = 1,000 microns. The human hair has a diameter of about 100 microns, and the smallest particle visible to the naked eye is about 15 microns. The smallest range of particles we need consider here is of the order of 0·01 to 0·1 micron, which is represented by smokes of various kinds, such as tobacco smoke. The upper range of particle size we need consider is about 15 microns.

Pollution in all its forms, and especially atmospheric pollution, is the subject of increasing public concern, though apart from smoke, fumes and soot it appears doubtful whether much of the other air-borne dusts and dirt are susceptible of reduction. Tests are regularly carried out and records kept of suspended black pollution material from which it is possible to deduce the following approximate table of contamination according to locality, expressed in milligrams per 100 cubic metres, as typical of winter conditions, when the greatest index occurs.

Typical average smoke index

Rural area - - - -	2
A very clean town - - -	4
Clean town - - - -	10
Average town - - -	20
Dirty town - - - -	30
Very dirty area - - -	40
Extreme pollution area - -	50

Necessity for air cleaning—If in a mechanically-ventilated or air-conditioned building air is blown in without some means for filtration, deposits of dust will be found to occur throughout the rooms, and the system of ducts will in itself become coated with solid matter. Heater batteries and fans will also become coated so that in time the efficiency of the system as a whole will fall off at an increasing rate.

Except in certain industrial applications, ventilation and air-conditioning systems therefore invariably include some means for filtration of the air.

The removal of the larger particles is, of course, a simple matter, since any mesh of fine enough aperture will arrest such particles. A plain mesh is however liable to become clogged very quickly, and hence is of little use for our purpose.

The finer material and the smokes are, however, much more difficult to arrest, and yet it is these which are largely responsible for the staining of decorations, the soiling of shirts and garments and, to some extent also, no doubt the bearing of harmful bacteria. Apart from fresh air, re-circulated air carries fluff from carpets, blankets and clothes, dust brought in on shoes and, in an industrial application, any dust resulting from the process.

The greater the degree of filtration, as a rule, the higher the cost of the equipment and the greater the space occupied. The selection of the best filter for a particular application therefore depends on whether great value is placed on a high degree of cleanliness or not.

Tests for Filters—*Weight test:* Under this method a carefully metered quantity of air is drawn through a filter paper from the unfiltered intake, and a similar quantity of air is drawn through another filter paper downstream from the filter. These are weighed on an accurate balance and a comparison of the two weights gives the gravimetric efficiency. The heavier particles, as explained, are the most easily collected and these constitute the greater part of the weight, hence even a poor filter will give a high efficiency of perhaps over 90 per cent by the gravimetric method.

Blackness test: Under this method equal quantities of air are drawn through filter papers in a similar manner to the above, and the resultant stains are viewed optically, and the reflected light measured by a light meter. A comparison of the relative reflected light values then gives the blackness test efficiency. This is a much more stringent test than the previous one, and many filters which may give over 90 per cent by the gravimetric method may well be less than 50 per cent on a blackness test.

It will thus be seen that the determination of the efficiency of a filter is by no means simple, as it depends on particle size and method of test. A test procedure is laid down in B.S. 2831 (1957) and certain standard dust sizes have been defined, being *British Standard Dusts*, numbers 1, 2 and 3; each of these has a range of particle sizes approximately as indicated at the base of Fig. 20.1. *Methylene Blue* in particle size closely resembles the distribution in atmospheric pollution, but this test can only be carried out in a laboratory.

In selecting a filter it is necessary to know by what method the maker's guarantee of test efficiency has been determined, and what type of dust was used, since it is unlikely that a determination was made under the particular conditions of atmospheric pollution obtaining in the case in question.

Practical Filters—Air filters fall into four main categories as follows:

(1) *Impact types.* Usually of some form of corrugated metal plates or metal coils or turnings or the like, in each case covered with a viscous oily liquid to arrest the particles on impingement. Other materials used instead of metal are glass fibres, similarly coated with a sticky fluid.

(2) *Fabric filters.* The material used in this type of filter is some form of textile, generally with a heavy nap, such as swansdown. The dirt particles are in this case arrested partly by being trapped in the interstices of the material, and partly by being caught on the fibres.

(3) *The Electrostatic filter.* In this system of filtration dust particles entering the filter are subjected to an electrostatic ionising charge and, on subsequently passing through the parallel plates which are alternately charged and earthed, the particles are repelled by the charged plates and adhere to the earthed plates. The electrical change is supplied from a power-pack containing the necessary transformers and rectifiers to produce the high voltage D.C. required.

(4) *Paper or Absolute filter.* This filter uses a special form of paper made usually from woven glass-fibre.

A further class of filter makes use of *activated charcoal* produced from cocoanut shell, and, by virtue of the high degree of porosity, is very active in absorbing odours, gases and the like.

Relative Efficiencies—Fig. 20.1 extracted from a paper by M. P. Penrose* indicates trends of efficiencies which may be expected from the

* *I.H.V.E. Journal* Vol. 29, 1961, p. 113.

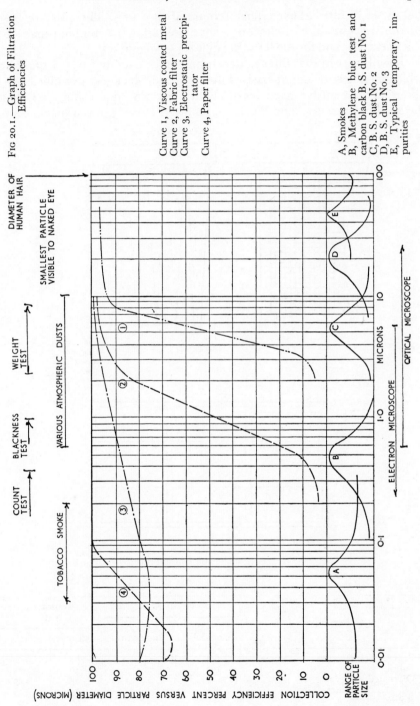

FIG 20.1.—Graph of Filtration Efficiencies

Curve 1, Viscous coated metal
Curve 2, Fabric filter
Curve 3, Electrostatic precipitator
Curve 4, Paper filter

A, Smokes
B, Methylene blue test and carbon black B. S. dust No. 1
C, B. S. dust No. 2
D, B. S. dust No. 3
E, Typical temporary impurities

four types of filter above referred to, numbers 1 to 4. The tests are based for the very small particles on a count test, for the middle range on a blackness test, and for the large particles on a weight test.

It will be seen that filter 1, depending on impact, has no arrestance on a blackness test or count basis. Filter 2 has a blackness test efficiency of about 50 per cent for particles of about 1 micron, diminishing rapidly for the smaller particles. The Electrostatic filter, curve 3, has a blackness test

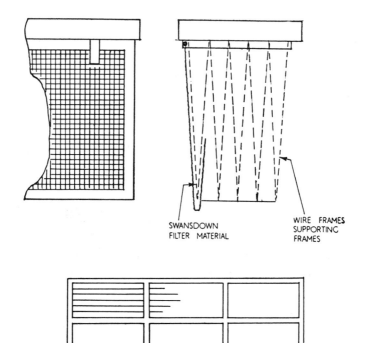

SWANSDOWN
FILTER MATERIAL

WIRE FRAMES
SUPPORTING
FRAMES

FRONT VIEW, SHOWING 9 CELLS

FIG. 20.2.—V Type Fabric Filter.

Filter Type	Velocity m/s		Resistance N/m²	
	Through material	Face velocity as made up	clean	dirty
Paper Absolute - - - -	0·1	5	250	—
Fabric - - - - - -	0·2	2	40	75
Oil coated, cell type - - -	—	2	75	125
„ „ automatic type - -	—	2	75	125
Roller, automatic - - - -	—	2·5	90	100
Electrostatic - - - - -	—	3	50	—

Plate XXII (above). Double inlet fans with vane control: Dual-Duct system (see pp. 362 and 431)

Plate XXIII (right). The dire results of poor water treatment in a refrigeration condenser (see p. 492)

Plate XXIV. Selectographic Control Panel (Honeywell) at Barclays Bank Head Office (see pp. 196 and 400)

efficiency of about 90 per cent whilst the paper filter 4 achieves virtually 100 per cent on a count basis for the smallest size of particles.

All filters vary in efficiency according to velocity of air through them. Velocities and resistances of various filters are given in the Table on page 464.

In the case of fabric and paper filters, owing to the large surface areas involved, designs are usually based on forming the material into a zig-zag formation.

Filter Cleaning—When heavily charged with dirt, the resistance of most filters rises sharply, thus reducing air flow, and hence ventilation rate. The cleaning of filters is achieved in a variety of ways as follows:

Paper Filters: These are discarded when dirty and replaced when new.

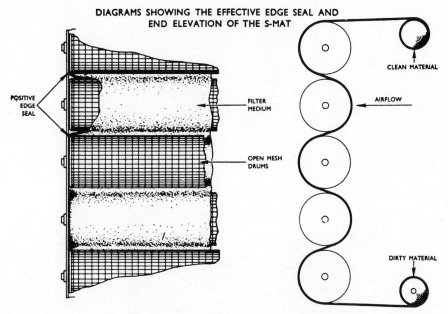

FIG. 20.3.—'S' Automatic Roller Filter (Ozonair Ltd.).

Fabric Filters: In one form, as in Fig. 20.2, these are mounted in frames and when dirty are thrown away. In another type the filter material is of a glass fibre or other suitable base, and is supplied in rolls. The roll is gradually wound off one spool and on to another on the principle of a camera film, see Fig. 20.3, and this movement is achieved by electric motor drive, controlled from a pressure differential switch across the filter, or from a timing device. Fresh filtering medium is unrolled only as it is required. This type of filter therefore requires no labour for changing, excepting at the long intervals of possibly three to six months for the changing of a complete roll.

Viscous Coated Metal Filters: These may be in the form of cells which are removed for cleaning by hand, see Fig. 20.4, being washed, re-oiled and replaced. They have also been developed on a self-cleaning principle. In one form, the zig-zag plates are vertical, and oil is caused to flow over

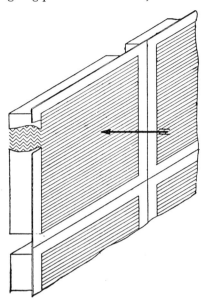

Fig. 20.4.—Oil-coated Filter, 'Ventex'.

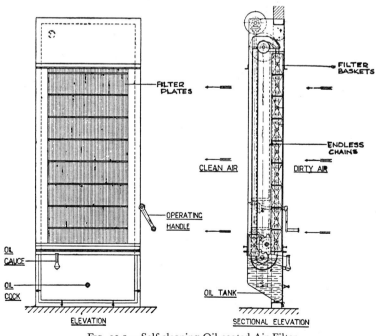

FILTER PLATES

FILTER BASKETS

ENDLESS CHAINS

CLEAN AIR

DIRTY AIR

OPERATING HANDLE

OIL GAUGE

OIL COCK

OIL TANK

ELEVATION

SECTIONAL ELEVATION

Fig. 20.5.—Self-cleaning Oil-coated Air Filter.

them from a pump drawing from the base tank at intervals. In another form the cells are on an endless chain dipping into the oil in the base tank and returning for re-use, see Fig. 20.5.

Electrostatic Filter: In this type of filter (Fig. 20.6), the dirt collects on the earthed plates and its removal is accomplished by washing with hot-water jets. This may be done by hand or automatically. After cleaning, a period of drying is necessary before the filter can be put back into use.

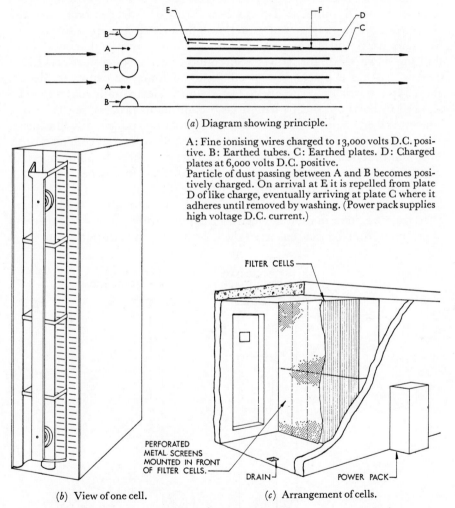

(*a*) Diagram showing principle.

A: Fine ionising wires charged to 13,000 volts D.C. positive. B: Earthed tubes. C: Earthed plates. D: Charged plates at 6,000 volts D.C. positive.
Particle of dust passing between A and B becomes positively charged. On arrival at E it is repelled from plate D of like charge, eventually arriving at plate C where it adheres until removed by washing. (Power pack supplies high voltage D.C. current.)

(*b*) View of one cell. (*c*) Arrangement of cells.

FIG. 20.6.—Electrostatic Precipitator.

Combination of Filters—Advantage may be taken of the properties of different types of filter in combination to achieve a high degree of filtration efficiency, with a minimum of labour in cleaning.

One combination is to use a coarse filter of impact type, such as the oil-coated one, which will remove the larger particles constituting the great bulk of material to be collected, and to follow it with an absolute filter. The latter will take out the fine particles which have passed through the coarse filter, hence renewal would be at long intervals.

Another combination is to use as a primary filter a roller type as already described, and to follow this with an electrostatic filter.

There is a good case for placing a reasonably good filter downstream of an electrostatic filter. This is to guard against short periods when the electrostatic filter may be out of action, and to trap any quantity of dust which may blow off the main filter in the event of power pack trip-out or in shutting down the plant. It has been found that if a roller curtain filter is used which has a blackness test efficiency of some 30 per cent the agglomerated particles which are collected on the electrostatic plates may be allowed to blow off and be caught as a continuous process, see Fig. 20.7. The overall efficiency is thus maintained at about 90 per cent by blackness. The roller curtain filter acts as the storage section for the collected dust. This combination may be found advantageous where a plant operates twenty-four hours a day and washing and drying time is not permitted, also where washing and draining facilities are difficult.

For clearance of fog only, the electrostatic and absolute filters are effective, or a combination of electrostatic plus fabric. Excessive collection of moisture from fog may cause an electrostatic filter to cut-out due to short

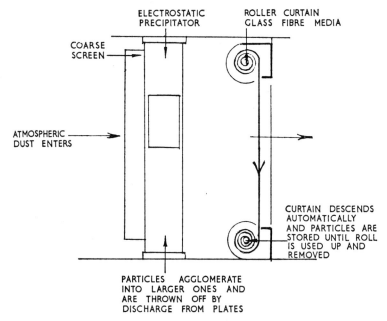

FIG. 20.7.—Combined Filter.

circuit. A special heater may be used ahead of the filter to ensure dry air entering the filter and avoiding this trouble.

Resin Wool—A further type of filter of electrostatic type makes use of the property of resin-coated woollen fabric to generate a static charge when air is passed through the material. This naturally operates without external electric supply. On a blackness test basis this material is believed to have an efficiency of 90 per cent when new, tending to drop with age. A velocity through the material of about 0·1 m/s is recommended.

AIR WASHERS

The treatment of air by water sprays might be thought to be an effective way of air-cleaning. Unfortunately this is not the case, owing presumably to the greasy nature of soot and other contaminants. The washer will remove heavier and gritty material but, on a blackness test, its efficiency is low even when wetting agents are tried. The air washer's chief function is therefore in humidification and de-humidification.

An air washer, see Fig. 20.8, consists of a casing with tank formed in the base to contain water. Spray nozzles mounted on vertical pipes connected to a header deliver water in the form of a fine mist. The spray is projected

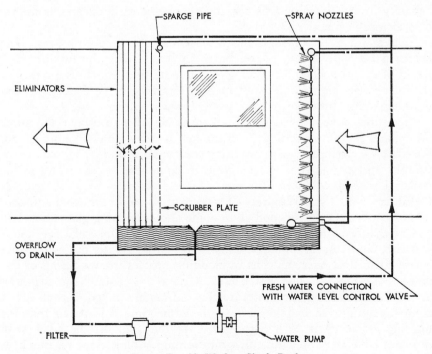

FIG. 20.8.—Air Washer, Single Bank.

either with or against the air current, or, where two banks of sprays are provided, in both directions, one with and one against the air stream. The water for the sprays is delivered by a pump under pressure at 2 to 3 bar gauge, the water being drawn from the tank through a filter. The casing has an access door and internal illumination, and may be of galvanized steel, or constructed of brick or concrete asphalted inside.

As the mist-laden air is drawn through the chamber it requires to have the free moisture removed, and for this purpose eliminator plates of zig-zag formation are provided to arrest the water droplets. Some makes precede these with scrubber plates, down which water is caused to run by a spray pipe at the top in order to flush down dirt which has collected from the air.

Galvanized eliminator plates are liable to rapid deterioration; a better material is copper. Glass plates with serrated ribs have been used successfully, and they are permanent.

A single bank air washer will not saturate the air more than that corresponding to about 70 per cent of the wet-bulb depression. A double bank washer may reach 90 per cent. For full air-conditioning and dehumidification, where it is desired to saturate at the dew-point, a two-bank washer is generally necessary with large spray nozzles to pass the necessary quantity of water for the temperature rise allowed. Spray nozzles vary in capacity from 0·05 to 0·25 litre/s, and should be easily cleanable and of non-corrodible metal.

The speed of air through an air washer is usually 2·5 to 3 m/s. The length is normally about 2·5 m, but is increased with the two or more banks of sprays needed for cooling—sometimes to as much as 4·0 m. On the inlet side straightening vanes, or a perforated grille, are to be advised in order to distribute the air evenly over the whole area.

The tank is kept filled by a ball-cock, with a hand valve for quick filling, and there is in addition a drain and overflow pipe. The pump is of normal type, preferably with insulated connections where noise may cause trouble, and arranged as to level so as to be flooded by the water in the tank, otherwise it will need priming.

Capillary Cell Type Washer—This type is shown in Fig. 20.9 and consists of a series of cells inclined at an angle and containing glass fibres laid in straight formation to a depth of some 200 mm. One cell, size 500 × 500 mm, passes an air quantity of about 550 litre/s. Water at low pressure is caused to flow over the cells by flooding nozzles from a pump of low power consumption. The water and air have to negotiate the striations of the glass fibres together, and are thus intimately mixed so that nearly complete saturation is achieved. At the same time it is stated that the filtering efficiency is of a high order. It may equally be used with refrigerated water for a full air-conditioning system, since as much as 0·75 litre/s per cell may be passed through; the normal rate is about 0·15 litre/s.

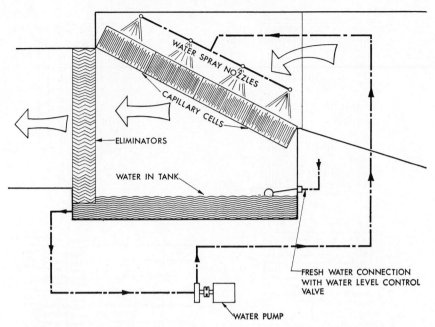

FIG. 20.9.—Diagram of Air-Conditioning Plant using Capillary Cell Washer.

HEATERS

Air heaters when originally introduced were commonly of plain tubing, as shown in Fig. 20.10 (*a*). In order to economise in space and achieve greater output from a given amount of metal, finned heaters are now generally used as in Fig. 20.10 (*b*). Plain tubes are less likely to become choked with dirt than finned—a fate all too common. Plain tube heaters are suitable for use in fresh air intakes to prevent wet fog collection on fabric or electrostatic filters. They should in this case be of galvanised steel or other non-corrodible metal, but not copper.

Heaters are arranged in stacks or batteries with automatic controls preferably of modulating type to give a steady temperature of output.

If the heater is warmed by hot water, a constant temperature supply is required from the boiler or calorifier with pump circulation. The flow required being relatively large, a control valve of diverter type will preserve the circulation irrespective of the demands of the heater. If the heater is fed by steam, it will require the usual stop valve and steam trap and a means of balancing the pressures so that the condense may not be held up by a vacuum caused by control valve shut-off.

Heaters may be enclosed in sheet-steel casings, 'packaged' with other elements into a prefabricated unit, or built into builders' work enclosures.

The number of rows of tubes depends on temperature rise, temperature and nature of heating medium, i.e. whether steam or hot water, and air

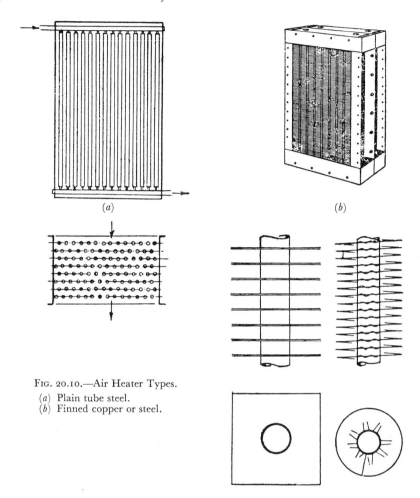

FIG. 20.10.—Air Heater Types.
(a) Plain tube steel.
(b) Finned copper or steel.

speed. The face area depends on air volume and free area between tubes. This is again determined by the velocity, and it is usual to fix this arbitrarily beforehand, generally between 4 and 6 m/s through free area, or 2·5 to 3 m/s face velocity. For sizing of heaters reference is necessary to makers' data.

Air Cooling Coils—Cooling coil surfaces as used in air-conditioning generally perform two functions—to remove sensible heat, and to remove moisture or latent heat.

The sizing and temperature of operation depend on the sensible : latent ratio. If low relative humidities are required, the dew-point will be low, and hence the water temperature will be low. If sensible cooling only is required, the coil surface temperature will be kept above dew-point.

In any coil a certain proportion of air fails to come in contact with the cold surfaces, and thus a part may be chilled and dehumidified and a part

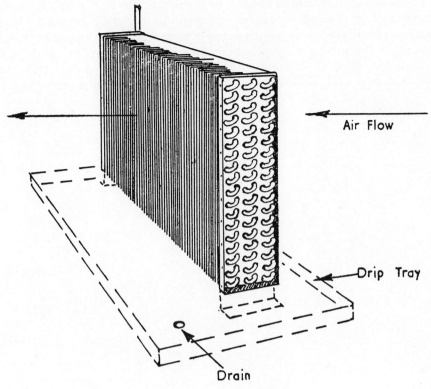

FIG. 20.11.—Cooling Coil.

remain unchanged. By correct selection of coil form the desired ratio may be obtained.

In direct-expansion systems, the actual refrigerant gas from the compressor is passed direct into the coils, which, in effect, form the evaporator of the refrigerating plant.

In a chilled-water system, chilled water is circulated through the coil, and this may be used down to about $3 \cdot 3°$ C. Below this an anti-freezing mixture such as calcium chloride brine becomes necessary.

In cooling by refrigeration, the higher the evaporator temperature the less power is consumed for a given duty, hence it is desirable to keep the cooling coil temperature as high as possible consistent with the final air temperature required to meet design conditions. This applies whether the coil is used for direct expansion or with chilled water or brine.

Cooling coil surface is therefore usually designed for small temperature differences between water and air; for instance, for air cooled from $24°$ to $13°$ C, water may be at $10°$ inlet and $14 \cdot 5°$ C outlet, (i.e. leaving above the air-outlet temperature). An air washer cannot give a comparable performance to this.

Coolers are generally of finned or block type (as in Fig. 20.11), comprising banks of small bore tubes threaded through plates between which the air passes, the whole being of copper tinned after fabrication or, more normally in present day practice, made up with copper tubes and aluminium fins. Coils are frequently arranged horizontally, with fins vertical, so as to facilitate drainage of condensation when dehumidifying.

CHAPTER 21

Refrigeration for Air-Conditioning

THE COOLING NECESSARY FOR full air-conditioning is in nearly all cases effected by means of a refrigerating machine. Such a machine may be similar to the usual types of plant used for cold storage work, ice-making, etc., except that the temperature to be produced is not so low as in such applications.

Mechanical Refrigeration depends on the principle that a liquid can be made to boil at a low temperature if its pressure is reduced enough. To boil, it must be supplied with heat from an outside body or fluid. This outside fluid thus loses its heat and becomes cool. By choosing the right liquid, the heat for boiling can be extracted from the surroundings at low temperatures, without unduly low pressures: such liquids are known as refrigerants. The vapour given off on boiling is compressed (which makes it hot), is liquefied by removing this heat whilst still under pressure, and its pressure is then suddenly reduced, which brings it back to the state in which it started, whereby it can be made to boil at a low temperature.

A refrigeration plant (see Fig. 21.1), therefore, consists essentially of:

(a) A *compressor* to compress the refrigerating medium.

(b) A *condenser* to receive the compressed gas and liquefy it; the latent heat is taken out of the circuit by some external means.

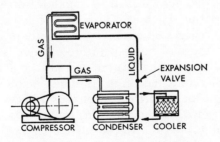

FIG. 21.1.—Diagram of Refrigeration Cycle.

One method is to cool the condenser with circulating water, which may then be either run to waste or passed to a cooling tower for re-use. Alternatively, the cooling may be by air.

(c) An *expansion valve* in which the pressure on the liquid medium is reduced.

(*d*) An *evaporator* in which the medium re-evaporates, extracting
heat from the surrounding material, i.e. from the cooling
water or air in an air-conditioning plant, or from the brine
where temperatures below freezing point of water are needed.

Application of Refrigeration—For use with air-conditioning using a
washer and dehumidifier, the water from the latter is returned to the
evaporator of the refrigerating plant generally at between 7 and 10° C
depending on the dew-point to be maintained: in passing through the
plant this is lowered by about 4 to 5° C. For the necessary heat transfer
to take place, the refrigerant must be at some temperature below that of
the water, but at the same time it must generally be slightly above freezing
point. Thus, in a typical example, the following conditions might obtain:

Apparatus dew-point   -   -   -   - 12° C
Washer outlet   -   -   -   -   - 9°
 „ inlet -   -   -   -   -   - 6°
Water at evaporator outlet   -   -   - 5·5°

The refrigerant in the evaporator would in this case be maintained at
about 1° C, giving 4·5° C differential for heat transfer. This small temperature potential means a very large cooling surface in the evaporator,
and various devices have been developed to augment the transfer rate.

Where chilled water coils are used, water is circulated from the cooling
plant in a closed system by a pump, and water temperatures down to
about 3·3° C flow and 7° C return may be used.

Where brine is used in cooling coils to enable lower temperatures to be
obtained (for example to achieve a low dew-point condition), temperatures
may be taken down to – 7° C flow and – 3° C return, or lower as desired.
The temperature to which the brine may be cooled is dependent on the
strength of solution, data for which are obtainable from standard tables.

When a refrigerating machine is used in a direct-expansion air-conditioning system, the refrigerant is conveyed directly to the cooling coils in
the airstream, and the surface temperature of these is dependent on the
air-conditions required, on the form of coil surface and on the air speed
over the coils. Refrigerant temperatures much below freezing point are
inadmissible owing to the risk of freezing on the surfaces when dehumidification is being performed, as such freezing would block the air flow. An
apparatus dew-point of 3° C is considered the minimum for direct
expansion coils, to avoid frosting.

Refrigerating Cycle—The refrigeration cycle may be considered on a
pressure-total heat (or enthalpy) diagram, as Fig. 21.2, which is drawn for
Refrigerant 12.

Inside the curved envelope the medium exists as a mixture of vapour
and liquid, and the increase of enthalpy from left to right on any pressure
line within the envelope represents the increase in latent heat. Within this
envelope lines of equal temperatures (isothermals) are horizontal.

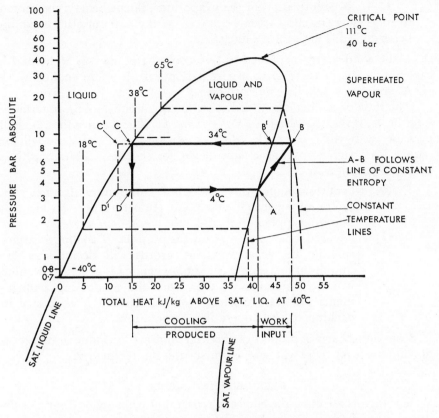

FIG. 21.2.—Pressure Enthalpy (Total Heat) Diagram for Refrigerant 12, showing Typical Refrigeration Cycle.

Outside the envelope to the left the liquid exists at a temperature below its saturation temperature, and isothermals are almost vertical. To the right of the saturated vapour curve, the vapour exists in a superheated form, and isothermals curve downwards.

The *critical point* is that at which latent heat ceases to exist. It is not possible to liquefy a gas by pressure alone if it is above its critical temperature.

The refrigerating cycle may be represented by the figure ABCD;

A—B the gas is compressed causing rise in pressure and total heat which equals the energy put into the gas by the compressor—all in the superheat region. This takes place at constant entropy.

B—B' is cooling the superheated gas in the condenser down to saturated vapour temperature.

B'—C is removing latent heat and condensing the gas to liquid in a the condenser.

C—D is pressure drop in the expansion valve, without change in total heat (adiabatic).

D—A is vaporization in the evaporator, latent heat—represented by increasing enthalpy—being drawn from the water or other medium being cooled. This is the cooling effect.

If the condenser is arranged to 'sub-cool' the liquid, say to point C', each unit mass of refrigerant in circulation will do more cooling work (D'A) and the cycle will be more efficient.

The ratio of cooling effect (as DA) to energy input (as AB), in terms of total heats, is termed *the coefficient of performance*. The smaller the range of pressure over which the cycle operates, obviously the less work will be required for a given cooling effect: hence, for economy of running, it is desirable to design for

(a) evaporator temperatures as high as consistent with other considerations (such as dew-point temperature in an air-conditioning application), and

(b) condenser temperature as low as possible. If cooling is atmospheric, the local weather records will decide the safe minimum to assume. Maximum cooling is generally wanted in hottest summer weather when the condensing arrangements are least efficient, so that caution is necessary in deciding on these operating temperatures.

Complete charts of properties of refrigerants are available in refrigeration literature for those who wish to pursue this matter further.

REFRIGERATING MEDIA

The factors affecting the choice of refrigerant will now be clear. A substance is required which can be liquefied at moderate pressures and which has a high latent heat of evaporation. The size of the compressor will then be kept down, and a relatively small amount of refrigerant need be circulated for a given amount of cooling. Such gases include ammonia, carbon dioxide, sulphur dioxide, and various organic gases.

Ammonia, whilst high in efficiency, and cheap, is ruled out in most cases for air-conditioning by the serious results which may follow a burst or leak. CO^2 calls for higher power input for a given capacity, much higher pressures and requires a skilled engineer for its operation. It is no longer used in air-conditioning work. SO^2 and methyl chloride are only suitable for small plants, but are not now used in air-conditioning applications.

Freon is a group name for a range of organic gases in which fluorine and chlorine are in combination with carbon in a number of different proportions. They are 'safe', non-toxic, colourless, non-inflammable and non-corrodible to most metals. Three of the forms in common use are:

R. 11 (trichloromonofluoromethane) CCl_3F, mainly used in centrifugal compressors for air-conditioning.

R. 12 (dichlorodifluoromethane) CCl_2F_2, used in piston compressors for air-conditioning and a wide range of other applications.

R. 22 (monochlorodifluoromethane) $CHClF_2$, used for low temperature work, as well as normal air-conditioning.

Air also may be used as a refrigerating medium. One method is to compress it to about 14 bar abs. and then, after removing the heat of compression, allow it to expand through a valve. At this pressure the air is not, of course, liquefied, and the latent heat is not therefore available. The specific heat being small, very large volumes of air must be handled to achieve the desired amount of cooling. In aircraft, the air cooling cycle is used in quite small turbo equipment running at very high speeds taking advantage of the extremely low temperature of the surrounding ambient for use in the condensing side. Another method is to use pressures high enough to liquefy the air; at least 200 bar is required. With either

TABLE 21.1

PROPERTIES OF REFRIGERANTS

	Refrigerant				
	Ammonia	R. 11	R. 12	R. 22	Water
Symbol	NH^2	CCl_3F	CCl_2F_2	$CHClF_2$	H_2O
Pressures (bar. g) Condenser (30° C) Evaporator (− 15° C) Evaporator (5° C)	+ 10·6 + 1·55 + 4·04	+ 0·248 − 0·806 − 0·525	+ 6·42 + 0·812 + 2·54	+ 11·0 + 1·95 + 4·27	− 0·97 — − 1·00
Boiling Point (°C) (Standard Pressure)	− 33·3	+ 23·8	− 29·7	− 40·6	+ 100
Critical Temperature (°C)	1333	198	111	96	373
Volume of Vapour at − 15° C (m³/kg)	0.509	0·766	0·093	0·078	616
Latent Heat of Evaporation at 15° C (kJ/kg)	1320	198	162	218	2530
Theoretical unit Energy input per unit Energy output (kW/kW	0.211	0·200	0·213	0·216	0·238
Coefficient of Performance (27° C to − 15° C)	4·75	5·00	4·69	4·65	4·20
Characteristics	Strong irritant. Forms explosive mixture under certain conditions	Odourless Non-toxic Non-irritant Non-inflammable Innocuous			Wet
Used in	Large Plants. Piston Machines.	Centrifugal Plants. All sizes of Piston Machines.			Centrifugal Plants. Steam Jet Plants.

method expensive plant is necessary, but the weight penalty in aircraft applications is of course the principal criterion. For normal land use, bearing in mind the relative cost of plant required, air is not a practical choice as a refrigerant for air-conditioning.

Water is also used as a refrigerant in a manner described later.

The refrigerant most suitable for direct expansion into coils in the air-way is R.11, R.12, or R.22, the others all being objectionable owing to their smell, inflammability, or inefficiency. Table 21.1 gives the properties of R.11, R.12 and R.22, with ammonia and water for comparison.

TYPES OF REFRIGERATION PLANTS

Types of refrigerating plants are:
Piston or reciprocating.
Rotary (which is another form of piston but without valves and mainly used in small sizes).
Helical Screw.
Centrifugal.
Absorption.
Steam jet.

Piston Compressors—The reciprocating type of machine may consist of one, two, three or four cylinders, according to the load, suction and discharge valves being operated by suction and pressure only and not mechanically.

Fig. 21.3 shows a diagram of a piston-compression unit of the older slow speed type occupying considerable space for a given duty.

Suction

Water Jacket

Safety Head

Discharge

Valves in head
& Piston

Driving Pulley or
Motor Coupling

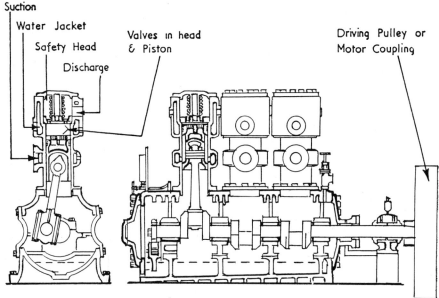

FIG. 21.3.—Old type 3-Cylinder Piston Compressor, for Ammonia or Freon.

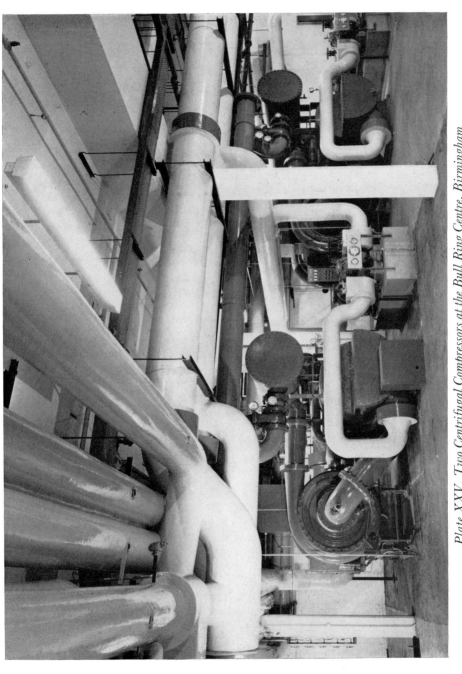

Plate XXV. Two Centrifugal Compressors at the Bull Ring Centre, Birmingham
(see p. 484)

Plate XXVI. A Centrifugal Compressor for air-conditioning hoisted to a roof-top plant chamber (see p. 482)

Plate XXVII. Evaporative Coolers on the roof of a building in the City of London (see p. 492)

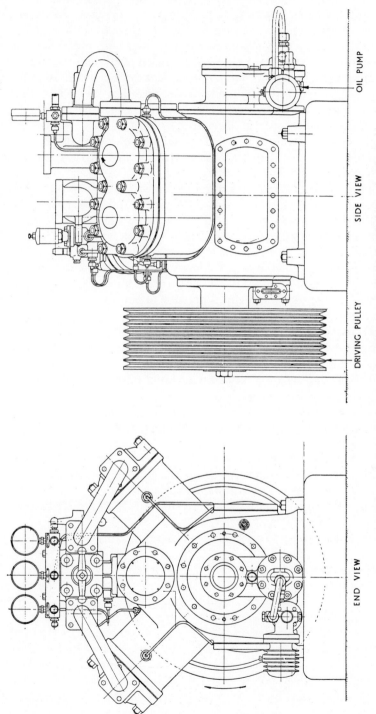

OIL PUMP

SIDE VIEW

DRIVING PULLEY

END VIEW

FIG. 21.4.—4-Cylinder V Type Freon Compressor, 140 kW Capacity (Lightfoot Refrigeration Co. Ltd.).

The gland where the crank shaft passes through the casing is of rotary type and various systems are used for keeping this gas-tight. The cylinders are water-cooled. An oil separator is a necessity with ammonia machines to prevent oil from the crank case being carried over into the coils.

A more modern type of reciprocating compressor running at a higher speed is one in which the cylinders are arranged in V or W formation, as shown in Fig. 21.4. Four, six or eight cylinders may be arranged in one block, thus enabling a large duty to be obtained in a small space with marked freedom from vibration. Two such machines may be coupled to one motor on a common centre line, and in this way capacities up to 700 kW may be obtained in one set.

Totally enclosed Freon compressors are available in which the motor is contained within the compressor casing, so dispensing with the need for a gland. These are termed 'Hermetic' (see Fig. 21.5). Such sets are usually quiet-running and particularly suitable for self-contained air-conditioning units.

Multi-cylinder machines are sometimes provided with a bye-pass, or unloading valve, on one or more cylinders, enabling the load to be varied either by hand or automatically. Variation of output may also be achieved by speed regulation.

Helical Screw Compressor (Dunham Bush)—Compression is achieved by two rotating helical screws, the seal being achieved by oil. A retract-able vane enables load variation in a simple manner over a wide range. This form of compressor combines in a way the advantages of a centrifugal type with those of a reciprocating type.

Centrifugal Compressors—The centrifugal compressor is used chiefly where large duties are required. The advantages are:

(1) Saving of space as compared with reciprocating machines.
(2) Absence of vibration (thus suitable for a roof-top plant chamber, see Plate XXVI, facing p. 481).
(3) Reduced maintenance due to there being no wearing or reciprocating parts.
(4) Throttling to suit load gives corresponding power reduction.

Refrigerants used in centrifugal compressors are usually one of the Freon group, such as R. 11, which for air-conditioning temperatures operate at low pressures and over a small pressure range, thus reducing slip losses between blades and casing.

An important consideration is the turn-down range, i.e. the ability to follow load variations. The two-stage machine shown in Fig. 21.6 has a turn-down to 10 per cent of full load, thus making it highly flexible in operation.

In the diagram (Fig. 21.6) of this centrifugal plant, it will be noted that the condenser and evaporator are close-coupled to the compressor and

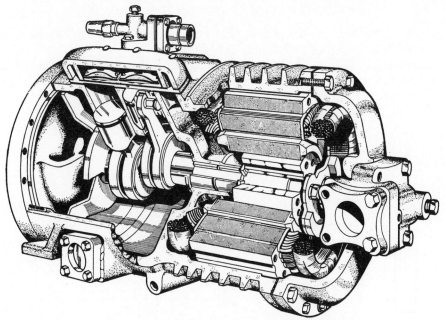

Fig. 21.5.—Hermetic Compressor (Dunham-Bush).

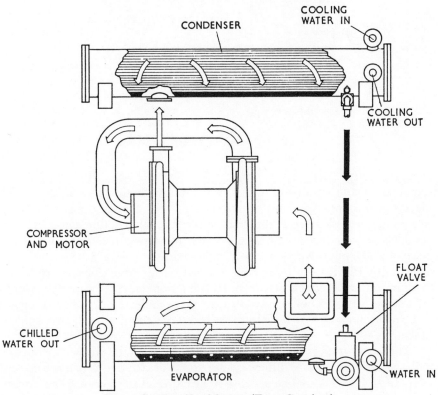

Fig. 21.6.—Centrifugal System (Trane Centrivac).

usually form one complete unit. Plate XXV, facing p. 480, shows two units. The capacity of each unit is 2000 kW.

Absorption Plant—One form of absorption plant is shown in Fig. 21.7. It uses lithium bromide. It has no moving parts except pumps, the source

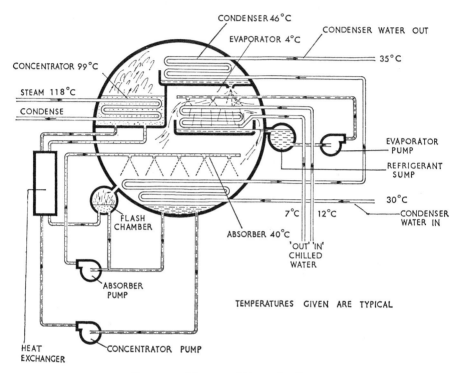

FIG. 21.7.—Diagram of Absorption System.

of heat energy being either steam, hot water or gas. The heat to be removed by the condenser water is greater with this system than with mechanical plant. As a unit of nominal 350 kW capacity uses a nominal steam quantity of 220 g/s, the heat rejected to the cooling tower is some 900 kW, as compared with 540 kW for mechanical compression refrigeration. The plant works under a high vacuum.

Steam-Jet Plant—An interesting type of refrigerator using water as the medium is that shown in Fig. 21.8. Its operation depends on the possibility of causing water to boil at low temperatures under high vacua. Thus, at 7° C, water boils at 0·01 bar abs.: i.e., about one hundredth of atmospheric pressure.

The water used is the same as that circulated to the washer or cooling coil, so that temperature difference due to heat exchange as in a refrigerant evaporator is avoided. The absence of any special refrigerant is an advantage and an economy.

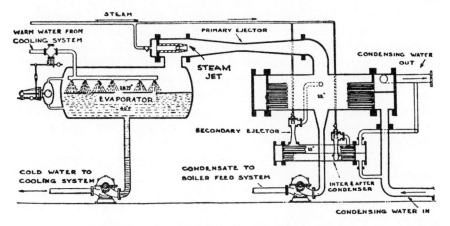

Fig. 21.8.—Diagram of Steam-Jet Refrigeration Plant.

The heat input with this equipment is much greater than with the positive compression types owing to the inefficiency of jet compression. All the steam used has to be condensed and the heat rejected at the cooling tower. The condensing requirement is about five times that of a mechanical compression plant.

A variation of the same system, but using a centrifugal compressor in place of the steam-jet compressor, has also been developed. It avoids the above mentioned disadvantage of high heat input. The great difficulty with both types is the maintenance of the extraordinarily high vacuum for long periods.

Choice of Refrigeration Plant—The selection of refrigeration plant will depend on a number of factors, among which are: available space and location, whether water is available, whether noise is important and whether condenser heat must be rejected at a distance from the compressor.

Hermetic piston-type compressors are widely available up to 30 kW and are used in combination to make larger capacities. Such machines are obtainable built as a weatherproof unit complete with an air-cooled condenser. A low silhouette arrangement is available for installation on flat roofs. The refrigerant lines must be run to the evaporator located inside the building at a not too distant point.

Refrigeration units of all sizes may be water-cooled and they must be so arranged if there is no free access to open air. Up to about 15 kW they are sometimes cooled by fresh water run to drain, but larger sizes are connected to cooling towers. The cooling tower may be located a considerable distance from the refrigeration plant, for instance a plant in a basement may be connected to a cooling tower on the roof.

Evaporative condensers are available in sizes up to 1000 kW. They give an increase in efficiency because one exchange of heat is eliminated as compared with a water-cooled condenser connected to a cooling tower. An evaporative condenser must be installed fairly close to the compressor, either inside the plant room or outside in the open. Its design must be carefully considered where capacity control is required and where winter operation is envisaged. It is used only with reciprocating compressors.

Open type or non-hermetic reciprocating compressors are available in sizes up to 750 kW and beyond. They may be connected to direct expansion air coolers or water chillers on the suction side and to air-cooled, water-cooled, or evaporative condensers on the discharge side. The large air-cooled condenser would only be used in dry or desert climates.

Centrifugal or turbo compression refrigeration plant is used in sizes from 500 to 5000 kW and has been made up to 14,000 kW. This plant invariably includes a water or brine chiller and a water-cooled condenser using water from a cooling tower, or other economical source.

The absorption refrigeration unit has been developed for use in air-conditioning and is attractive where a supply of cheap heat or exhaust steam may be freely obtained. Sizes of 300 to 2000 kW are usual. It is finding a new field of development making use of natural gas as the heat source.

Steam jet refrigeration is not in wide use, but has been found economical in large industrial enterprises where exhaust steam is available.

REFRIGERATING PLANT COMPONENTS

Evaporators—The evaporator is tubular in form. The tubes may contain the refrigerant and the whole is immersed in the liquid to be cooled. In the *dry system* the evaporator coils are filled with vapour having little liquid present. When operated on the *flooded system* the liquid refrigerant discharges into a cylinder feeding the coils by gravity. As evaporation takes place the gas returns to the top of the cylinder and from there returns through the suction pipe to the compressor.

A more common form of evaporator for application to air-conditioning practice is the shell and tube type, used with a closed circuit system such as with cooling coil surface. It is necessary to include safety cut-outs to prevent icing up. The water is contained in the tubes and the refrigerant in the shell.

A development of the shell and tube type is the direct-expansion shell type evaporator in which the refrigerant is in the tubes and the water in the shell. The refrigerant tubes may be arranged in two or more circuits, each with its own expansion valve and magnetic valve on the liquid inlet to allow step control. Different arrangements of baffles in the shell control the water velocity over the tubes to improve heat transfer.

In the centrifugal machine the evaporator forms part of the unit, the water passing through the tubes and the refrigerant being dripped over the outside. Plate XXVIII, shows pumps associated with a plant this of type.

It is generally not necessary to resort to brine for air-conditioning purposes, as chilled water at about 4° C satisfies all normal requirements. Precautions against accidental freezing of the water in the evaporators include low suction pressure cut-outs and low water temperature cut-outs as well as water-flow switches.

Modern water-chilling plant does not generally require the addition of chilled-water storage to act as a 'flywheel', because it can be arranged in suitable steps of capacity control to suit the variations in load. Little complication is experienced with piston compressors having four to eight steps of control. Centrifugal units which turn down to 10 per cent can virtually run on pipeline 'losses'. In cases where these machines are limited to a small number of starts per hour, it is possible to fit a 'hot gas by-pass' valve which automatically passes hot gas from the condenser into the evaporator to provide some load. This can come in automatically at a predetermined point on the capacity controller travel and enables the machine to run continuously when there is no cooling load at all.

Chilled-water storage-tanks should be avoided in general because they require the chilled water to be produced at a lower temperature than is required to be used. This is because water at about 4° C does not stratify as in a hot-water calorifier but tends to mix. Some ingenious arrangements of weirs and baffles have been used in chilled-water tanks to overcome this difficulty, but they add complications to the system. At the present day a chilled-water tank would be justified in connection with perhaps two to four small single-step cooling units, so providing water chilling in steps.

Condensers—The evaporative condenser, consisting of coils of piping over a tank from which water is circulated and dripped over the pipes, is the simplest form of condenser. Its use is restricted to cases where the compressor can be near to the condenser, otherwise long lines of piping containing refrigerant under pressure are necessary.

More often in air-conditioning systems, the condenser takes the form of a heat exchanger of the shell and tube multi-pass type.

Circulating piping from the condenser to the water cooler with these types alone traverses the building, and all equipment containing refrigerant is then confined to the plant room.

Air-cooled condensers are used in small self-contained or split package systems and in duties of up to 350 kW in capacity. They are often found convenient to use and economical in first cost in this country in sizes up to say 100 kW. Air-cooled condensers are often used in the tropics, despite the fact that they are bulky, in circumstances where the outside air temperature is high. This apparent paradox arises from the simple fact that evaporative condensers and cooling towers require some measure of skilled maintenance, whereas the air-cooled condenser requires little attention other than simple basic cleaning and, of course, some lubrication. A further advantage is that it requires no water supply in those places where such may happen to be scarce.

Evaporative Coolers—The heat extracted by the refrigerating machine, together with the heat equivalent of the power input to the compressor, raises the temperature of the condenser water by an amount which is dependent upon the quantity of the water which is circulated through the condenser.

The lower the temperature of the condenser water the less power will be required to produce a given refrigerating effect; and it also follows, conversely, that with a given size of plant the greater will be the amount of cooling possible.

Water from a well or from the main supply will always be the coldest, the former at $12°$ C and the latter at about $18°$ C in summer. The quantity to be wasted, however, generally rules this method out. For instance, a 700 kW plant with power input of about 120 kW with a $10°$ C rise through the condenser would require

$$\frac{700 + 120}{10 \times 4\cdot2} = 20 \text{ litre/s}$$

This at 2p per 1000 litres would exceed the cost of current for running the compressor by about three times.

Applications exist outside the U.K., however, where well water is more freely and cheaply available, which produce economical solutions. In one known instance in Europe, such well water is first passed through a pre-cooling coil integral to the air-handling plant prior to use in the condenser. Similarly, in some parts of the Caribbean, clear sea water is available via fissures in the coral which may be pumped in quantities of up to 1000 litre/s through specially designed condensers.

However, cooling the water by evaporation is more generally adopted. Evaporative coolers depend on the ability of water to evaporate freely when in a finely divided state, extracting the latent heat necessary for the process from the main body of water, which is then returned, cooled, to the condenser. In the case stated above, the consumption of water with an evaporative cooler (assuming no loss of spray by windage) would be only

$$\frac{820}{2258 \text{ (latent heat)}} = 0\cdot36 \text{ litre/s}$$

Evaporative coolers divide themselves into two categories, i.e., *Natural Draught* and *Fan Draught*. The former is represented by:

(a)   *The Spray Pond.* In this type the water to be cooled is discharged through sprays over a shallow pond in which the water is collected and returned to the plant. To prevent undue loss by windage the pond is usually surrounded by a louvred screen. Owing to the large area needed for the spray pond system and, moreover, the consequent probability of pollution, this type of equipment is seldom possible for air-conditioning applications.

(*b*) *Cooling Tower.* In this, the water is pumped to the top of a tower which contains a series of timber or specially designed plastic slats arranged so as to split up the water stream and present as large an area as possible to the air, which is drawn upwards due to the temperature difference, and by wind. The base of the tower is formed into a shallow tank to collect the water for return to the plant. Again, owing to its size and height, this type of cooler is not frequently used for air-conditioning.

(*c*) *Condenser Coil Type.* Where the refrigerating machine is near the point where an outdoor cooler may be used, the condenser heat exchanger may be dispensed with and the refrigerant is then delivered to coils outside, over which water is dripped by a pump. The water collects in a tank at the base and is recirculated. A louvred screen is usually necessary surrounding the coils.

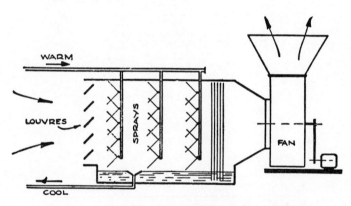

Fig. 21.9.—Fan Draught Cooler for Condenser Water.

Fan draught systems are more commonly used owing to the compact space into which they may be fitted. Where possible they are placed on the roof, but if this is impracticable they may be used indoors, or in a basement, with ducting connections for suction and discharge to outside. One system in Paris uses the air extracted from an underground car park for this purpose, thus economising in the overall power requirement of the engineering systems.

The following three methods are typical:

(*a*) *Air Washer Type*, Fig. 21.9. This uses a normal type air washer containing several banks of sprays, base tank, and eliminators in the normal way.

(*b*) *Forced Draught Cooling Tower*, Fig. 21.10. In this the water is delivered to the top by the condenser pump, and delivers over timber or plastic slats as with the natural draught type. A cased fan is arranged to blow air upwards over the slats at high velocity so that a much reduced area of contact is required.

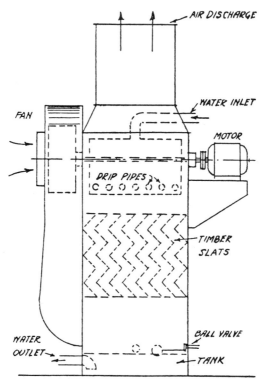

FIG. 21.10.—Forced Draught Cooling Tower.

(*c*) *Film Cooling Tower*, Fig. 21.11. The water is not sprayed or broken up in any way, but is allowed to fall from top header troughs down wooden or plastic slats arranged in egg-crate form. The water remains as a film on the surface of the slats. Higher air velocities than usual are permissible without risk of carry over of water, and yet air resistance is low due to the open nature of the surfaces.

Rating of Cooling Towers—The heat to be removed from the condenser cooling water by the cooling tower is equal to the sum of cooling load plus heat equivalent of power absorbed by the compressor. An approximate figure of 1·2 kW per kW of refrigeration may be used.

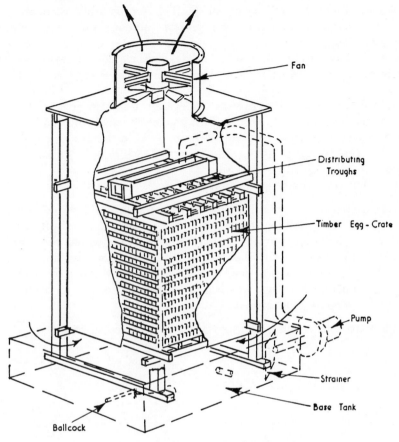

FIG. 21.11.—Film Type Cooling Tower.

The quantity of water to be passed through the tower is dependent on the temperature drop allowable between inlet and outlet. A usual figure is 5° C and the inlet water flow will then be

$$\frac{1\cdot 2}{5 \times 4\cdot 2} = 0\cdot 057 \text{ litre/s per kW of refrigeration.}$$

Part of this will be evaporated as above and made up by fresh water through the ballcock.

The temperature to which the cooling tower may be expected to cool the condenser water depends on the maximum wet bulb of the atmosphere and the design of the tower: the higher the efficiency the closer will be the water outlet to the wet-bulb temperature. A good efficiency will give a difference of about 3° C, so that, if the maximum wet bulb is taken at 21° (the highest in England except on rare occasions), the cooling water will be brought down to 24° C, and, with 5° temperature rise through the condenser, the outlet will be at 29° C. This temperature then, in turn, forms the basis for design of the compressor and condenser (e.g. condensing at about 38° C).

Certain new designs of cooling towers achieve considerable capacity in minimum space, an example of which is shown in Plate XXVII, facing p. 481.

Condenser Water Treatment—As evaporation proceeds, there is a continuous concentration of scale-forming solids which may build up to such a degree as to foul the condenser tubes (as in the example shown in Plate XXIII, facing p. 464). Similarly, the evaporative surfaces of the cooler suffers a build-up of deposit.

In order to overcome this problem:

(a) a constant bleed-off of the water is required from the base tank of the cooling tower, the rate of which bleed-off may be calculated from the known analysis of the water, the evaporation rate and the maximum concentration admissible;

(b) as is common practice, the water can be treated by a regular chemical dosage which also generally contains an additive to prevent algae growth.

These are matters on which specialist advice is desirable.

CHAPTER 22

Tall Buildings

THE PRESENT VOGUE OF building vertically to some 20 or 30 storeys poses some serious questions for the heating and ventilating engineer.

With the normal building height of 30 metres or so, some shelter from sun and wind may be expected from adjacent buildings, but this is not so with the isolated towers now becoming common.

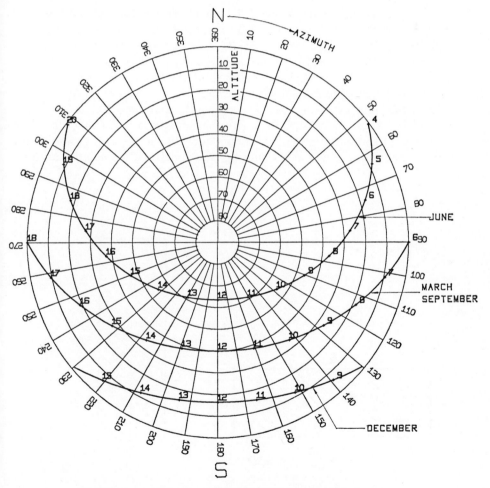

FIG. 22.1.—Diagram of Sun Positions at 50° North Latitude.

A block of 60 to 100 metres high is exposed to the sun on various sides throughout the day, whereby solar-heat gain can cause internal temperatures to vary considerably.

The force of wind at these higher levels is greater in strength than at ground level, rendering the opening of windows for ventilation well nigh impracticable under certain conditions. It has been found by experience that wind pressure on one side of a building and corresponding suction on the other is sufficient to cause undesirable pressure differences on doors to corridors, as well as discomfort in the rooms. It appears paradoxical that a tall building, which should have the freshest of ventilation from the upper atmosphere, cannot so avail itself as a general rule.

Solar Gains—Reference to the calculation of solar-heat gain was made in Chapter 17. The orientation is important in tall buildings as will be evident from the following notes:

Fig. 22.1 is an azimuth and altitude chart for Latitude 50° N for various seasons of the year plotted by a computer. It is assumed that a horizontal surface at this latitude at sea level (or up to 300 metres above) in June will receive at maximum peak 910 W/m² solar radiation with a clear sky. Intensities on surfaces other than the horizontal will be proportional to the cosine of angle of incidence.

Fig. 22.2 illustrates the plan of a rectangular slab block: Case 1 is with the main axis east and west, Case 2 with the main axis north and south.

Fig. 22.3 shows the plot of solar incidence on the east, south and west faces throughout the day at the summer solstice. It will be noted that due

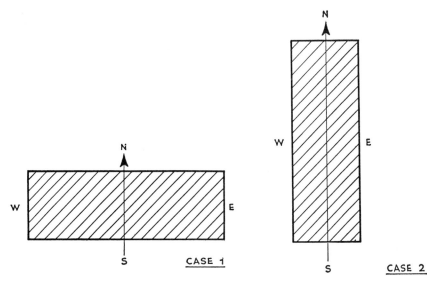

Fig. 22.2.—Plan of Slab Block of buildings.

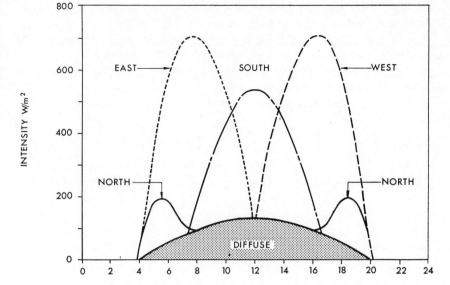

Fig. 22.3.—Graph of Solar Heat on a Vertical Face on June 21.
S and N Faces—24 hour mean value, 245 W/m²
E and W Faces—24 hour mean value, 390 W/m²

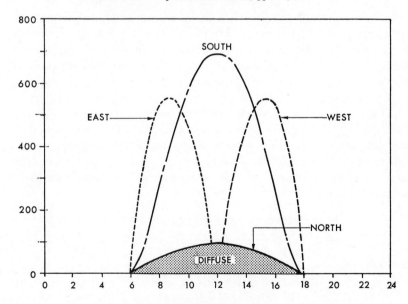

Fig. 22.4.—Graph of Solar Heat on a Vertical Face on March 22
and September 22.
S and N Faces—24 hour mean value, 241 W/m²
E and W Faces—24 hour mean value, 236 W/m²

to the high angle sun at midday the solar incidence is lower on the south face than on the east and west faces where the angle is less.

Fig. 22.4 is a similar plot for the spring and autumn equinoxes. The incidence on the south face now exceeds that on the other two.

A measure of the relative incidence of heat is given by the 24 hour mean values, and these are stated on the figures.

In Fig. 22.2 assume length = 3 × breadth.

Case 1, in June:

S and N faces 245 × 3 =	735
E and W faces 390 × 1 =	390
	1125

Case 2, in June:

S and N faces 245 × 1 =	245
E and W faces 390 × 3 =	1170
	1415

Case 1, in September:

S and N faces 241 × 3 =	723
E and W faces 236 × 1 =	236
	959

Case 2, in September:

S and N faces 241 × 1 =	241
E and W faces 236 × 3 =	708
	949

It is evident that at mid-summer the orientation of Case 1 produces lower heat gain than Case 2, but in spring and autumn they become roughly equal. In reality, in a slab block of this kind the narrow ends would be windowless which, when due allowance is made for gain through glass, shows Case 1 to even greater advantage. From a planning point of view, however, Case 1 is no doubt less attractive than Case 2, and the consequences of the additional heat gain of the latter might have to be accepted.

This example may serve to illustrate how other cases may be compared, and Fig. 22.5 shows the generalised solutions for a number of different building orientations in mid-summer, at the solstice and year round.

Glazing—U factors for single and double glazing are given in Chapter 2. Due to high exposure, we take external surface resistance as nil. The factors become:

Single glazing in metal frames - $U = 9.5$ W/m² °C
Double ,,   ,,   ,,   ,,   - $U = 4.9$ W/m² °C

This concerns us chiefly in calculating winter heat losses, as in summer winds may be assumed to be less and normal U values may be taken.

The reduction in solar-heat gain with double glazing, and also the reduction due to various forms of shading is referred to in Chapter 17; for instance, for double glazing with a venetian blind between:

factor = 0·6 with clear glass.

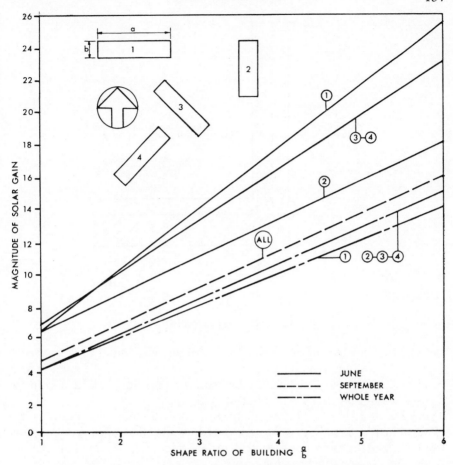

FIG. 22.5.—Solar Gains to Buildings of Different Shapes and Varying Orientations.
(All faces have been assumed to be glazed in equal proportion.)

Curtain Walling—The infilling between continuous fenestration in tall buildings is frequently some light-weight material, as in Fig. 22.6. The U factor for such a construction was considered in Chapter 2 (pages 20 and 21). If external surface resistance is ignored due to exposure, the example therein has a U value of 1·26 W/m² °C instead of 1·19. Other constructions may be similarly evaluated, but the importance of bridging by concrete or metal ribs should be noted (as in the U values given on page 22).

The effect of solar radiation on this form of construction is to raise the outer surface temperature rapidly. It can be shown that in this country the rise will be to about 50° C. The difference between this temperature and that to be maintained in the room in summer (say 21° C) multiplied by the U value gives the heat gain to be allowed per square metre in respect of this wall area.

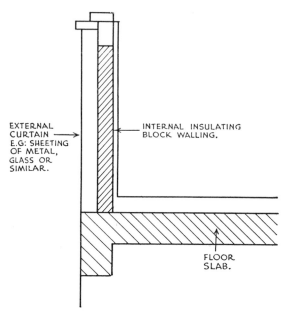

EXTERNAL
CURTAIN ——→
E.G: SHEETING
OF METAL,
GLASS OR
SIMILAR.

INTERNAL INSULATING
BLOCK WALLING.

FLOOR
SLAB.

FIG. 22.6.—Curtain Walling.

The time lag due to the mass of material for this kind of light construction may be about 1 hour; thus, if solar gain (say on a S.E. face) is at a peak at about 10.30 a.m., it may be assumed that the peak heat gain *via* curtain walling will arrive at about 11.30 a.m.

Time Lag—It has been shown* that for British climatic conditions the effect of heat storage in a building is to reduce the effect of peak heat gains from all sources. The paper postulates the factors given in Table 22.1, below, as a reasonable basis for design.

It is stated: 'They are based on 24 hour operation of air-conditioning plant during peak load conditions or pre-cooling. They should also be suitable for 12 hour operation providing an upward momentary swing in air temperature is allowed.'

TABLE 22.1
MULTIPLYING FACTORS FOR COOLING LOAD
(J. C. and J. L. Knight)

Instantaneous heat source	Multiplying factor for Cooling Load
(*a*) Peak solar radiation through bare glass	0·55
(*b*) Peak solar radiation but with venetian blinds on inside - - - -	0·75
(*c*) Peak solar radiation through double windows with venetian blinds in inter space - - - - - -	0·6
(*d*) Lighting and occupants - - -	0·6

* Paper by J. C. and J. L. Knight, *I.H.V.E. Journal*, Vol. 30, p. 1.

It will be clear that the effect of solar gain through glazing, in so far as effect on temperature rise of the structure and contents is conerned, is less than the peak, but this and the effect of other heat gains will depend on many factors, such as the nature of the materials of the building and its contents. Another approach to this same problem of the actual effect of the time lag on the cooling load is that referred to on page 368, where the choice of two factors is given depending on the mass of the internal construction. It would appear unsafe in the design stage to make too great an allowance for time lag, the actual details of construction and contents being then generally vague.

Diurnal Cycle—The diurnal cycle of the sun's motion subjects the various faces of a building to more or less solar incidence at the various hours of the day.

When considering the effect on a tall building, it will be apparent that changes can be quite rapid. Thus, although for some hours in the morning an easterly face may receive solar radiation, by about 10.30 a.m. it will begin to lose the benefit and, by 11.00 a.m., will be almost in shadow. If the building contains wings, such as an L, T or cruciform shape, the degree of solar gain is more complicated and more violent changes are possible, bearing in mind, also, angular incidence over roofs. A simple way to study these effects is to set up a small model of the building at the centre of a chart, such as Fig. 22.1. If external shading or sun-breaks in some form are being considered, such will be particularly necessary.

Wind Effects—The pattern of wind-flow over a tall building is of the form shown in Fig. 22.7.* The positive pressure on the windward side can

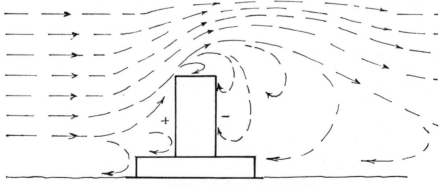

Fig. 22.7.—Wind Currents on a Tall Building.

cause peculiar and undesirable effects within the building if leakage through windows occurs. Corridor doors are found difficult to open due to pressure differences. Infiltration on the windward side may account for a high rate of air-change, rendering the maintenance of temperature difficult,

* See A. E. Wise, *The Architects' Journal*, 10 Feb., 1965, p. 330.

whilst on the leeward side air movement tends to be outward, the only ventilation being by second-hand air from the corridor.

Wind speeds up to 25 km per hour account for up to 30 per cent of frequency of winds in the prevailing direction. Higher speeds are a less frequent phenomenon. Wind speeds as published by the Meteorological Office are at 10 m above ground, but the greater the height the greater the wind speed. The Meteorological Office formula for correction for height is:

$$VH = V_{10}\left(\frac{H}{10}\right)^{\alpha}$$

where VH = Velocity in m/s at a height of H metres.

V_{10} = Velocity in m/s at 10 metres.

α = a function depending on location, 0·14 in flat open country and 0·5 in a city centre.

If 5 m/s is taken at 10 metres, it will be found that the velocities at other heights above ground in a city centre would be:

at 30 m	8·7 m/s
at 60 m	12·3 m/s
at 90 m	15·0 m/s

For a tall building, an upper limit may perhaps be taken at 20 m/s for winds of significant frequency, corresponding to a velocity pressure of about 240 N/m² or 24 mm w.g.

The leakage of air into and out of tall buildings through cracks around windows is dealt with in detail in the *I.H.V.E. Guide*, the basis being work by P. J. Jackman.* It is concluded that wind pressure is the principal factor, the pressure differences due to stack effect being small in comparison except in certain unusual circumstances. Some mention of this subject has already been made in Chapter 2 (page 24), but the following few comments are appropriate in the present context.

Air flow through cracks may be calculated from the formula

$$V = C(P_1 - P_2)^{0.63}$$

where V = volume of air in litre/s per metre run of window crack.

C = a constant depending on the window type, 0·25 where there is no weather-stripping and 0·05 or 0·13 for weather-stripped pivoted and sliding windows respectively.

$P_1 - P_2$ = pressure difference across the window in N/m².

Solutions for this equation are, however, strictly applicable only when the building is in open plan form and the wind entering on one side has free access to a similar escape route on the other. Where the building has many

---

* *A study of Natural Ventilation in Tall Office Blocks*. H.V.R.A. Lab. Report No. 53, 1969.

internal partitions which impede the cross air flow, then the total infiltration may be reduced to as much as 40 per cent of the calculated value. An average figure for normal cases might be 70 per cent.

For a twenty-storey building, 60 m high, the *average* infiltration rate (which is based upon the *total* height to allow for disturbed wind currents as shown in Fig. 22.7), with a prevailing wind of 3 m/s and windows without weather-stripping, would therefore be:

$$\text{Speed at 60 m height} \qquad 3 \times \left(\frac{60}{10}\right)^{0.5}$$

$$= 7\cdot35 \text{ m/s}$$

$$\text{Equivalent velocity pressure} = 0\cdot6v^2$$

$$\text{(see also Fig. 19.14)} \qquad = 0\cdot6 \times 7\cdot35^2$$

$$= 32\cdot5 \text{ N/m}^2$$

Thence,

$$V = 0\cdot25 \times 32\cdot5^{0\cdot63} = 2\cdot24 \text{ litre/s per metre run of crack}$$

The *Guide* proposes that, for the topmost floor, the average result should be increased by 8 per cent and thus, for application to a situation where average partitioning exists, the infiltration to a private office 5 m × 5 m × 3 m with opening windows manufactured to a one metre module would be:

$$\text{Crack length} = 5 \times 4 \text{ m} = 20 \text{ m}$$

$$\text{Infiltration} = 2.24 \times 1.08 \times 0.7 \times 20$$

$$= 33.8 \text{ litre/s}$$

Since the volume of the office is 75 m³, this infiltration rate *with the windows closed* amounts to 1·6 air changes per hour, or a ventilation loss of 0·53 W/m³ °C.

Stack Effect—A tall building, warm inside and cold outside, acts like a chimney. Cold air will enter through entrance doors and other apertures at ground level, rise through lift shafts, staircases and other vertical communications to the upper storeys whence it will tend to be exhausted through smoke vents, crackage and other openings at high level.

A theoretical approach to this problem is valueless on account of the unknowns—e.g., what is the resistance to air-flow through a complex of shafts and corridors? T. C. Min* gives the results of a useful survey of research work and draws some conclusions.

(a) At ground floor, wind pressure is found to be insignificant and temperature forces dominate.

(b) For higher levels wind effect is more important than stack effect.

* H.E.V.A.C. Conference, 1961. Theme.

(c) It is not economically feasible to use excessive ventilation to pressurize a tall building in an endeavour to eliminate infiltration due to wind.

(d) The criterion of tightness of the envelope of a building (such as curtain walling) has not been established for quantitative approach.

(e) Swinging door entrances—for a particular set of conditions — infiltrate about 25 m³ per person entering or leaving for a single door; 15 m³ per person for a vestibule-type entrance with air lock.

(f) Manual revolving doors under the same conditions infiltrate about 2 m³ per person, and motor driven doors about 1 m³ per person.

The conclusions of this study are:

(i) that the greater part of the stack effect problem concerns the ground floor entrances and vestibules where adequate heating in winter is required either by warm air curtain just inside the door, or by forced convectors augmented by floor panel heating.

(ii) that in summer the reverse stack effect is not serious, but some form of air-lock at entrances is necessary if control is to be maintained.

Perimeter and Core Areas—Tall buildings are often planned in depth. The perimeter areas will be subject to solar gains, and to exposure to wind. The inner and core areas will be subject to internal heat gains only and ventilation will be essential. In tall buildings, the intensity of daylight at upper levels is often such that shades are left down and artificial light kept on continuously. The intensity of lighting for inner and core areas which receive no daylight must be such as to compare favourably with the high daylight intensity, and this has led to some very high loadings—as much as 50 W/m².

CHOICE OF SYSTEM

To summarise the foregoing. The tall building is:

(a) subject to considerable solar gain varying in amount throughout the day on the various aspects of the building.

(b) subject to wind effects such that infiltration or exfiltration through openable lights is liable to cause excessive uncontrolled ventilation.

(c) subject to high internal heat gains from artificial light of high intensity.

Heating and Ventilation—Consider, first, heating for the perimeter. From the foregoing it will be clear that the system must be quickly responsive to change. Any system with large thermal mass will be liable to cause overheating when solar gains occur. Embedded systems, such as floor heating or panel heating, are unsuitable for this reason.

Convectors in any form, or heated metal acoustic ceilings having low thermal mass, are suitable provided control is adequately zoned or sectionalised room by room.

Reliance on opening of windows being inadvisable, a heating system needs to have a ventilation system in combination. This would introduce fresh air warmed in winter to room temperature, or slightly below, and means for extraction would be provided. This would suffice for that period of the year when outside temperatures are low enough to cause sizeable heat loss, say 7° C and below. Control of room temperature can then be achieved by adjustment of the amount of heat supplied by the heat emitters. At higher external temperatures, solar gain plus internal heat gains may cause overheating, even with heat shut off.

In the inner areas not subject to variations in weather, heating is not required, but ventilation only. The ventilation air will be the vehicle for the removal of heat gains and, when air can be introduced from outside some 5° C lower than room temperature, results will be satisfactory. At higher external temperatures the same 5° C rise will occur, so that, when the outside temperature is 21° C and over, temperatures of 27° C and over may be expected internally (see also page 367 with reference to this).

Air-Conditioning—For the tall building 'straight' heating and ventilation can therefore only be partially satisfactory. A means for cooling becomes a necessity.

Air-conditioning provides such a means and includes also the duty of heating. Thus, one system replaces two and, in addition, supplies the means for dehumidification in summer. Windows may be sealed if cleaned externally, or openable by key for cleaning from inside. Temperatures may be controlled individually room by room.

Air-conditioning is immediately responsive to control; air, being the medium, can be varied in temperature rapidly to deal with sudden changes of heat gains or heat losses.

Core areas may be maintained at a constant condition regardless of the season of the year.

Air-conditioning of a space may use less air supply than 'straight' ventilation in the kind of system where local recirculation is used, such as in the induction system and fan coil system. This leads to economy in duct sizes, plant size and filter renewal costs.

In view of the many advantages of air-conditioning, it is not surprising that it is being included more and more extensively in tall buildings in spite of its greater cost.

Systems of Air-Conditioning—Of the various systems described in Chapter 17, the most suitable for tall buildings are: the *Induction* system and the *Dual-duct* system. Variations such as *heat-recovery* and *panel-cooling*

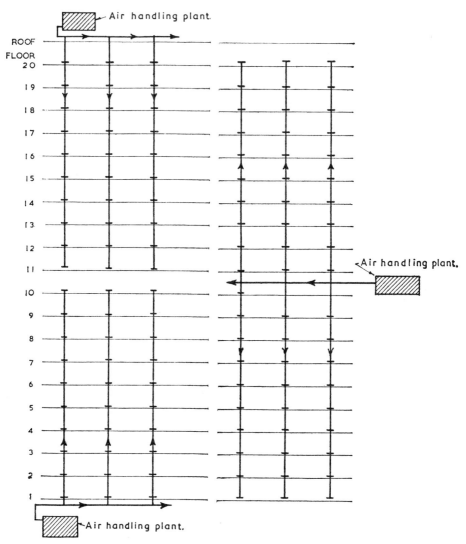

FIG.22.8.—Plant Rooms in Basement and on a Roof.

FIG. 22.9.—Plant Room at Intermediate Floor Level.

would no doubt be considered. *Fan coil,* or other systems depending on puncturing the wall face, would not be suitable due to wind pressure effect.

An induction system would serve the perimeter areas, but the inner areas would require conventional central station air-conditioning plant or plants. The system of ducting for the primary air and water circulation to the induction units would be determined by the plant location. Figs. 22.8, 22.9 and 22.10 show some alternatives which will be self-explanatory: the arrangements of Figs. 22.8 and 22.9 are to be preferred where possible.

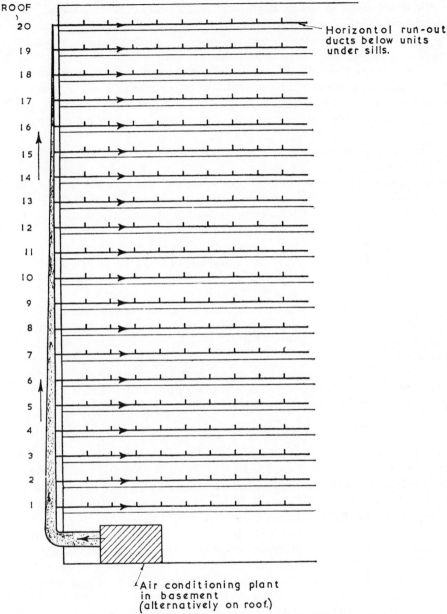

FIG. 22.10.—Arrangement of Ducts from Main Trunk at end of Building.

The perimeter areas may be regarded as about 5 metres depth from the windows, and certainly no more than 6 metres.

The dual-duct system would serve both perimeter and inner areas from the one set of ducts and plant. The perimeter would be dealt with by blender boxes beneath windows or, alternatively, in zones with single ducts taken to sound attenuator boxes beneath the windows. The blender box feeding the zone would then be remote. The inner area would be fed from blender boxes serving ceiling diffusers and the like, room by room, or by a variable volume single-duct system.

The arrangement of ducts might follow one of the patterns shown for the induction system so far as the feeding of the perimeter goes, but it would be more usual to concentrate the main risers at one or two points in the core, and to run out floor by floor over the corridor in a ring main fashion, the two ducts being side by side, as in Fig. 22.11.

The ring main system has the advantage that the air handling capacity of the ducts is reduced, due to the fact that exposure on all sides of a building does not occur simultaneously, as the cooling load follows the sun. Similarly, the heating load is greatest on the windward side. Hence the ring need only handle the average load, whereas individual ducts to each section must handle 100 per cent.

As has been explained in Chapter 17, the dual-duct system circulates more air than the induction system and needs ducts for return air; duct spaces are therefore larger. But to set against this is the saving of separate plants for core areas and avoidance of water-circulating piping.

The *water-chilling plant* would be common to all systems and could be located in one or other of the plant spaces indicated, such as the basement, roof or intermediately, due regard being paid to reduction of noise and vibration.

The *air heater batteries* would be served from boilers either at the bottom of the building, or on the roof. The latter avoids a flue, which is an expensive item in a tall building as well as taking up valuable space.

REDUCTION OF STATIC PRESSURE

Apart from the areas served by air-conditioning, there will be staircases, entrance halls, lavatories and the like where direct heating in some form is necessary. In order to avoid undue static pressures on apparatus such as radiators or convectors, valves and fittings, a tall building of over 50 metres height should preferably be divided into two.

The same applies to hot and cold water services where a pressure of little over 30 metres is the desirable maximum on taps and other fittings.

An intermediate plant room half way up the building, as shown in Fig. 22.12 is thus an advantage. The lower part of the building is served from tankage at this level, with calorifiers in the basement. In the intermediate plant room are accommodated calorifiers serving the upper half of the building. The tankage for the upper half is of course on the roof, as shown in Fig. 22.13.

A building 100 metres high would require two such intermediate plant rooms, and so on. The same intermediate level(s) would also be convenient for air plants serving up and down, as in Fig. 22.9.

The chilled-water system must perforce be designed for the full head of

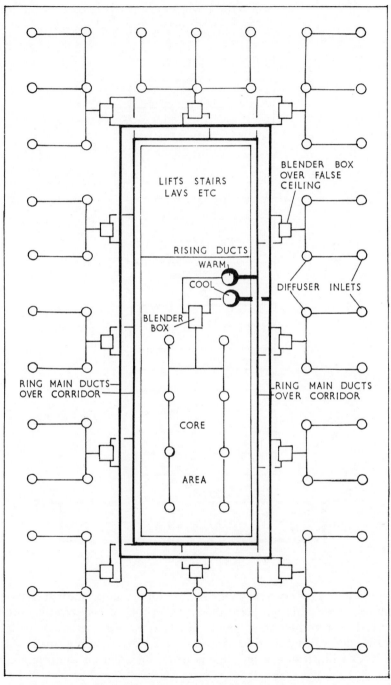

FIG. 22.11.—Plan of Building showing ring main system of Dual Ducts,
serving Zone Blenders. (The return air system is not shown.)

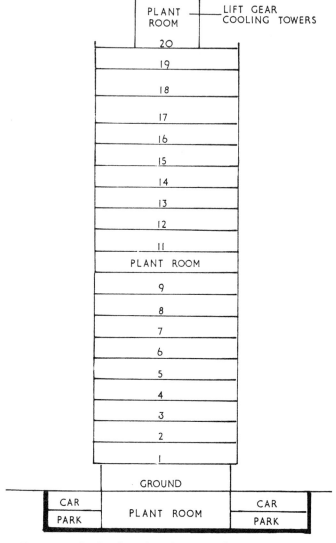

FIG. 22.12.—Section through Tall Building showing Plant Spaces.

the building, as any heat-exchange surface for breaking static head would introduce undesirable temperature drops probably necessitating the use of brine. The introduction of water chillers at an intermediate floor level, or on the roof, can be considered as a means of overcoming this problem. Plate XXVI illustrates water chillers being installed on the roof of a tall block.

Turbine and Absorption Plant in Combination—One interesting development of the last named method is to provide in the basement a

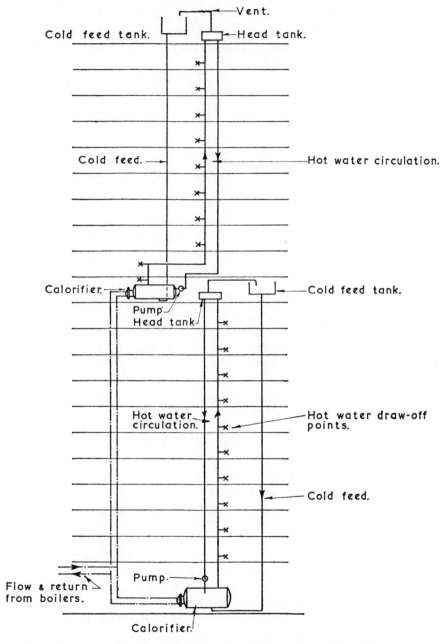

FIG. 22.13.—Diagram of Hot-Water Supply System for a Tall Building.

centrifugal compressor plant driven by steam turbine, and to provide at the upper plant room an absorption cooling plant. The absorption system of refrigeration can make use of the exhaust steam from the turbine, so that a highly efficient thermal balance is obtained. To obtain the optimum efficiency, it is necessary for the two plants to be in a similar state of load, although in practice this may not always be possible. The steam is obtained from boiler plant serving heating, in the winter, as well as other requirements. It is probably true to say that this system has arisen chiefly where district steam is available, as in N. America.

AIR-CONDITIONING AND HEAT RECOVERY

The tall building contains within itself a problem inherently favouring the use of some form of heat-recovery system (see p. 265). The inner and core areas of such buildings are heat producing on a massive scale due to high lighting and other electrical loads, whilst the perimeter areas are, in winter, heat losers. Thus, if the system of ventilation is such that the exhaust air from the centre can be used as the low grade heat supply to a heat-recovery system, a great deal of the heating of the perimeter can be had for nothing.

For air-conditioning in summer, cooling plant is required in any event, and thus it is obvious that the same plant can be put to good use, with the cycle reversed in the winter, without extra capital costs for such plant.

This principle is used in the specialised system referred to on page 356, but may equally be applied to conventional central plant, provided the additional complications are accepted. It is obvious that running costs for the year-round conditioning of the building as a whole should be considerably less than with a conventional approach.

CONCLUSION

Tall buildings, as will be seen, present a number of unusual problems. experience of which for conditions obtaining in this country is not yet fully digested. Further research work already in progress will no doubt in due course throw further light on these matters, and more precise forecasts will be possible, particularly as to wind effects, exposure factors, sun gains and so on.

Other aspects not relevant to the present treatise concern water supplies, sanitation, lifts, electrical distribution, and fire protection, all of which fall within the province of the building services engineer and require the most careful co-ordination if the structure as a whole is to provide a tolerable environment for the occupants and a satisfactory investment in both capital and running costs to the building owner.

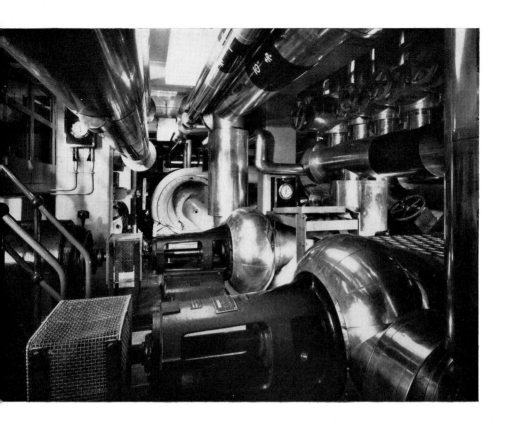

ate XXVIII (above). Chilled Water
mps for an air-conditioning installa-
ion in a City building (see p. 486)

ate XXIX (right). District Heating:
'pipe within a pipe' during installation (see p. 515)

Plate XXX. Above: Vekos coal-
fired boilers serving a District-
Heating System. Left: the pumping plant associated therewith
(see p. 513)

CHAPTER 23

District Heating

DISTRICT HEATING, PIONEERED IN THE USA and on the Continent on a small scale late in the last century, was later developed to such an extent that over 400 schemes now exist in Denmark alone. In Great Britain, however, probably due to the more equable climate and the traditional plentiful supplies of cheap fuel, little interest was shown in such schemes until the period immediately following the Second World War, although quite large institutional and factory sites were served from a single central boilerhouse. Fig. 23.1 shows the layout of such a heat distribution scheme for a university, having a site area of just over 1·1 km², the boiler capacity being 40 MW.

With the development of the New Towns and the intensive building programme after 1945, there was a growing awareness in this country of

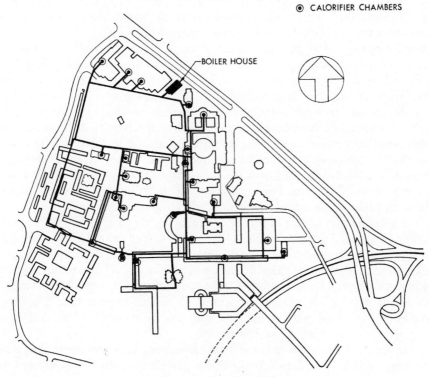

⊚ CALORIFIER CHAMBERS

BOILER HOUSE

FIG. 23.1.—Site plan of a University.

the advantages of district heating and many design projects were initiated, although regrettably few came to fruition. The publication in 1953 of *Post War Building Studies, Nos. 31 and 32*, provided a mass of detailed information but, by this date and in the prevailing economic climate, interest was receding.

Over the last ten years, with the tendency to plan for compact multi-storey building complexes, many heating schemes serving building groups —some of considerable size—have been constructed. But these do not fall strictly within the definition of 'district heating', since the groups are usually of common ownership and are often so arranged that the pipe-distribution system runs totally (and economically) in basements, car parks, interlinking corridors or other structural elements, thus avoiding such problems as underground networks and road crossings.

The advantages inherent in district-heating systems lie in the economic use of the World's fuel resources, the avoidance of atmospheric pollution with the accompanying hazards to health, a reduction in the road traffic inherent to retail fuel delivery and the provision to building owners and occupiers of a convenient and trouble-free utility service on attractive terms.

FIRST CONSIDERATIONS

The primary elements of a district-heating system are a source of heat supply, a system of distribution and a means of utilisation.

Heat Supply—The first and most obvious source of heat supply is the electrical generating station which, at best, operates at only about 35 per cent efficiency. The second is the municipal refuse incineration plant and the third is the conventional boiler plant burning one fuel or another.

Difficulties of load balancing (heat supply to electrical supply) and some technical complications have tended in the past to militate against the use of thermal electric-generating plant; in fact, less than two per cent of the European systems are so arranged. One notable exception exists, however, in the centre of London where part of Pimlico is heated by waste energy from Battersea Power Station. Also, Aldershot has a district-heating plant using waste heat from a diesel generator station.

Refuse incineration as a source of heat is widely used on the Continent; and new plants operated by the GLC and by Nottingham City are about to come into service, so that British experience should soon be available. Domestic refuse has a calorific value of about 12 MJ/kg, about half of which can be made available by combustion. Since refuse output is rising to about 300 kg per person per annum it would appear that dwellings could be provided with about ten per cent or so of their annual heat requirements by 'fuel' otherwise wasted.

In context with the present time, however, it would seem that the conventional boiler plant (coal-, oil- or gas-fired) must be considered as the most usual heat source for district-heating plants. Such installations have been discussed in some detail in earlier chapters and it is necessary only to emphasise that district-heating systems, being large by their very nature, can be equipped to burn the lower and cheaper grades of coal and oil with least disadvantage to the environment, and can take advantage of the most favourable tariffs offered by Gas Boards. Plate XXX (facing p. 511) shows an automatic coal-fired boiler plant serving a district-heating system. Plate I (*frontis.*) also shows a plant serving a like system.

Heat Distribution—Generation of heat in a supply station is, in many ways, the least part of the design problem associated with the inception of a district-heating scheme.

The heat there produced must be distributed to the consumers in the most economical manner possible, from the viewpoint of both capital and recurrent costs. As is often the case in engineering economics, these two aspects are fundamentally opposed—the requirement for minimum capital costs militating against the need for excellence in choice of materials for pipework, techniques for heat insulation and structures for underground containment of both.

Traditional British methods of pipe insulation and enclosure have of recent years come under scrutiny, particularly since practice in the USA and on the Continent is known to be at variance with them. Following an expected pattern, the North American developments have been towards the production of relatively complex factory-built units in order to reduce site works to a minimum; conversely, European tendencies have been towards simplification in the choice of materials and their containing structures whilst retaining meticulous attention to detail. In each case, however, the result of these advancing practices has been to reduce capital costs.

So much information has recently been published on this subject that it would be out of place here to do other than list briefly the principal methods used, all as shown in Fig. 23.2:

(a) The 'traditional', where excavation is made and an in-situ brick or concrete passage formed, in which pipes are installed and subsequently insulated with glass fibre or like material. A concrete cover is then fixed and the ground surface put back.

(b) The 'semi-traditional', where the in-situ passage is replaced by preformed channel, circular or semi-circular concrete sections.

(c) The 'enclosed loose fill', where the segmental insulation of the previous methods is replaced by a loose granular material packed around the pipes.*

(d) The 'solid fill', where a preformed channel section is laid in an excavation and the pipes are insulated by means of a light-weight aerated concrete material poured round them to fill the channel completely.*

* This method and also those under (d) and (e) are, reported here to complete the list of past practices, but they are not recommended.

2L   K.H.A.C.

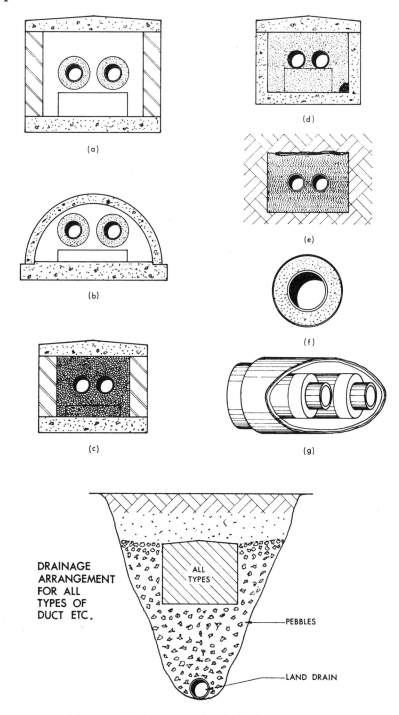

FIG. 23.2.—Typical constructions for Underground Mains.

(e) The 'open loose fill', where bare pipe lengths are laid in an excavation prior to insulation by means of a powdered mass of hydrophobic material, which cures on the application of heat to form a protective skin on the outer surface of the pipe, no structural enclosure of any type being provided.*

(f) The 'simple preformed', where single steel pipe lengths are factory-set concentric within a larger plastic tube and the annulus filled with a hard-setting plastic foam, prior to burying directly in an excavation without further protection or structural surround. (See Plate XXIX, facing p. 510).

(g) The 'complex preformed', where single or multiple steel pipe lengths are factory-insulated and enclosed within an outer steel tube (protected by up to ten laminates of various materials) or an asbestos cement tube, prior to burying directly in an excavation without any structural surround.

Long term experience has shown that in this country, despite much ingenuity and the most careful design, small size in-situ or preformed structural ducts, as described in (a) and (b) above, are sooner or later subject to leakage such that ground water enters. Provision for drainage can in many instances prove disastrous, since at times of heavy rainfall surface water will actually enter the ducts *via* the drains, which are not normally allowed to be connected to a deep outfall system but only to soakaways. Water entry having been accepted as an unfortunate fact of life has resulted in attempts being made to keep the insulation dry by wrapping. It has been found, however, that tiny failures occur in due time and that the insulation in the wrapping becomes poulticed on to the pipe by moisture unable to escape and quickly causes corrosion. The attempted cures have thus often proved more fatal than the diseases.

Therefore, unless generously dimensioned walk-in tunnels and subways can be formed (an unusual circumstance in terms of modern building economics), one of the preformed systems is to be preferred, and these may be associated with preformed access chambers for inspecting valves etc., as shown in Fig. 23.3. For these, many novel methods have been proposed to monitor possible failures: an air pressure may be maintained within the protecting enclosure tube, or tracer cables laid so that moisture

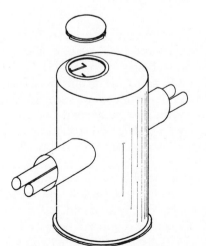

FIG. 23.3.—Factory-built access chamber.

* Not recommended. See footnote, page 513.

causes current flow to provide both indication and fault location. Mr A. E. Haseler, first chairman of the recently formed District Heating Association, has written extensively on this subject,* and the Heating and Ventilating Research Association are actively initiating developments.

Table 23.1 lists data regarding dimensions and heat losses from one proprietary type of preformed system ('pipe-in-pipe'—to use a common description). Reference should be made to the *I.H.V.E. Guide* for further information on heat loss from buried pipes.

TABLE 23.1

HEAT LOSS FROM PREFORMED 'PERMAPIPE'
BURIED 600 mm BELOW GROUND LEVEL

Size of service pipes (mm)	Insulation thickness on pipes (mm)	Diameter of enclosing conduit (mm)	Heat loss, W/m run of enclosing conduit for following mean water temperatures		
			75 °C	100 °C	125 °C
Two × 20	25	250	39	53	63
Two × 25	25	250	42	58	73
Two × 32	25	300	50	65	83
Two × 40	25	300	54	71	89
Two × 50	25	300	58	79	99
Two × 65	25	350	69	92	117
Two × 80	37	350	62	84	108
Two × 100	37	400	73	98	123
Two × 125	37	450	83	110	140
Two × 150	37	600	96	130	165

These data have been obtained by converting manufacturers published figures into SI units. The published values were calculated according to the methods of BS 4508, Part 1.

In recent City Centre and housing developments, there has been a tendency towards separation of pedestrian and vehicular traffic by elevation of the former to bridges, overhead walkways or upper levels of a podium. These proposals offer further solutions to the problem of heat distribution since it is often possible in such instances to suspend the necessary pipework below the pedestrian ways and thus avoid the problems inherent in any form of excavation and burying; protection against vandalism is all that is necessary. Fig. 23.4 shows two instances of application in this way.

Heat Utilisation—Heat energy utilisation, at the consumer end, can follow any of the methods described in previous chapters. Initially, of course, much will depend upon the medium used for distribution since if steam or high-temperature water be employed, these may require to be fed into a 'transformer' station to reduce temperature potential to that which may be

* *I.H.V.E. Journal*, Vol. 38, December 1970: *New Heat Mains Techniques for Telethermics.*

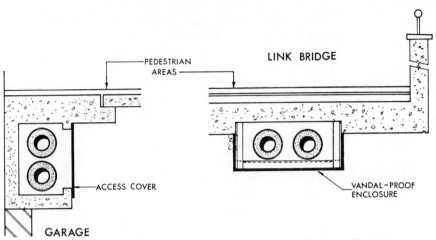

FIG. 23.4.—Heating Mains suspended below elevated pedestrian walkways.

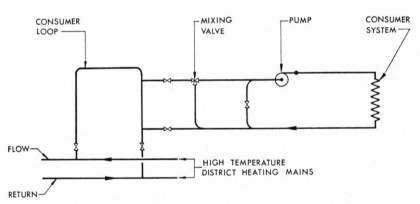

FIG. 23.5.—High-temperature/low-temperature mixing arrangement.

employed safely by the consumer. Such a transformer station would incorporate heat exchangers—storage or non-storage according to the service required—or, in the particular case of high-temperature water, some form of mixing arrangement, as illustrated in Fig. 23.5.

HEATING MEDIA

Basic design decisions for a district-heating system are determined by the nature of the load to be served. Domestic and commercial premises rarely require other than water at low to medium temperatures, whereas steam or some other high-temperature energy source may be necessary for industrial processes. Each case must be considered on its merits, some of the alternatives being as follows:

Mixed Areas

(1) Primary distribution at high temperature (water at say 180°C or steam at 10 to 12 bar absolute):

Industrial buildings would be connected directly to the primary mains.

Domestic and commercial building systems would be connected either individually or in groups to the primary mains but served thence, *via* transformer stations, at low temperature.

(2) Dual primary distribution, at high temperature, as above, and with low temperature (water at say 90° to 100°C) in parallel:

Industrial buildings would be connected to the high temperature mains.

Domestic and commercial building systems would be connected to the low temperature mains.

(3) Primary distribution at low temperature (90° to 100°C as above):

Industrial processes would be served by some local heat generating plant.

Industrial, domestic and commercial building systems would be connected to the primary mains.

Domestic and Commercial Areas

(1) Primary distribution at medium temperature (water at say 120 to 150°C:

Building systems would be connected in groups to the primary mains but served thence, via transformer stations, at low temperature.

Building systems, if of warm air type, would be connected directly to the primary mains.

(2) Primary distribution at low temperature as above:

Building systems of any type would be connected directly to the primary mains.

Obviously, there are any number of combinations of these principles which could be adopted and space will not permit further discussion. It must suffice to say that the essence of the matter lies in the selection of the most economic arrangement in both capital and operational cost, since these are the penalties which have to be set against the advantages which may be gained from centralised heat generation.

PIPING DISTRIBUTION SYSTEMS

Where district heating is to serve domestic and commercial premises only, there are four principal methods which may be used in arranging the distribution system, as shown in Fig. 23.6. These are:

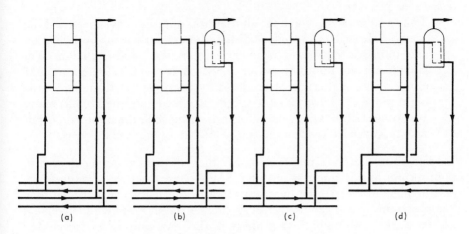

FIG. 23.6.—Alternative methods for distribution.

The 4-Pipe Primary/Secondary System (a)

In this case, similar in many ways to that which might occur in a single large building, one pair of distribution mains serves the various heating appliances in the buildings connected, and a parallel pair of distribution mains provides domestic hot water to draw-off fittings from a central storage vessel or vessels. There are obvious cost disadvantages in that the domestic water mains may well have to be in copper and, in any event, will be subject to scaling and the other varieties of attack inherent to any pipe distribution network conveying a changing water content. Further, and perhaps more important, over-use or wastage in any building connected will limit the supply of hot water available to other buildings.

The 4-Pipe Primary System (b)

Here, one pair of distribution mains serves the various heating systems in the buildings connected and the parallel pair serves the primary coils of hot-water storage vessels, also, in the various buildings. The latter mains may now be in steel and are not subject to scaling or internal attack. Any building requiring year-round service could demand *heating* connection to the *hot-water* primary mains.

The 3-Pipe Primary System (c)

This is a refinement of the last system whereby two flow mains, one for heating and one for hot water primaries, have a common return. The advantages of the 4-pipe primary system are retained and there is some saving in capital cost.

The 2-Pipe System (d)

This arrangement, which for all practical purposes has superseded the previous three, has one pair of distribution mains serving both heating and domestic hot-water primary coils in the buildings connected. Heat losses and pumping power are saved during the summer months either by use of alternative smaller-duty pumps or by arranging for the main pumps to have variable speed. The latter alternative has many advantages since, with reasonable operating skill, the pumped output may be governed to meet the actual load at any time of the year.

The four methods mentioned are those which have been, or are, appropriate to this country. In the USSR, where distances are immense, and in Iceland, where the heat source is geothermal springs, single pipe 'open' systems have been used.* In such cases the single flow pipe conveys heated water to the buildings connected, the maximum usage is made of the energy potential by balancing heating and domestic hot-water demands where possible, and the residual is sent to the cold-water mains, to swimming pools or to waste.

As emphasised previously, there are many possible combinations of piping systems. It may be desirable, for instance, to use a 2-pipe distribution for sixty per cent of the connected load and a 4-pipe primary/secondary system for groups of buildings making up the remaining forty per cent. Once again, it must be stressed that the most economic solution for a given set of circumstances will determine the 'right' choice of system.

DOMESTIC BUILDINGS

District heating, or often Group heating as applied to large housing developments, is currently in the public view. This presents certain very special problems insofar as heat requirements and heat consumption become a matter for the householder's pocket. Many such schemes are initiated by local authorities who obtain guidance from the *Housing Subsidies Manual* (Appendix XIII: *District Heating*) published by HMSO for the Department of the Environment.

Standards for Internal Temperatures—The Parker Morris Report,† in advance of contemporary thought but now outdated, included the following recommendations with regard to standards:

'*Minimums*'

'The heating installation should be one capable of heating the kitchen and the areas used for circulation to 55°F (13°C) and the living and dining areas to 65°F (18°C), both temperatures attainable when the outside temperature is 30°F (− 1°C)'

---

* J. C. Knight. *I.H.V.E. Journal*: Vol. 36, p. 107/108.
† *Housing for Today and Tomorrow*. HMSO: 1961.

'Recommended additions'
'Where daytime and evening use of bedrooms is likely to be considerable, it will be worthwhile to adopt a higher standard and provide an
installation capable of heating the bedrooms to 65°F (18°C) as well.'

These recommendations have been variously interpreted by local
and other authorities, as shown in Table 23.2 Dwellings for old persons
are required to be heated to 21°C throughout.

TABLE 23.2
STANDARDS FOR INTERNAL TEMPERATURE

Authority	'Parker Morris standard'	'Full Parker Morris standard'
A	18°C in all living areas (not bedrooms!)	—
B	18°C in living room 13°C in hall. No heat in bed-rooms	—
C	18°C in living room, 13°C in hall, kitchen and one bedroom	—
D	—	21°C in living room, 16°C in all other rooms
E	—	21°C in living room, 18°C in all other rooms

Undesirable Standards—It is in context to make particular mention
here of the fallacy which exists in attempting to warm only the living room
and hall to temperatures of 18°C and 13°C respectively, leaving the
bedrooms unheated. Fig. 23.7 shows a typical flat, the shaded areas being
those directly heated. The sinuous arrows show the route of the heat loss
from the flat to the outside air, whereas the straight arrows show the heat
transfer between the heated and unheated rooms of the flat.

The figures circled in each room show the temperature levels which
would be achieved in steady state conditions, with the outside temperature

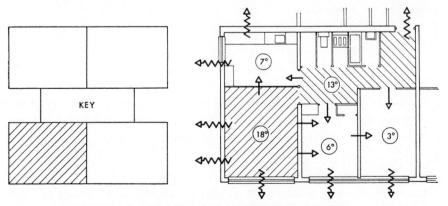

FIG. 23.7.—Heat Transfer in a partially heated dwelling.

at $-1°C$; and it will be noted that although kitchen and bedrooms are excluded, these are in fact heated to some extent, second-hand. The point that requires emphasis here, of course, is that this condition is only achieved by putting excess heat into the living room and hall with the inevitable result that, if kitchen and bedrooms are provided with local heaters, such as electric fires, the living room and hall will be *overheated* by the fixed plant.

Quantitatively, for the case considered, the following apply:

Living Room
Heat required by that room -   -   -   -   1·7 kW
Heat added there to lose to kitchen   -   -   0·3 kW
Heat added there to lose to bedrooms   -   -   0·4 kW

Hall
Heat required by that room -   -   -   -   0·7 kW
Heat added there to lose to kitchen   -   -   0·1 kW
Heat added there to lose to bedroom   -   -   0·5 kW

The apparent paradox exists, therefore, that

the nett heat required to heat the living room
    and hall $(1·7 + 0·7)$ is   -   -   -   -   2·4 kW
the actual heat required to serve these rooms,
    including spill heat to bedrooms and
    kitchen, is   -   -   -   -   -   -   3·7 kW
Whereas the heat required to heat the whole
    flat might be only   -   -   -   -   -   5·4 kW

This goes towards proving that it is more reasonable to design for whole-dwelling heating than to use any half-baked, half-way measure.

Standards for Consumption of Domestic Hot Water—The more important of the standards laid down with regard to provision for domestic hot-water consumption are:

The Parker Morris Report (5-person house)
'All systems should allow for a supply of 250 gallons (1150 litres) of hot water per week at 140°F (60°C).'

The I.H.V.E. Guide (1965)
For low rental flats, consumption is quoted as 15 gallons (70 litres) per person on the day of heaviest demand. This is usually interpreted to mean 60 gallons (280 litres) per person per week.

Records exist which indicate that the recommendations set out above are too high if the system incorporates some in-built means for the prevention of water wastage. With district heating, the provision of a local storage vessel in each dwelling, with a limited facility for re-heating, has been found to produce such economies. From data available from several sources, it is considered that, if separate facilities are provided for bulk laundry use, hot-water consumption may be limited to approximately 180 to 230 litres per person per week, on the basis of nominal design occupancy.

Typical Consumptions of Heat Energy—For domestic premises, the peak requirement for space heating will vary considerably according to the construction, size and isolation, or otherwise, of the dwelling. A detached four-bedroom house might require about 18 kW, whereas similar accommodation within a multi-storey block would require only about 10 kW. Most available statistics relate to local authority housing and suggest a range between 5 and 8 kW, with an average of about 6·5 kW for a two- to three-bedroom (four-person) dwelling in a low rise block, fully heated.

Annual consumptions of heat energy vary considerably according to the method of heating and the fuel used. Electricity Boards are able to produce evidence that annual usage for electric warm-air heating in a dwelling is much less than that which would normally be calculated for the connected load. The aspects of ease of metering and the relatively high fuel cost are important factors here.

Again, for local authority dwellings, statistics show that, where district heating is available, a total annual consumption of between 53 GJ and 80 GJ may be anticipated, with an average of about 65 GJ. Of this total, some 13 GJ per annum may be considered as usage for domestic hot water, leaving about 52 GJ per annum for space heating. Current rising standards in demand for comfort in the home suggest that future total requirements may well be of the order of 65 GJ per annum for space heating plus 15 GJ for domestic hot water, making a total of 80 GJ per annum for the 'average' dwelling.

CHARGES FOR HEAT SUPPLY

Individual occupiers may be charged for the supply of district heat to their premises in four principal ways:

By a flat-rate charge included in the total of the rental where appropriate.

By a separate flat-rate charge for heating service.

By a unit charge derived from metering.

By a fixed charge plus a unit charge derived from metering.

It is usually thought to be important that an impression should not be created that the individual can wholly 'contract out' of receiving service, since the economics of the case overall are at a maximum with universal supply.

The *Parker Morris Report* mentioned previously states that, for dwellings, some form of heat metering is desirable. This conclusion is no longer accepted, since metering inevitably introduces the necessity for meter reading, preparation of accounts and collection of debts.

Metering—In the early days of district heating in the USA, heat metering was comparatively easily achieved. Energy was distributed by steam mains and a water meter on the condense line provided a measure of the heat consumed.

With the advent of water distribution, matters became more complex since it was then necessary to meter not only the water quantity flowing to each consumer but to integrate this with the temperature difference between the supply and return mains.

Heat Meters—Complex Btu (now presumably GJ) meters have been produced and proved to be successful. They depend for their operation upon some means which will measure water flow and integrate this with temperature difference. Three principal types have been evolved:

(a) *Mechanical*. In this case (the principle of which is shown in Fig. 23.8 (a)), the heat-energy reading is produced by biassing the output drive from a water meter mechanically, by means of temperature sensitive elements in the supply and return mains.

(b) *Electrical*. Here, as shown in Fig. 23.8 (b), both flow rate and temperature difference are measured electrically and are integrated by an electronic device.

(c) *Shunt*. In this case (Fig. 23.8 (c)), a known small proportion of return water is electrically heated back to the supply temperature, as sensed by a thermostat. The heating current is recorded by an ordinary kWh meter to provide an inferential measure of the energy abstracted from the circuit proper.

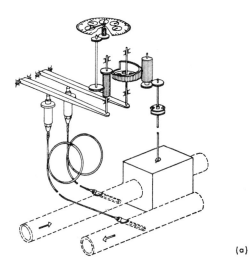

(a)

FIG. 23.8. (a)—Mechanical Heat Meter.

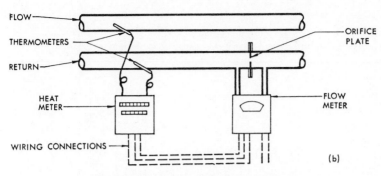

Fig. 23.8. (*b*)—Electrical Heat Meter.

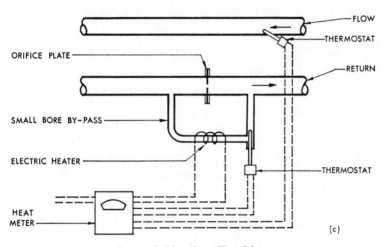

Fig. 23.8. (*c*)—Shunt Heat Meter.

Such meters have the disadvantage of being quite large, relatively expensive and subject to error if not regularly maintained and re-calibrated. Recourse has thus been made to simpler devices, but these meter by apportionment rather than directly: i.e. they provide data whereby the total operating costs of the system may be divided between the users without actually measuring the energy consumed.

Inferential Meters—For radiator systems, the pattern shown in Fig. 23.9 (*a*) has been evolved. This consists of a housing bearing a scale which is clamped to each radiator and which conceals a removable open phial containing a volatile liquid. The amount of liquid evaporated by radiator heat is a measure of usage, and this may be read against the scale each season or other period. The scale readings are then totalled and the cost per radiator apportioned. An obvious disadvantage exists here in that access to individual radiators by meter readers is necessary, and this is not always easy to achieve in domestic premises. The manufacturers, however, offer an all-in meter reading and accounting service.

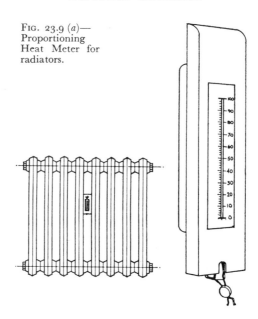

Fig. 23.9 (*a*)—
Proportioning
Heat Meter for
radiators.

For warm-air systems, inferential readings are easily achieved. A cyclometer type 'hours run' meter may be wired in parallel with the fan control, as Fig. 23.9 (*b*), and read from outside the premises. The products of hours run and unit size may be totalled and then used to apportion the cost to users.

Flat-rate Charges—Some system of producing flat-rate charges—per m² of dwelling area, for example—is an obvious alternative to metering, and this, for domestic premises at any rate, seems to be the preferred method on the Continent. For local authority developments, this approach has advantages in that the tenant who is unfortunate enough to live on the twenty-fifth floor of a tower block, facing north east, is not penalised *vis-à-vis* his near neighbour living in an identical flat on the ground floor, facing south. The disadvantage, of course, is that no incentive exists to be economical in heat usage and that the frugal tenant is penalised at the expense of his extravagant equivalent, who opens the windows rather than turning the radiators down and leaves all equipment running at full blast whilst he is away on holiday.

Flat-rate plus Unit Charges, etc.—A system of flat-rate plus unit charges would seem to have advantages over both the alternatives: the economics of the system would be assured and equity would be preserved.

It would seem reasonable that charges should fall into three categories:

(*a*) Connection.
(*b*) Floor area, cube or loading.
(*c*) Usage.

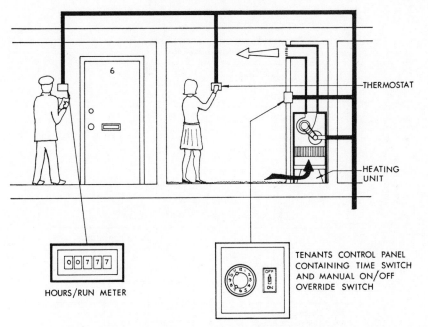

HOURS/RUN METER

TENANTS CONTROL PANEL
CONTAINING TIME SWITCH
AND MANUAL ON/OFF
OVERRIDE SWITCH

FIG. 23.9. (b)—A Proportioning Heat Meter for warm-air systems.

The first of these would, not unreasonably, cover the actual cost of service provision, and the second would be related to the physical aspects of each building or dwelling connected. The last component, heat metering aside, could be related quite simply to the product of the connected load and the hours of operation: a sealed time-switch could provide a simple means of measuring the latter, or even a 'white-meter' assessment related to the operation of local circulating pumps or thermostatic-control devices.

SOCIOLOGICAL ASPECTS

Much has been written upon the incidence of ill health and even death, in the case of elderly persons, due to hypothermia, a matter directly associated with the availability of low-cost district heating. One Lloyds insurance firm has, in fact, offered a ten per cent reduction in premiums to persons living in centrally-heated houses. The social costs of the 'four damp cold walls which are some peoples environment' is not inconsiderable in terms of loss of production and drain on the Health Service.

With particular respect to the viability of district-heating schemes, however, and specifically to those which provide a metered supply, it is necessary to consider the occupation incidence of homes in Britain today. Statistics collected and published in relation to one housing estate show that the occupancy set out in Table 23.3 may be anticipated for a local authority housing development. This leads, *via* certain empirical assumptions, to the conclusions listed in Table 23.4.

TABLE 23.3

Typical Population Distribution in a
Housing Estate

Dwelling		Occupancy	
Type	No.	Adults	Children
Bed-Sitting Rooms	6	6	—
One-Bedroom Flats	112	200	12
Two-Bedroom Flats	84	169	76
Three-Bedroom Houses	27	77	51

TABLE 23.4

Occupancy during Working Week

Dwelling	Assumed condition 9.00 to 1700 hrs (%)		
	Empty	Occupied	
		In part	In full
Bed-Sitting Rooms	100	—	—
One-Bedroom Flats	75	10	15
Two-Bedroom Flats	55	20	25
Three-Bedroom Houses	25	40	35

Such data (as shown in the above Tables) are very pertinent when it becomes a matter of considering the annual heat consumption which may be expected from a district-heating scheme (particularly when the heat supply may be metered), in parallel with the following assumptions for hours of operation:

Unoccupied during working hours
5 days from 7 a.m. to 9 a.m.
5 days from 5 p.m. to 11 p.m.
2 days from 9 a.m. to 11 p.m.
Total = 68 hours/week

Part occupied during working hours
5 days 7 a.m. to 9 a.m.
5 days 1 p.m. to 11 p.m.
2 days 9 a.m. to 11 p.m.
Total = 88 hours/week

Fully occupied
5 days from 7 a.m. to 11 p.m.
2 days from 9 a.m. to 11 p.m.
Total = 108 hours/week

TYPICAL HOUSING DEVELOPMENT

In order to bring the subject into perspective, a housing development with ancillary amenity buildings (as shown in Fig. 23.10) may be considered. There are 1200 dwellings with a notional occupancy of 4500 persons.

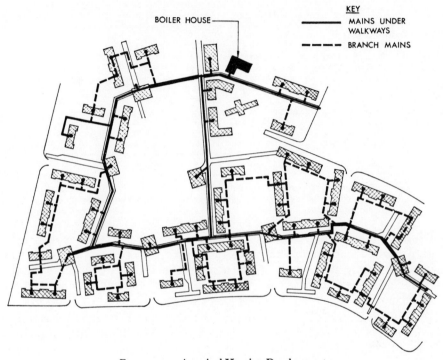

FIG. 23.10.—A typical Housing Development.

Assuming the average peak heat requirement per dwelling to be 6·0 kW, the annual requirements per average dwelling might be:

Space heating	56 GJ
Hot-water supply	14 GJ
Total	= 70 GJ

It would be necessary to consider the comparative capital costs of equipment and plant within and outside the dwellings, as set out in the first two columns of Table 23.5, for district and local heating. These costs would then be amortised over varying periods, using current rates of interest, to produce the annual owning costs listed in the last two columns of the same Table.

2M

K.H.A.C.

TABLE 23.5

CAPITAL AND ANNUAL COSTS OF EQUIPMENT FOR ALTERNATIVE
HEATING METHODS

Item	Capital Cost £ per dwelling*		Annual Cost† £ per dwelling*	
	District Heating	Individual Unit	District Heating	Individual Unit
Equipment				
Boilers, Pumps, etc.	60	—	5·9	—
External Mains	80	—	7·8	—
Internal Mains	60	—	5·5	—
In dwellings	180	330	16·4	38·2
Builders Work				
Boilerhouse and Flue	30	—	2·7	—
External Mains	5	—	0·5	—
Internal Mains	5	—	0·5	—
In Dwellings	10	20	1·0	1·8
Total	430	350	40·3	40·0

* Based on a 4 person 2 bedroom flat.
† Interest rate taken at 9% over terms of 10, 20, 30 and 60 years as
appropriate.

District Heating—Assuming that the fuel used is either coal, a light
grade oil or gas on an 'interruptible' tariff, the operating cost of the
district-heating system would be calculated as set out in Table 23.6, the
raw fuel cost in each case being about 25p per GJ. It will be noted that the
ancillary buildings are included here as being equivalent to an additional
100 dwellings and that, fortuitously, the addition of an item for sundries
brings the total annual cost to a round £100,000. This addition enables
the individual costs to be viewed in terms of percentage of the total. It
will be seen that the charges for interest and repayment of capital make up
more than half of the annual cost—which situation reflects the high
interest rates currently in force.

Individual Heating Units—For the individual local units, assuming
usage of fuel purchased at a 'retail' price of say 100p per GJ (approximately typical of off-peak electricity and domestic-tariff gas burnt at
80 per cent efficiency), an annual cost-in-use could then be as follows:

	£
Owning cost	40·0
Standing charge (say)	6·0
Maintenance	4·0
70 GJ × 1·0	70·0
	£120·0

or £2·3 per week approximately.

TABLE 23.6

COST-IN-USE FOR DISTRICT HEATING

Item	£	£
Fuel		
Heat used/annum		
$= 1300 \times 70 = 91\ 000$ GJ		
Heat losses, fuel pre-		
heating (where appro-		
priate) etc. $= 19\ 000$ GJ		
Heat generated $= 110\ 000$ GJ		
Fuel used at 80% eff. $= 137\ 000$ GJ		
Cost $= 137\ 000 \times 25$p	34,250	
Electricity to Plant		
Approximately 10% of fuel	3,425	37,675
Salaries of Operators, Spares etc.		
Operators	3,600	
Insurance and holidays (10%)	360	
Management	1,000	
Spares, 0·5% of equipment	2,470	7,430
Insurance and Rates		
Insurance, 0·1% of equipment cost	494	
Rates, 0·25% of equipment cost	1,235	1,729
Interest and Repayment of Capital		
Boiler Plant	7,670	
External Mains	10,140	
Internal Mains	7,150	
Dwelling Equipment	21,320	
Builders Work	6,110	52,390
Sundry Items		
Say		776
Total cost-in-use/annum		£100,000
Cost-in-use/dwelling per annum		£77
Cost-in-use/dwelling per week		£ 1·5

However, if it were assumed—and this is the situation reflected by statistics quoted by the supply authorities—that the occupier economises in energy usage at the inevitable expense of comfort conditions in the dwelling, then the consumption might only be 40 GJ per annum. The resultant cost-in-use would then be £90 per annum and the weekly charge would fall to £1·7 approximately.

Comparison—It will thus be seen that for comparable service, district heating is 80p per week cheaper than individual units. Parity of cost is only achieved by a greatly diminished service from the individual units.

Other Factors—It will be appreciated by the reader that this fiscal examination is necessarily superficial and refers to a hypothetical development: fuel costs are constantly varying, as are interest rates, so that any precise comparisons would quickly be out-dated. The picture presented is, however, fairly typical of the results of a number of surveys completed during the period 1968 to 1971 in relation to real situations.

As examples of circumstances which may confuse matters, the following are typical:

Some electrical supply authorities claim, not unreasonably, that in order to provide a supply to a new housing development it is necessary for a considerable capital investment to be made by way of transformer stations, distribution networks and so on. If district heating is provided, then a remunerative component of the potential total load is denied to the electricity supply, and Boards in some instances have required that the developing authority make a contribution towards the capital cost: figures of up to £50 per dwelling have been mentioned. Such a sum if capitalised over fifty years could be considered as *adding* about 9p per week to the cost of district heating per dwelling, a sum well within the cost advantage of the service over the alternative. Gas-supply authorities have been known to suggest that, if a district heating boilerhouse uses that fuel, then a reduction in the domestic cooking tariff will be made available to the individual dwellings. This situation might produce a saving of about 3p per week for each dwelling, which could be considered as *reducing* the cost of district heating.

HEAT SERVICE OPERATORS

A comparatively recent development in this country has been the formation of heat-service organisations which will undertake the operation of district-heating plants and will offer expertise imported from the Continent, where such arrangements have been in being for some decades. In some instances, these organisations will also provide capital assistance towards the initial cost of the installation.

Such arrangements relieve the owning authority of operational responsibilities for heat generation and include the cost of fuel, insurance, maintenance and spares, etc., in the boilerhouse. In return, the authority will be required to enter into a long term contract with the service organisation. Terms of payment vary: that most common in the immediate past being a standing charge plus a unit charge for heat as metered at the 'boilerhouse wall', the owning authority retaining all responsibility for external and internal mains, dwelling equipment, etc.

There are current indications that operators are now prepared to accept responsibility for the operation and maintenance in good order of external mains up to the entrance of each building, thus placing themselves on all fours with other public utilities.

CONCLUSION

This brief survey of district heating has necessarily omitted mention of many technical aspects implicit in the design of installations, pressure-

and temperature-limiting controls, volume regulators and the like. These are beyond the scope of this volume. But it is worthy of note that among the most obvious features of successful plants visited on the Continent is the severe simplicity of the arrangements overall: maintenance skills for complex equipment are at a premium in present-day circumstances, and this would seem to dictate the avoidance of unnecessary 'gadgets' all too prone to failure.

CHAPTER 24

Total Energy

THE FIRST, SECOND AND THIRD EDITIONS of this book, published in 1936, 1943 and 1957 respectively, included a chapter entitled *Combined Electrical Generating Stations*. The subject matter of these earlier inclusions now reappears, suitably modernised, under a new title: *Total Energy**.

In 1967, the *Journal of the Institute of Fuel* proposed a definition for the total energy concept as being 'a scheme for the energy supply to a site in which the whole of the electricity and heat requirements are derived from only one form of energy'. This definition has been accepted in principle for the present chapter, with the rider that the equipment concerned should be an assembly of heat-producing, heat-accepting and heat-rejecting units arranged in a manner which derive the maximum conversion of energy in a fuel to useful energy forms over extended periods. In short, with a minimum of waste.

BASIC CONSIDERATIONS

Heat generated in a conventional boiler plant is normally obtained at high thermal efficiency, whilst the generation of electrical power from heat is seldom obtained at overall thermal efficiencies exceeding 35 per cent. The intentional wastage of heat through cooling towers is a feature of generating stations: some small attempts, such as the Pimlico scheme referred to in Chapter 23, have been made to use the cooling water for heating, but nothing in this country compares with the co-ordinated use of energy resources by other countries. It is sufficient to say that the least thermally efficient of the existing British power stations could be made the most efficient if use were made of the waste heat. Further, it is obvious that, when an appreciable proportion of the nation's electricity supply has to be generated in stations which have efficiencies as low as 20 per cent, the cost to the consumer will inevitably be high. Fig. 24.1 illustrates the comparative efficiency of various forms of electrical, thermal and thermal-electric energy production.

A total-energy installation will necessarily comprise a fuel supply, a prime mover, an electrical power generator, heat-recovery equipment and

* Much of the content of this chapter has been taken from a paper read by one of the Authors, in conjunction with Mr D. G. Axford, to the 1970 Conference of the Combustion Engineering Association.

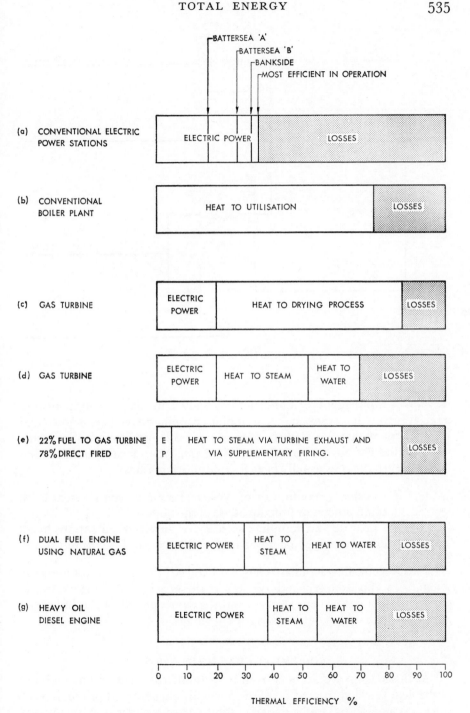

Fig. 24.1.—Comparative Efficiencies for various methods of energy production.

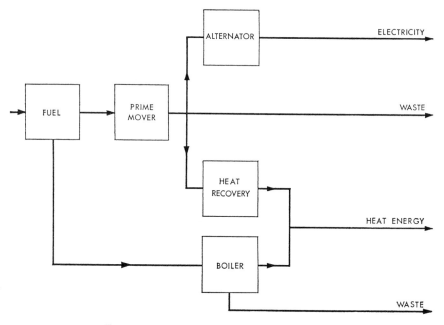

FIG. 24.2.—Components of a Total-Energy System.

perhaps a supplementary boiler—all as shown in diagram form in Fig. 24.2. It is true that components in similar groupings have been known and built for many decades, and the idea of independence from public electricity supplies has always appealed to planners. When local generation has been employed, it has usually been for one of two reasons:

(*a*) To ensure a continuity of electrical supply where essential to some process or functional use of the building.

(*b*) To provide the complete security of an electrical supply in all circumstances, including national disaster or insurrection.

Two recent developments have advanced the use of total-energy schemes in this country and the USA, these being the discovery and exploitation of natural gas reserves and the development of the industrial gas turbine, for which natural gas is an ideal fuel.

EQUIPMENT CHARACTERISTICS

At this stage it would be as well to examine the form and availability of the heat rejected from some of the normally encountered prime movers, since it is an essential to the design of total-energy systems that the marked differences in the characteristics of prime movers are not only appreciated but are deliberately put to use. The more important of these character-istics are:

(a) the heat-to-power ratio at both full and partial load;
(b) the level or levels of heat rejected;
(c) the heat input rate throughout the operating range.

The first of these has the greatest influence on equipment selection, since the exercise will be centred upon the matching of the heat-to-power ratio of the total-energy installation to the building energy-demand profiles. However, it is not sufficient to ensure a match at full or at any one loading, but rather to maintain the balance as far as possible through-out the year. The object, ultimately, is not to create a match of maximum energy rates but to transfer the greatest possible quantity of fuel energy into useful energy forms.

PRIME MOVERS

Prime movers fall into three general classifications:

1. Steam turbines.
2. Gas turbines.
3. Reciprocating engines.

Of these, the steam turbine has the highest heat-to-power ratio at some ten or more to one. It follows that installations using such equipment are most suited to applications which have large thermal demands in relation to electric-power requirements: these will normally lie in the industrial field where a high process heat load may be accompanied by an associated electric-power demand in a ratio particularly suited to these machines. Gas turbines have heat-to-power ratios of the order of three to one, but the manner of application can materially increase this ratio by adding supplementary direct firing to the waste-heat boiler. The third group includes gas ignition engines, diesel engines and dual-fuel compression engines having heat-to-power ratios of approximately one to one.

On occasions, of course, the thermal and electrical demands of the building complex can be matched with the heat-to-power characteristics of a total-energy system by employing a combination of prime movers.

Steam Turbine Generator Sets—High efficiency electricity generating stations on a national scale employ high-pressure boilers and sophisticated techniques to obtain the maximum electric power from the fuel input. However, thermal-electric stations of comparable generating capacity, designed to produce useful heat as well as electric power, would be arranged for quite different pressure and operating conditions and would, in general, tend to employ less complex and cheaper steam-raising plant. In such cases, back-pressure turbines are most likely to be used or, where some proportion of higher pressure steam is required for process use, this may be obtained either directly from the boiler or by pass-out from the

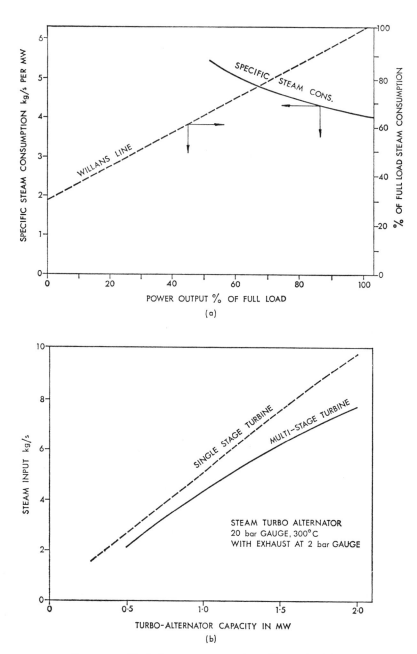

Fig. 24.3.—Typical energy consumptions of Steam Turbines.

turbine at an intermediate pressure. Where the generator load is less than say 700 kW, consideration would be given to reciprocating engines which in this range may be more economical of steam.

Turbine steam consumption per kilowatt produced varies widely according to the load and capacity of the machines. Fig. 24.3 (*a*) shows the variation in steam consumption for a 1 MW set under a varying load. Typically, 4 kg/s of steam per megawatt would be available at a back pressure of say 2 bar gauge. This represents a heat-to-power ratio of ten to one. Fig. 24.3 (*b*) gives full-load data for other sizes of machines and for single and multi-stage turbines.

Gas-Turbine Generator Sets—The full-load thermal efficiencies of typical open-cycle gas-turbine sets are illustrated in Figs. 24.1 (*c*) and (*d*). This shows that the generation of electricity without subsequent heat recovery produces an efficiency of only 19 per cent. In fact, industrial gas-turbine sets have full-load heat ratios varying from about 5 to 7, as shown in Fig. 24.4. These represent an efficiency range of between 14 per cent and 20 per cent and hence, as straight generators of electricity, such machines are not very efficient units.

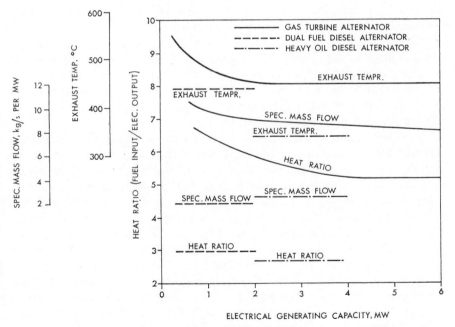

FIG. 24.4.—Comparative full-load characteristics of Gas Turbines and Diesel Engines.

However, the exhaust gas temperature at full load will normally be between 450° C and 550° C and will thus permit relatively simple heat exchange from a single unit; further, the thermal efficiency can be

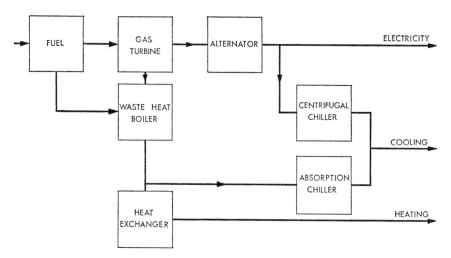

FIG. 24.5.—Line diagram of a Gas Turbine Total-Energy System.

increased by arranging two or more stages of heat transfer. Fig. 24.5 represents a flow diagram of a typical arrangement.

Occasionally, very high efficiency can be achieved by allowing the exhaust gases to fall to quite low temperatures: such an application might involve a direct drying process as shown in Fig. 24.1 (c). In practice, there are remarkably few problems associated with low final gas temperatures when the fuel used is good quality natural gas and the inlet air supply to the turbine has been adequately filtered.

Another characteristic of gas turbines has an important bearing on overall efficiency, this being the requirement that large volumes of air are handled by the turbine in order that the temperature of the turbine blades is limited. The limitation is usually about 850° C for industrial machines. This results in substantial quantities of excess air which pass through the turbine and into the heat exchanger, which large volumes of preheated air clearly invite supplementary firing. Fig. 24.1 (e) shows the increase in overall efficiency of gas-turbine-boiler combinations which can be achieved by adopting this arrangement.

When considering gas turbines, it is important to take account of the condition under which the machine is to produce maximum electrical output. Machines are normally rated at N.T.P. and deviations from this with respect to inlet-air temperature affect the rating, as illustrated in Fig. 24.6. Similarly, exhaust temperature, inlet pressure loss and exhaust-system pressure must be considered when assessing the actual output of a gas-turbine set under specific conditions. Of particular importance is an appreciation of how the heat input rate varies with load. Below 50 per cent of full load this rate increases sharply, as shown also in Fig. 24.6.

VARIATION IN FULL LOAD RATING

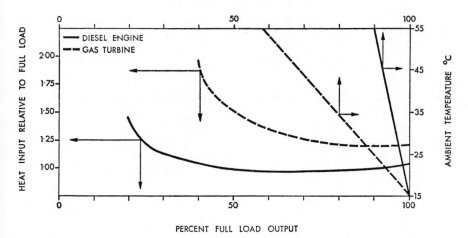

FIG. 24.6.—Variation in Gas Turbine and Diesel Engine Performance at part-load and with ambient temperature.

Reciprocating Engines—Whilst the open cycle gas turbine has the great merit of rejecting waste heat in a single fluid stream and at a relatively high temperature, it has a high proportion of fuel energy rejected as heat, and the shaft power for generation of electricity is comparatively low.

The various forms of reciprocating engine, on the other hand, can convert an equivalent amount of fuel into approximately double the shaft power obtained with gas turbines. Among the reciprocating machines generally considered, and available in Britain for total-energy systems, are:

(*a*) Diesel engines using light diesel oil. These may be either naturally aspirated or turbo-pressure charged.

(*b*) Diesel engines using heavy fuel oil, up to 70 centistokes (3500 secs., Redwood No. 1, at 38° C). These engines are normally turbo-pressure charged.

(*c*) Dual fuel engines capable of using either diesel fuel oil or natural gas. These may be naturally aspirated or turbo-pressure charged.

(*d*) Spark ignition gas engines.

Heat recovery from reciprocating machines may be better appreciated by reference to Fig. 24.7, which is a typical flow diagram for an installation capable of utilising heat at the various thermal levels at which it is available. Low grade heat may be obtained from the engine jacket, lubricating oil cooler and, where fitted, the inter-cooler. In addition, a higher grade heat is available from the exhaust gas *via* a waste-heat boiler. British practice is to take the waste heat from the engine jacket in the form of low-temperature hot water, whilst the exhaust gas is used to generate steam or

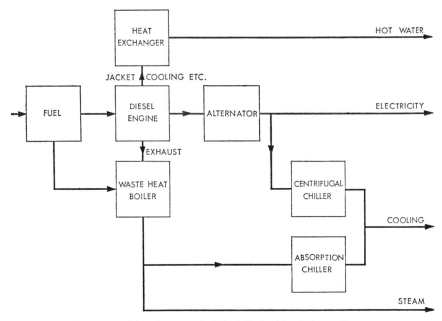

FIG. 24.7.—Line diagram of a Diesel Engine Total-Energy System.

to heat water to higher temperatures. In the United States, however, it is more usual to employ ebullient cooling which virtually generates steam within the cylinder jacket. This steam is not allowed to collect in the waterways but is removed by a steam separator above the engine. It may be that, in due course, this practice will extend to this country.

The various waste-heat sources not only have different heat potential but also have differing limiting factors. For example, the inter-cooler water-outlet temperature is usually limited to 30° C, and hence it is often connected in series flow with the oil cooler which can allow the oil temperature to approach 90° C and so raise water to a temperature of say 60° C.

Full-load thermal efficiencies of diesel and dual-fuel engines are shown diagrammatically in Fig. 24.1 (f) and (g). The heavy-oil diesel is shown to convert some 38 per cent of the heat input into shaft power for producing electricity and, provided that the grades of heat transferred to water and low-pressure steam are acceptable, then overall efficiencies of nearly 80 per cent are obtainable. The dual-fuel engine heat recovery is similar to that of the heavy-oil diesel and, although heat input conversion to shaft power is a little lower, higher overall efficiencies can be obtained. Notwithstanding the lower potential of the heat available, as compared with the gas turbine, satisfactory and economically sound installations are quite possible by using separately fired boilers to provide higher pressure steam or heat demands in excess of the heat recoverable from the engine.

Further, the diesel or dual-fuel engine has the merit of a substantially constant heat ratio over a wide range of operating loads, as shown in Fig. 24.6, and suffers much smaller changes in output in relation to intake-air temperature as compared with the gas turbine. Fig. 24.6 also illustrates the variation in output for diesel engines in relation to air-intake temperature. Changes in heat-input rate with varying load conditions are also shown in this Figure.

BUILDING LOAD CHARACTERISTICS

To ensure maximum utilisation of the potential energy in the fuel, the heat and power demands of the building load must be matched by the output of the prime movers quantitatively and in an acceptable ratio of heat to power. This match must be considered not only at full load but in accordance with load variations.

Electrical Supply—Total-energy systems set out to provide all the on-site energy requirements, including electrical power: i.e. including an electrical supply which would otherwise be obtained from the national network or public utility. It follows, therefore, that the on-site power supply should provide no less security than that obtained from public supplies. In some circumstances, of course, a guaranteed minimum supply may be the major reason for considering on-site generation. This security factor will affect the number, sizes and, on occasion, the type of prime mover and generator combination.

All machinery requires periodic maintenance, and it may not always be convenient to arrange for a machine to receive the service during light-load periods of the year. Therefore, where the load is substantially constant all the year round, provision should be made for at least one set in excess of maximum requirements. Such an arrangement, however, would do no more than allow a spare machine to be itemised on the maintenance schedule and, whilst this would reduce the possibility of breakdown, it could not be depended upon to obviate it. It is quite possible, then, that security of supply at full load may demand two machines in excess of maximum requirements.

It is on this aspect that the economic viability can turn so easily. The choice will lie between a small number of large machines with consequent large and costly excess capacity, or a larger number of small machines with lower capital cost for provision of standby but with higher cost per installed kilowatt of capacity. Frequently, too, the smaller machines would provide flexibility in load matching which would allow longer periods of running at part total-load at relatively low heat-rate and without by-pass and associated heat wastage. Reference to Fig. 24.6 will show that the heat rate increases as the power output is reduced, and this characteristic is particularly pronounced with gas turbines.

Electrical Load—The electrical load will be made up from components of power and lighting and will vary considerably between buildings put to different usage. In commercial buildings, offices and the like, the power element (apart from a small amount of power to socket outlets) is likely to be mainly for motors driving mechanical building-systems plant, such as pumps, fans, compressors and lifts. An industrial complex, on the other hand, may have a similar type of load confined to the office section of the site combined with a predominant process-power load elsewhere.

The lighting, too, can vary considerably, even when comparing buildings of similar usage. Deep planning will tend to show that the lighting load is substantially constant the year round, although the peak demand may be close to that of a shallow plan building of equal area and lit to comparable standards. Lighting may also be deliberately employed as a heating element during the pre-warm periods of an intermittently heated building.

Space and Process-Heat Load—Space- and process-heating loads for any one complex having such a dual demand may not only vary considerably quantitatively but are frequently of different character. For example, heated water at the relatively low temperature of 95° C will ordinarily satisfy space heating and domestic hot-water requirements. A process, on the other hand may well require higher temperature water or steam at a level such as 200° C for moulding presses. The type of demand will therefore influence the choice of prime movers and supplementary plant, but process applications are not considered in this Chapter.

Air-Conditioning Load—A desirable feature of any installed plant is that it shall have high utilization and this aspect will considerably influence feasibility in a total-energy study. It follows that substantially even demands, sustained over long periods, for both heat and electrical power have the highest acceptability. By its very nature, space heating will be seasonal, so that in Britain, between May until late September, the requirement is likely to reduce progressively to zero and then re-build similarly. However, an air-conditioned building will have a cooling demand which can be expected to increase inversely with movements in outside temperature. Cooling will not of course be entirely related to outside temperature, but this variable will nevertheless influence changes in demand considerably.

But a cooling requirement can become a heat demand if considered in conjunction with absorption refrigeration. This type of refrigeration plant not only accepts heat directly but at a relatively low level. Steam at 1 bar gauge is more usual, but hot water at the equivalent temperature of about 110° to 120° C is also used. This temperature requirement is significant since there is no restriction in choice of type of prime mover, all of which can easily be arranged to transfer rejected heat at this level.

Total Load—Fig. 24.8 illustrates how load factors, expressed there in heat-to-power ratios, may vary for two typical applications. The matching of these with the characteristics of the prime movers is essential to the success of any total-energy plant. The process of carrying out a detailed

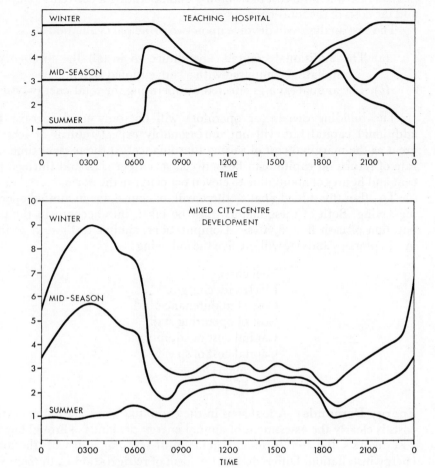

FIG. 24.8.—Heat-to-power demands for typical applications.

examination of the electrical and thermal loads, hour-by-hour and season-by-season, is tedious, but the economic viability and practicality of the plant depends upon such an examination in great detail. A superficial approach and a cursory treatment of load profiles can lead to highly misleading conclusions.

ECONOMIC FEASIBILITY

Where a total-energy installation is to be considered as an alternative to conventional methods of obtaining power and heat, the *feasibility* must be founded on economic viability and not on the practicability of installing the equipment and meeting the energy demands. Thus a comparison must be made between the cost of installing and operating a conventional plant and the costs of the total-energy system.

The cost analysis will devolve upon two principal evaluations:

(a) The additional capital expenditure to install the total-energy installation compared with a conventional system.

(b) The annual savings effected by operating the total-energy system.

The building owner (or operator) will not only need to raise the additional capital but will not unreasonably expect returns at least as great as the marginal cost of raising that finance. At the present time, the rate of return on capital can lie between seven per cent and thirteen per cent and figures of about nine to eleven per cent are the norm.

To offset this added finance there will, of course, be an annual operating saving. Both of these factors must be taken into account in the preparation of cash flow analysis. Computations, similar to those described in Chapters 15 and 23, will involve the following:

> Fuel costs.
> Electricity charges.
> Cost of maintenance.
> Cost of operating staff.
> Capital cost of equipment.
> Capital cost of space.
> Insurance charges.
> Cost of finance.

Energy Demands—A first step in the computation of annual operating costs is clearly the assessment of annual energy demands. Throughout the cost analysis, the conventional scheme must be compared with the total-energy installation. Differences in the form of refrigeration equipment will affect the energy demands of the two installations. For a conventional scheme it may be assumed that electrically driven compressors would be employed: the total-energy installation on the other hand may use absorption type water-chilling apparatus.

The conventional plant would incur energy costs in two parts. Firstly, the provision of heat by gas-fired or oil-fired boilers and, secondly, the cost of bought-in electricity to supply power and lighting together with the power to drive the refrigeration plant. The total-energy system, conversely, would buy-in fuel to enable the prime movers to meet the electrical requirements and, in addition, fuel would be required to fire any supplementary boiler plant.

Fuel—Since the case for total energy may largely turn upon the cost of fuel it is imperative that the annual energy demands are accurately assessed; it is important, too, that the timing of these demands is carefully related to fuel costs where more than one fuel may be involved. For example, if advantage is to be taken of an interruptible supply tariff for natural gas, it is not sufficient to adjust running costs by the number of hours of gas cut-off. Full consideration must be given to the loading during those hours.

Bought-in Electricity—These outgoings will only apply to the conventional plant, although, on occasion, a case can be made for *partial energy* schemes. Negotiations with the Area Supply Authority for the bought-in portion of electric power must, however, be firmly concluded before carrying out the feasibility analysis. The supply of power may sometimes involve a contribution towards the cost of switchgear and transformers or the laying of cables, and conventional system costs should not fail to include such charges where these occur.

Capital—Since the total-energy system sets out to generate electric power on site rather than to buy-in this commodity, it follows that space requirements will be greater than those for a conventional plant. Space inevitably costs money and in most cases there will be the added cost of enclosing that space, and comparative costings must take full account of this. Equipment costs, too, must be carefully examined so that every item which is not common to both types of installation is properly listed in the analysis.

Maintenance and Staff—Whilst maintenance will generally not be a major component of total cost, there will be marked differences between the different types of prime mover. In general, reciprocating engines with their relatively large number of moving parts will require more frequent maintenance routines than turbines—which latter will normally run for 8000 to 20 000 hours between overhauls on good quality fuel. The maintenance costs will not only include the wages of operating staff and overheads but also the replacement of components, small expendable tools, lubricating oil, consumable stores and administration costs.

INVESTMENT APPRAISAL

There are several methods of appraising the economic soundness of any project, and difficulties in comparing the different methods have been shown to lead to wrong decisions. It is for this reason that the *discounted cash flow* method is recommended in examining the case for total energy. This process takes account of variation in earnings over the expected life of a project and allows investment incentives and taxation to be taken fully into account.

The various means of setting out the economic case must in one way or another recognise that:

(a) equipment, and to a lesser extent buildings, will deteriorate and depreciate in value;

(b) any additional capital expenditure involved in installing a total-energy system must not only be met but must show realistic returns.

K.H.A.C.

Notwithstanding a high yield when correctly assessed, commercial circles may be unwilling to countenance pay-out periods in excess of five years. Very often in practice, therefore, the paramount issue becomes return on capital expenditure within periods which tend to swamp the effect of depreciation of equipment. Space for housing the installation must also be financed, but the returns for this are usually long term. It is not unusual, for example, to amortise the cost of additional space over periods of forty to sixty years.

CONCLUSIONS

Recent feasibility studies (the results of three of which are summarised in Table 24.1), have shown some quite remarkable returns on capital; and other studies, whilst less exciting, have demonstrated that total energy, if given the right conditions, can not only be shown to be economically viable for the managing authority using thoroughly realistic commercial

TABLE 24.1

COMPARATIVE TOTAL RECURRENT COSTS FOR
CONVENTIONAL AND TOTAL ENERGY SYSTEMS

Central Plant	Cost Ratios for following cases		
	Case A	Case B	Case C
Conventional System Boiler plant plus bought-in electricity for all power and lighting etc.	100	100	100
Total Energy Steam Turbines	94	—	—
Gas Turbines	76	66	81
Diesel Engines	73	—	—
Dual Fuel Engines	62	—	72

Note
Cases A and B include major air-conditioning components.

rates of return, but can be considered as a step towards reducing the enormous waste of energy resources throughout the World and, in consequence, reducing the pollution of the atmosphere by waste products from unnecessary fuel consumption.

This book may seem to have come a long way from the first chapters dealing with the simple elements of heat, water, and air; and yet, as forecast at the outset, the basic principles of all engineering design in this field depend just on a full understanding of the phenomena related to these elements.

The problems posed by modern life and by building techniques demand increasing sophistication in engineering systems, as well as their integration with other services as a whole. Nevertheless, in spite of such complications, the Revisers hope that in this work the reader can gain a sufficient clue to whatever may be the problem in hand to enable a path to be seen by which it may be solved.

SI UNIT SYMBOLS

Quantity	Unit	Symbol	Common multiples or sub-multiples	Multiplier	Symbol
Length	metre	m	kilometre	$m \times 10^3$	km
			millimetre	$m \times 10^{-3}$	mm
Area	square metre	m^2	hectare	$m^2 \times 10^5$	ha
			sq. millimetre	$m^2 \times 10^{-9}$	mm^2
Volume	cubic metre	m^3	litre	$m^3 \times 10^{-3}$	(litre)*
			cu. millimetre	$m^3 \times 10^{-9}$	mm^3
Time	second	s	hour	$s \times 3600$	h
			minute	$s \times 60$	
Velocity	metre per second	m/s			
Acceleration	metre/second/second	m/s^2			
Frequency	hertz (cycle per sec)	Hz			
Rotational frequency	revolutions per sec	s^{-1}			
Mass	kilogramme	kg	tonne	$kg \times 10^3$	t
			gramme	$kg \times 10^{-3}$	g
			milligramme	$kg \times 10^{-6}$	mg
Density		kg/m^3			
Specific volume		m^3/kg			
Mass flow rate		kg/s	litre per sec	$(m^3/s) \times 10^{-3}$	litre/s
Volume flow rate		m^3/s			
Momentum		kg m/s			
Force	newton	N	meganewton	$N \times 10^6$	MN
			kilonewton	$N \times 10^3$	kN
Torque		Nm			
Pressure (and stress)	newton per square metre	N/m^2	meganewton per square metre	$(N/m^2) \times 10^6$	MN/m^2
			kilonewton per square metre	$(N/m^2) \times 10^3$	kN/m^2
			bar	$(N/m^2) \times 10^5$	b
			millibar	$(N/m^2) \times 10^2$	mb
Viscosity					
Dynamic		Ns/m^2	centipoise	$(Ns/m^2) \times 10^{-3}$	cP
Kinematic		m^2/s	centistokes	$(m^2/s) \times 10^{-6}$	cSt
Temperature	degree Kelvin	°K			
	degree Celsius	°C			
Heat					
Energy			gigajoule	$J \times 10^9$	GJ
Work	joule	J	megajoule	$J \times 10^6$	MJ
Quantity of heat			kilojoule	$J \times 10^3$	kJ
Heat flow rate (power)	watt	W	gigawatt	$W \times 10^9$	GW
			megawatt	$W \times 10^6$	MW
			kilowatt	$W \times 10^3$	kW
Thermal conductivity		W/m °C			
Thermal resistivity		m °C/W			
Specific heat capacity		kJ/kg °C			
Latent heat		kJ/kg			

NOTE. For conversion factors, Imperial units to SI units, see Table 1.4 (page 15).

* This book does not accept *l* as a symbol (see the *Introduction*).

LIST OF TABLES

I N D E X

554